This book describes the manifestations of chaos in atoms and molecules.

About ten years ago, atomic physics received a rejuvenating jolt from chaos theory with far-reaching implications. The study of chaos is today one of the most active and prolific areas in atomic physics. This is the first attempt to provide a coherent introduction to this fascinating area. In line with its scope, the book is divided into two parts. The first part (chapters 1–5) deals with the theory and principles of classical chaos. The ideas developed here are then applied to actual atomic and molecular physics systems in the second part of the book (chapters 6–10) covering microwave driven surface state electrons, the hydrogen atom in a strong microwave field, the kicked hydrogen atom, chaotic scattering with CsI molecules and the helium atom. The book contains many diagrams and a detailed reference list.

The book will be of interest to graduate students and researchers in atomic, molecular and optical physics.

CAMBRIDGE MONOGRAPHS ON ATOMIC, MOLECULAR AND CHEMICAL PHYSICS

General Editors: A. Dalgarno, P. L. Knight, F. H. Read, R. N. Zare

CHAOS IN ATOMIC PHYSICS

CAMBRIDGE MONOGRAPHS ON ATOMIC, MOLECULAR AND CHEMICAL PHYSICS

Chaos in Atomic Physics

R. BLÜMEL
Albert-Ludwigs–Universität, Freiburg

W. P. REINHARDT
University of Washington, Seattle

CAMBRIDGE
UNIVERSITY PRESS

CAMBRIDGE UNIVERSITY PRESS
Cambridge, New York, Melbourne, Madrid, Cape Town, Singapore, São Paulo

Cambridge University Press
The Edinburgh Building, Cambridge CB2 2RU, UK

Published in the United States of America by Cambridge University Press, New York

www.cambridge.org
Information on this title: www.cambridge.org/9780521455022

First published 1997
This digitally printed first paperback version 2005

A catalogue record for this publication is available from the British Library

Library of Congress Cataloguing in Publication data

Blümel, R. (Reinhold)
Chaos in atomic physics / R. Blumel, W. P. Reinhardt.
p.   cm.   (Cambridge monographs on atomic, molecular, and chemical physics: 10)
Includes bibliographical references.
ISBN 0 521 45502 2
1. Chaotic behavior in systems.   2. Atoms.   3. Molecules.
I. Reinhardt, William P.   II. Title.   III. Series.
QC174 17.C45B47   1997
539.7d–c21   96–39292 CIP

ISBN-13 978-0-521-45502-2 hardback
ISBN-10 0-521-45502-2 hardback

ISBN-13 978-0-521-01790-9 paperback
ISBN-10 0-521-01790-4 paperback

To Lynn and Tina

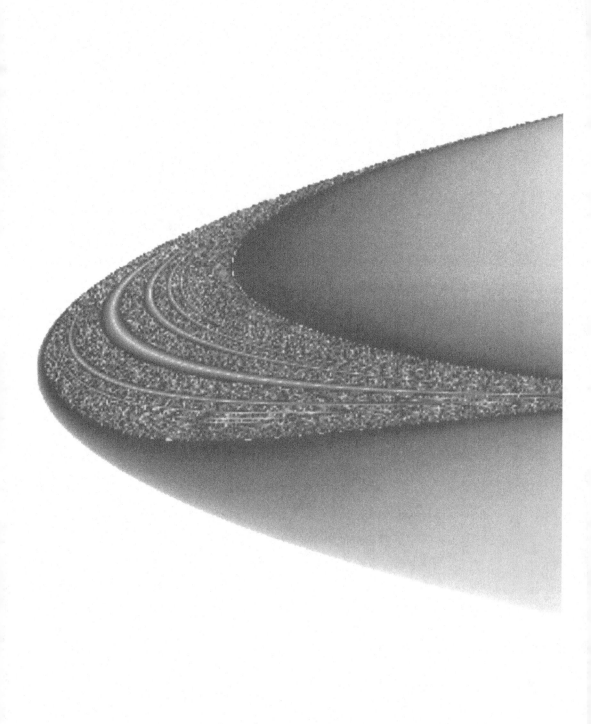

Contents

Preface

Atomic physics is one of the oldest fields of physics. A barren and "academic" discipline? Not at all! About ten years ago, atomic physics received a rejuvenating jolt from chaos theory with far reaching implications. Chaos in atomic physics is today one of the most active and prolific areas in atomic physics. This book, addressed at interested students and practitioners alike, is a first attempt to provide a coherent introduction into this fascinating area of contemporary research. In line with its scope, the book is essentially divided into two parts. The first part of the book (Chapters 1 – 5) deals with the theory and philosophy of classical chaos. The ideas and concepts developed here are then applied to actual atomic and molecular physics systems in the second part of the book (Chapters 6 – 10).

When compiling the material for the first part of the book we profited immensely from a number of excellent tutorials on classical and quantum chaos. We mention the books by Lichtenberg and Lieberman (1983), Zaslavsky (1985), Schuster (1988), Sagdeev *et al.* (1988), Tabor (1989), Gutzwiller (1990), Haake (1991), Devaney (1992) and Reichl (1992).

The illustrative examples for the second part of the book were mostly taken from our own research work on the manifestations of chaos in atomic and molecular physics. We apologize at this point to all the numerous researchers whose work is not represented in this book. This has nothing to do with the quality of their work and is due only to the fact that we had to make a selection. We were not even able to devote a separate chapter to every one of the most important classically chaotic systems currently under active investigation. The most important omission is probably the hydrogen atom in a strong magnetic field, although we took care to mention it several times in this book, and we discuss its potential for future research in Chapter 11.

For the experienced practitioner of atomic physics there appears to be an enigma right at this point. What does *nonlinear* chaos theory have to do with *linear* quantum mechanics, so successful in the classification of atomic states and the description of atomic dynamics? The answer, interestingly, is the enormous advances in atomic physics itself. Modern day experiments are able to control essentially isolated atoms and molecules to unprecedented precision at very high quantum numbers. Key elements here are the development of atomic beam techniques and the revolutionary effect of lasers. Given the high quantum numbers, Bohr's correspondence principle tells us that atoms are best understood on the basis of classical mechanics. The classical counterpart of most atoms and molecules, however, is chaotic. Hence the importance of understanding chaos in atomic physics.

During the past ten years the area of chaos in atomic physics has matured to an established field with a rapidly expanding literature of hundreds of published papers. Therefore, we feel that the time has come where it makes sense to include this field in the set of courses routinely offered to students at the graduate level. In fact, this book is based on a course on chaos in atomic physics delivered at the University of Freiburg in the Winter Semester 1995/1996.

Excellent reviews exist addressing the topic of chaos in atomic physics. The most comprehensive attempt at a review of the entire subject of quantum chaology and applications is a recent collection of reprints and original articles edited by Casati and Chirikov (1995). Its section on atoms in strong fields especially is highly relevant for the topic of this book. Additional material on chaos and irregularity in atomic physics can be found, e.g., in a collection of articles edited by Gay (1992).

Many of the subjects discussed in this book were investigated and developed in collaboration with Prof. Uzy Smilansky. We owe many thanks to Uzy for sharing his continuous stream of ideas with us, and for hosting uncounted extended visits at the Weizmann Institute. This fruitful collaboration started in 1983 and was always generously supported by the MINERVA foundation.

The topic of Chapter 5 was originally investigated in collaboration with Prof. Shmuel Fishman. His hospitality on the occasion of many visits to the Technion is very much appreciated.

Prof. Peter Koch made it possible for one of us (R.B.) to visit his lab on several occasions for extended periods of time. This was a unique opportunity to see the inner workings of one of the most productive modern atomic physics labs. Many thanks!

Many thanks are due to our colleagues at the University of Maryland, Professors Tom Antonsen, Celso Grebogi, Ed Ott and Richard Prange for a very productive collaboration in 1995.

Last but not least we would like to thank our colleagues at Freiburg, especially Dr G. Alber, Prof. J. S. Briggs, Dr A. Bürgers, Dr F. Großmann and Dr J.-M. Rost for valuable discussions and suggestions. We owe special thanks to T. R. Neal, our science editor at Cambridge University Press for continued encouragement.

1

Introduction

By now the "chaos revolution" has reached nearly every branch of the natural sciences. In fact, chaos is everywhere. To name but a few examples, we talk about chaotic weather patterns, chaotic chemical reactions and the chaotic evolution of insect populations. Atomic and molecular physics are no exceptions. At first glance this is surprising since atoms and molecules are well described by the *linear* laws of quantum mechanics, while an essential ingredient of chaos is *nonlinearity* in the dynamic equations. Thus, chaos and atomic physics seem to have little to do with each other. But recently, atomic and molecular physicists have pushed the limits of their experiments to such high quantum numbers that it starts to make sense, in the spirit of Bohr's correspondence principle, to compare the results of atomic physics experiments with the predictions of classical mechanics, which, for the most part, show complexity and chaos. The most striking observation in recent years has been that quantum systems seem to "know" whether their classical counterparts display regular or chaotic motion. This fact can be understood intuitively on the basis of Feynman's version of quantum mechanics. In 1948 Feynman showed that quantum mechanics can be formulated on the basis of classical mechanics with the help of path integrals. Therefore it is expected that the quantum mechanics of an atom or molecule is profoundly influenced, but of course not completely determined, by the qualitative behaviour of its underlying classical mechanics. To be specific, we expect to see qualitatively different quantum behaviour in an atom or molecule depending on whether its classical mechanics is regular or chaotic.

   The central theme of classical deterministic chaos is the occurrence of the most astonishing complexity in the simplest systems (see, e.g., Schuster (1988), Ott (1993)). And what could be simpler than, e.g., a hydrogen atom in a strong magnetic field? On second thought, however, we realize that on the classical level this system possesses the crucial

1

ingredient for the emergence of chaos: nonlinearity. In this case the nonlinearity consists in the combined interactions of the electron with the proton and the applied magnetic field. Sure enough, it was shown by many scientists (see, e.g., Friedrich (1990) and references therein) that chaos does indeed occur in the classical version of the hydrogen atom in a magnetic field. Therefore, we expect complex behaviour to appear in the quantum mechanics of this system. While complexity may reveal itself in many quantum observables, the most familiar observable is the energy. Fig. 1.1(a) shows the energy spectrum of the hydrogen atom as a function of the magnetic field in a range of principal quantum numbers around $n = 40$. At first glance the spectrum appears "wild" and "chaotic". On second thought one might argue that Fig. 1.1(a) is merely "congested", i.e. an assembly of many hydrogen lines, all in principle quite regular, but producing an irregular effect simply by plotting many of them in one single figure. To demonstrate that this is not so, i.e. that something fundamentally new is going on in Fig. 1.1(a), we show a magnification of the framed area of Fig. 1.1(a) in Fig. 1.1(b). The magnification reveals that Fig. 1.1(a) actually consists of hundreds of avoided crossings which force initially "regular" energy levels to bend in "erratic" ways as a function of the magnetic field strength producing considerable complexity in the energy spectrum. The bending of the energy levels makes it very hard, if not impossible, to assign quantum numbers to the individual energy levels at fixed magnetic field strength. In fact, it was argued by Percival in 1973 that one of the key quantum signatures of chaos is the loss of our ability to assign quantum numbers.

Chaos does not only wreak havoc in otherwise orderly atomic spectra, it also provides a natural framework, indeed a common language, in which one can discuss such seemingly unrelated systems as, e.g., ballistic electrons in mesoscopic semiconductor structures, the helium atom, and Rydberg atoms in strong external fields. All these systems have one feature in common: their classical counterparts are chaotic. Chaos imprints its presence on their spectra and manifests itself in spectral features which are very similar for all these systems (universality).

Especially in the highly excited semiclassical regime the quantum properties and dynamics of atomic and molecular systems are most naturally discussed within the framework of chaos. Not only does chaos theory help to characterize spectra and wave functions, it also makes specific predictions about the existence of new quantum dynamical regimes and hitherto unknown exotic states. Examples are the discovery of "frozen planet" states in the helium atom by Richter and Wintgen (1990a) and

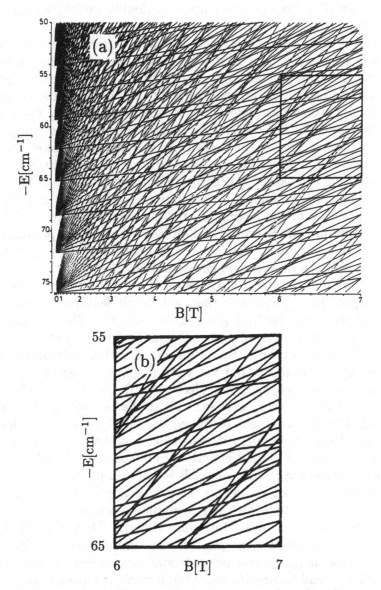

Fig. 1.1. Energy levels of the hydrogen atom as a function of magnetic field strength B. (a) Transition from order to "chaos" for increasing magnetic field strength (Friedrich and Wintgen (1989)). (b) A magnification of the framed detail in (a) resolves the complex behaviour of the energy levels in the $E - B$ plane for high magnetic field strengths.

the prediction of a "gas of resonances" in doubly excited helium (see Chapter 10).

But chaos is more than a tool. There are as yet unsolved philosophical problems in its wake. While relativity and quantum mechanics necessitated – and in fact originated from – a careful analysis of the concepts of space, time and measurement, chaos, already on the classical level, forces us to re-think the concepts of determinism and predictability. Thus, classical mechanics could not be further removed from the dusty subject it is usually portrayed as. On the contrary: it is at the forefront of modern scientific research. Since path integrals provide a link between classical and quantum mechanics, conceptual and philosophical problems with classical mechanics are bound to manifest themselves on the quantum level. We are only at the beginning of a thorough exploration of these questions. But one fact is established already: chaos has a profound influence on the quantum mechanics of atoms and molecules. This book presents some of the most prominent examples.

Among the atomic systems discussed in this book we find the helium atom (Chapter 10). This topic rings a bell. It featured prominently in the final stages of the "old quantum mechanics". The old pre-1925 quantum theory was based on classical notions such as periodic orbits. Periodic orbit quantization of the helium atom within the framework of the old quantum mechanics did not work out and presented an insurmountable problem (see, e.g., Van Vleck (1922)). But all the difficulties with the old quantum theory were solved with one blow when Heisenberg, Born, Jordan and Schrödinger developed the "new" quantum mechanics in 1925 and 1926. Heisenberg in particular was very proud of having eliminated the concept of classical orbits from the structure of quantum mechanics (Heisenberg (1969)). But his joy proved premature. Periodic orbits are very much alive! In fact, they currently enjoy a key role as one of the few known tools for the systematic semiclassical quantization of classically chaotic systems (Gutzwiller (1990)). Thus, the fall of the old quantum mechanics was not primarily due to the use of classical concepts, but to the inappropriate use of classical procedures, such as adding probabilities instead of amplitudes. Also, it was not known then how to incorporate properly the intricacies of classical mechanics, epitomized in the phenomenon of chaos. We note that the difficulties of incorporating the ideas of chaos into the old quantum theory were well appreciated by Paul Ehrenfest's student Burgers (1916), Einstein (1917) and Dirac (1925).

In order to develop the mind set and methods needed to understand and use the fingerprints of chaos in quantum mechanics, we must set to work. Our journey through chaos in atomic physics begins head-on with a schematic, but physical, example of chaos in Section 1.1. The remaining

sections of Chapter 1 present a general discussion of some of the philosophical implications of the existence of chaos in the classical world. Formal tools and concepts, indispensable for a deeper understanding of chaos,
are presented in Chapter 2. Chapter 3 is essentially an elementary review of Lagrangian and Hamiltonian mechanics with special emphasis on
the role of chaos in classical mechanics. In Section 3.2, e.g., we present a
simple physical system (the double pendulum) which shows many generic
features of a chaotic Hamiltonian system. Chaos in quantum mechanics
is discussed in Chapter 4. The main point here is to eliminate some of
the confusion surrounding the topic of "quantum chaos" by proposing a
classification of quantum systems into three categories: (I) systems whose
quantum dynamics is not chaotic, but which "feel" the underlying classical chaos ("quantized chaos"); (II) systems whose quantum dynamics
is fully chaotic, but which are coupled to at least one classical degree
of freedom in the sense of a dynamic Born-Oppenheimer approximation
("semi-quantum chaos"); and (III) systems that are fully quantized and
show fully developed chaos. While the existence of type I and type II systems is confirmed and their usefulness in atomic and molecular physics
established, the very existence of type III systems is still much debated.
Therefore, it may well be established by future research that category
III, no doubt the most interesting of the three, is empty. With Chapter
4 we finish the introductory part of the book, whose main purpose is to
provide the reader with the necessary tools and concepts for a thorough
understanding of the remaining chapters of the book, which deal with the
manifestations of chaos in specific atomic and molecular physics systems.
Chapter 5, a study of the physics of an impulsively driven rotor, builds a
bridge between the more formal introductory parts of the book and the
applications. It links with the introductory chapters by providing further tools needed for the discussion of driven atomic physics systems, but
also connects with the following more "applied" chapters by proposing
a laboratory experiment with diatomic molecules driven into chaos by
the application of a sequence of strong electric field pulses. In Chapters
6 – 10 the ideas of chaos and nonlinear systems developed in Chapters
1 – 5 are applied to actual atomic physics systems. Two different types of
systems are discussed: driven and time independent (autonomous). Representatives of driven systems are discussed in Chapters 6 – 8. Chapter
6 presents the classical and quantum dynamics of surface state electrons.
Surface state electrons are an essentially one-dimensional system whose
physics is not encumbered with the presence of additional degrees of freedom. Moreover surface state electrons provide an excellent model for
microwave-driven hydrogen Rydberg atoms discussed in Chapter 7. The
importance of phase-space fractals for the properties of atomic decay is
discussed in Chapter 8. With Chapter 8 we conclude the discussion of

time dependent atomic systems. Representatives of not explicitly time
dependent atomic and molecular systems are discussed in Chapters 9 and
10. Chapter 9 is on chaotic scattering theory applied to molecular scattering in external fields. It prepares for a discussion of the helium atom
(Chapter 10), which may be classified as an autonomous chaotic scattering system. The book concludes with Chapter 11, a discussion of status,
trends and developments of chaos in atomic physics.

1.1 Chaos: a physical example

The difference between regular and chaotic motion is best explained with
the help of a physical model. The model also illustrates one of the central messages of chaos theory: the possibility of complex motion in the
simplest physical systems.

We assume Newtonian mechanics to be valid and consider a mass
point M bouncing in a two-dimensional square box of side length 1 (see
Fig. 1.2). The box shown in Fig. 1.2 is used to illustrate regular motion. Therefore, we call it "R". Another box is shown in Fig. 1.3. It is
equipped with a hard stationary disk of radius $r = 1/4$ at its centre. It
serves to illustrate chaotic motion. Therefore, we call it "C". We assume
that inside the boxes the mass point M travels on straight line trajectories subject only to specular reflection whenever it hits the walls of R or
C, or the central disk of box C. Since the motion of M is free between
bounces, the mass of M is irrelevant for the kinematics of M. Therefore,
M's velocity can be normalized to 1.

The walls of the boxes are labelled a, b, c and d as shown in Figs. 1.2
and 1.3. The mass point M is injected into the box at a launch point L
located at $x = 0, y = 1/2$. The launch velocity of M is $\vec{v} = (v_x, v_y) = (\cos(\varphi), \sin(\varphi))$ where φ is the initial inclination of the trajectory of M

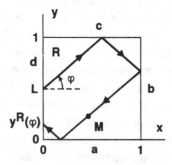

Fig. 1.2.   Regular motion of a mass point M inside an empty square box labelled
R. A trajectory launched at L with $\varphi = 0.69$ returns to wall d after three
bounces.

(see Figs. 1.2 and 1.3). In order for $\vec{v}$ to point into the interior of the box, the angle φ may range from $-\pi/2$ to $\pi/2$. Due to symmetry, and without loss of generality, φ may be restricted to the interval $0 \le \varphi < \pi/2$.

At first sight the box systems may look dry and abstract. But adding a simple modification immediately converts the boxes into models for molecular reactions. We declare the wall d to be "sticky". By this we mean that whenever the mass point M hits the wall d it is "absorbed". This way its final position on the y axis, denoted by y^R (y^C, respectively), can be determined as a function of the initial angle φ. In analogy to the theory of molecular reactions, the functions $y^R(\varphi)$, $y^C(\varphi)$ are called *reaction functions*.

First, we discuss the motion of M in box R. Following its departure from L, the mass point M ricochets around inside R. Returning to Fig. 1.2, a sample trajectory launched at L with $\varphi = 0.69$ is shown. The resulting motion is simple. The successive points of impact at the walls can be computed using only elementary geometry. As a result, the reaction function $y^R(\varphi)$ can be calculated analytically. We obtain:

$$y^R(\varphi) = 1 - \left| 1 - \left[\frac{1}{2} + 2\tan(\varphi) \right] \bmod 2 \right|. \qquad (1.1.1)$$

It is important to note that $y^R(\varphi)$ can be expressed analytically with the help of elementary functions. This feature is characteristic for simple systems that do not show chaos. The simplicity of the reaction function (1.1.1) is reflected in the graph of $y^R(\varphi)$, shown in Fig. 1.4(a).

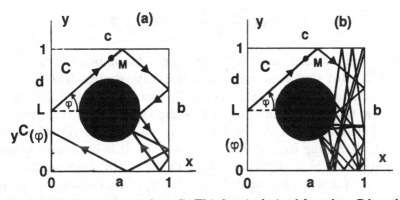

Fig. 1.3. Irregular motion in box C. This box is derived from box R by adding a totally reflecting disk at the centre of R. (a) A complicated but exiting trajectory is produced for the launch angle $\varphi = 0.69$. (b) Dynamically trapped trajectory for a launch angle close to $\varphi \approx 0.692$.

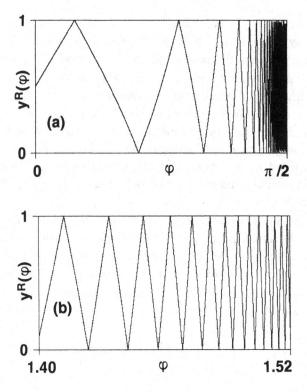

Fig. 1.4.　Reaction function $y^R(\varphi)$ for box R. (a) Full range of angles $0 < \varphi < \pi/2$. (b) Magnification of a detail of (a) in the interval $1.4 \leq \varphi \leq 1.52$.

The graph of $y^R(\varphi)$ displays a regular, saw-tooth like behaviour. The cusps of the reaction function $y^R(\varphi)$ occur at

$$\varphi_n = \arctan\left(\frac{2n-1}{4}\right), \qquad n = 1, 2, \dots. \qquad (1.1.2)$$

The only "complication" in $y^R(\varphi)$ is an accumulation point of "zig-zags" at $\varphi = \pi/2$. But this accumulation is perfectly regular. The positions of the accumulating cusps are predicted accurately by (1.1.2). In order to show that indeed nothing interesting is hidden in the "blur" in Fig. 1.4(a) close to $\varphi = \pi/2$, Fig. 1.4(b) shows a magnification of a detail of Fig. 1.4(a) ranging from $\varphi = 1.4$ to $\varphi = 1.52$. It is important to note that any details which might be hidden in Fig. 1.4(a) close to $\varphi = \pi/2$ can be *resolved* in any interval $[a, b]$, $a < b < \pi/2$. No matter how closely we look, i.e. no matter how large the magnification factor, nothing interesting will ever be revealed about the functional behaviour of $y^R(\varphi)$. On the analytical level this is already clear from the simple analytical form (1.1.1) of $y^R(\varphi)$ and the orderly position of its cusps according to (1.1.2).

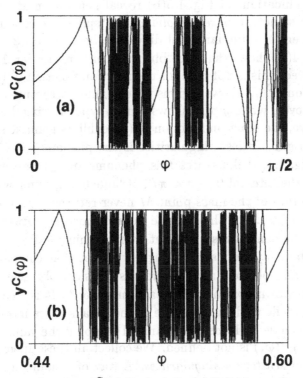

Fig. 1.5. Reaction function $y^C(\varphi)$ for box C. (a) Full range of angles $0 < \varphi < \pi/2$. (b) Magnification of a detail of (a) in the range $0.44 < \varphi < 0.60$.

This situation changes drastically in the case of box C. The new reaction function $y^C(\varphi)$ is shown in Fig. 1.5(a). This time the reaction function looks very complicated. The question is whether this complicatedness is only in degree or in quality. The answer is that the simple addition of a scattering disk at the centre of box R results in a profound qualitative change in the motion of M, so drastic, indeed, that the resulting motion can only be described as "chaotic".

There are various indicators for this qualitatively new type of motion. First of all, an analytical formula for $y^C(\varphi)$ is not known. By "analytical formula" we mean that $y^C(\varphi)$ cannot be written down in finitely many steps with the help of the known special functions of mathematical physics. Furthermore, the unresolved structures apparent in the reaction function displayed in Fig. 1.5(a) *cannot be resolved in principle*, no matter how large the magnification factor. This is illustrated in Fig. 1.5(b), which shows a magnification of a detail of Fig. 1.5(a) in the range $0.44 < \varphi < 0.60$. Indeed, instead of appearing simpler, as was the case with magnifications of unresolved structures in $y^R(\varphi)$, the magnification shown in Fig. 1.5(b) appears to be even more complicated.

Further magnifications of Fig. 1.5(b) reveal generation after generation of additional structure *never resolving* the "spiky" behaviour of $y^C(\varphi)$. Obviously there is a fundamental difference between $y^R(\varphi)$ and $y^C(\varphi)$. While Fig. 1.4(b) shows that the "blurry" structures in Fig. 1.4(a) can in fact be resolved, this is not true for the reaction function $y^C(\varphi)$. The reaction function $y^C(\varphi)$ shows regions which exhibit structure on all length scales. Moreover, while $y^R(\varphi)$ shows a mild type of "singularity", called "cusps" above, the reaction function $y^C(\varphi)$ exhibits genuine singularities in φ. The origin of these singularities is the phenomenon of *dynamical trapping*. Fig. 1.3(b) illustrates this phenomenon. There exist certain angles φ_s in the interval $0 < \varphi < \pi/2$ of launch angles for which the resulting trajectory of the mass point M never returns to wall d of box C. As shown in Fig. 1.3(b) the particle bounces forever between the walls a, b, c and the central disk without ever reaching wall d. The important point about dynamical trapping is the fact that the particle is not hindered by any obvious means, such as physical obstacles, or energy considerations, from exiting the "reaction region". It is free to leave at any time, but fails to do so, because of the details of its motion.

If a launch angle φ_s leads to dynamical trapping the value of the reaction function $y^C(\varphi_s)$ is not defined. We collect all these special angles φ_s into a set S of *scattering singularities*. A way of visualizing the set S is to plot the number of bounces of M against the walls versus the launch angle φ. We call the number of bounces the *lifetime l* of a trajectory with launch angle φ. Fig. 1.6 shows the lifetime $l(\varphi)$ as a function of φ. Smooth regions alternate with regions that contain "spikes" representing scattering singularities characterized by $l = \infty$. The singularities in l correspond to the exceptional angles φ_s collected in the set S. Since the singularities in l occur precisely at the places where y^C is not defined, we conclude that in analogy to Fig. 1.5 the spiky regions of Fig. 1.6 are not resolvable. This implies that the set S of scattering singularities has a very complicated structure. Moreover, since no degree of magnification yields an end to the hierarchy of scattering singularities, we conclude that S must have infinitely many members φ_s, i.e. S is an infinite set. Even more astonishing: it turns out that the elements of S are not countable. This means that no scheme exists according to which the singularities in $y^C(\varphi)$ can be listed one by one. Therefore, paradoxically, the singularities of $y^C(\varphi)$ ($l(\varphi)$, respectively) are just as numerous as the initial angles φ in the interval $0 < \varphi < \pi/2$. This fact is truly counterintuitive. Its precise meaning is explained in Chapter 2. We will return to a more detailed study of scattering singularities in Chapter 9, where we study a molecular physics example of chaotic scattering. In that chapter we

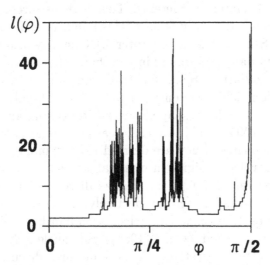

Fig. 1.6. Lifetime function $l(\varphi)$ for box C. The spikes in l indicate the presence of scattering singularities.

will also see that the reaction function y^C is by no means academic, but occurs in similar form frequently in atomic and molecular physics.

"Strange" sets such as the set of singularities S were first studied by the mathematician Georg Cantor toward the end of the last century. Cantor was understood only by very few of his peers (see, e.g., Dauben (1979), Purkert and Ilgauds (1987)). But Cantor's theories, especially his theory of sets, are nowadays recognized as among the most profound developments of 19th century mathematics. Cantor's thoughts are fundamental for the theory of the number continuum, and since physics is mostly concerned with mapping physical phenomena to convenient number spaces (physical modelling), Cantor's ideas are ultimately of the utmost importance for physics. Just as quantum mechanics requires an essential use of complex, rather than just real, numbers, chaos requires a far more detailed understanding of the real number system than is required in traditional physical modelling. We will see in Chapter 2 that the theory of chaos has profited immensely from Cantor's ideas.

The box models R and C discussed above were chosen for various reasons. (i) The two models are "physical" in the sense that actual realizations of the box systems can be built in the lab. Two billiard tables, one with and one without a scattering disk at the centre, are excellent first approximations. One may even think of manufacturing quasi-two-dimensional solid state structures in the shape of the boxes R and C, and studying the quantized states of electrons in these structures. In fact, very similar systems have already been built in the lab (see, e.g., Baranger *et al.* (1993a,b) and references therein). This leads us directly to the topic of "quantum chaos", the main focus of this book. (For an introduction to this subject see Section 1.4 and Chapter 4.) One may also imagine filling super-conducting flat metal boxes in the shape of R and C, respectively, with microwave radiation. Such structures can very effectively model the quantum situation. This way energy spectra have already been obtained experimentally for similarly shaped boxes (Stöckmann and Stein (1990), Kudrolli *et al.* (1994)). (ii) Another reason for the choice of the box models is the following. They are derived from "Sinai's billiard", which is one of the central paradigms in classical and quantum chaos research (see, e.g., Gutzwiller (1990)). (iii) It was discovered recently that the behaviour of $y^R(\varphi)$ and $y^C(\varphi)$, respectively, is generic for the simplest scattering systems (see, e.g., Blümel and Smilansky (1988)) including molecular (see, e.g., Noid *et al.* (1986)) and nuclear (see, e.g., Baldo *et al.* (1993)) scattering systems. A more detailed discussion of chaotic scattering and its quantum manifestations is presented in Chapter 9.

So far, the radius of the scattering disk of box C has been thought of as constant and set to $r = 1/4$. But there is nothing special about $r = 1/4$. Indeed, an important question is how the chaotic set S of scattering singularities changes with a change of r. Thus, the radius r of the scattering disk becomes a *control parameter* of the system. We have encountered control parameters in the context of atomic physics systems before. In the case of the hydrogen atom in a strong magnetic field (see Fig. 1.1), the magnetic field strength is a control parameter. In the case of Rydberg atoms in a strong radiation field, it is the strength of the microwave field that plays the role of a control parameter. Another example of a control parameter relevant in atomic physics is the nuclear charge of helium-like ions. Changing this control parameter in discrete steps results in the helium iso-electronic series H$^-$, He, Li$^+$, Be^{++},.... . Changing the control parameter r of box C results in a change in the chaotic properties of the system. For $r \to 0$, e.g., the chaotic properties of box C are progressively switched off. For $r = 0$ we return to the regular behaviour of box R. Progressively regular behaviour is also observed for $r \to 1/2$ for the majority of scattering trajectories launched at L. But we note that for $r = 1/2$ two disconnected regions exist in C not dynamically connected with the "exit channel". One of these regions consists of the upper right corner of C bordered by b, c and the first quadrant of the unit circle; the other region consists of the lower right corner of C bordered by a, b and the fourth quadrant of the unit circle. If M is launched within one of these disconnected regions, it is trivially prohibited from exiting. Its trajectory is permanently chaotic for almost any initial direction of the velocity.

A particularly important question is the exact nature of the transition to chaos at $r \sim 0$ ($r \sim 1/2$, respectively). For the box model discussed in this section, the transition to chaos is "sudden" (Bleher *et al.* (1990)). This means that at least some trajectories in box C are chaotic for any r with $0 < r < 1/2$. Box C is regular only in two cases: (i) for $r = 0$ (in which case C is identical to R), and (ii) for $r = 1/2$ (apart from the chaotic disconnected regions discussed above). No doubt this "route" to chaos is important, but rather abrupt. Other systems show more slowly developing, and thus more interesting, routes to chaos as a control parameter is varied. A particularly important route to chaos, the *period doubling route to chaos* is discussed in the following section.

1.2 The transition to chaos

The period doubling route to chaos is best illustrated with the help of the *logistic map*

$$x_{n+1} = f_r(x_n) = rx_n(1 - x_n). \tag{1.2.1}$$

This is a difference equation widely used as a model in ecology and population dynamics (May (1974, 1987), Gleick (1987), Devaney (1992), Ott (1993)). Let x_n be the (normalized) number of individuals of some biological species present in year n. Then, the prescription (1.2.1) predicts the number of individuals in the following year $n + 1$. The logistic map (1.2.1) has a long history and served as a model for chaos for the past two decades. For $r = 4$ this mapping provided one of the earliest pseudo-random-number generators (Ulam and von Neumann (1947)) for applications in game theory and statistical modelling. The logistic map is a prime example of the most valuable lesson learned from chaos research: the possibility of complex and irregular behaviour in the simplest dynamical systems. Unlike in the box example discussed in Section 1.1, chaos appears at some nonzero value r^* of the control parameter r in (1.2.1), and the route to chaos is interesting in its own right. We discuss the logistic map (1.2.1) in order to introduce some technical terms frequently used in connection with chaotic systems. The logistic map is also an excellent example to introduce some graphical techniques generally used to illustrate and to exhibit chaos in dynamical systems.

Figs. 1.7(a) – (c) show the first 100 iterates of (1.2.1) for $x_0 = 0.1$ and $r = 1.5, 3.3$ and 4.0, respectively. Fig. 1.7(a) shows that for $r = 1.5$ the iterates of x_0 (marked by the plot symbols) quickly converge to the asymptotic value $x_\infty = 1/3$. Except for some special cases, this asymptotic value is independent of the value of x_0. Fig. 1.7(b) ($r = 3.3$) shows that (1.2.1) is capable of more complicated behaviour. After a short transient,

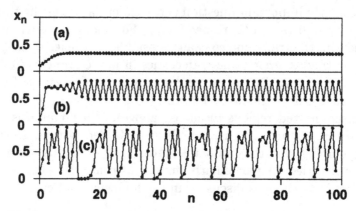

Fig. 1.7. The first 100 iterates (plot symbols) of the logistic mapping for $x_0 = 0.1$ and (a) $r = 1.5$, (b) $r = 3.3$, (c) $r = 4.0$. The full lines between the plot symbols are drawn to guide the eye.

the iterates of x_0 settle into a stationary pattern, jumping alternately between the two values 0.4794... and 0.8236.... Cyclic behaviour of the type shown in Fig. 1.7(b) actually does occur in nature. An example is the $10 - 11$ year hare-lynx cycle which can be inferred from the sales records of the Hudson Bay Trading Company (May (1980)). Cyclic behaviour of populations can also be seen in single-species laboratory-grown biological systems (May (1986)), which are more accurately modelled by a single logistic equation of the type (1.2.1) than are predator-prey systems which require two coupled equations (Beddington *et al.* (1975)).

Yet a third type of behaviour is displayed in Fig. 1.7(c). The iterates of x_0 never settle into any pattern but keep jumping irregularly in the interval $[0, 1]$. Thus, depending on the value of the control parameter r, the iterates of (1.2.1) can display three qualitatively different kinds of asymptotic behaviour: (i) convergent, (ii) cyclic, and (iii) chaotic.

In many applications, and especially in population dynamics, one is not so much interested in the transient behaviour of (1.2.1), but rather in its asymptotic behaviour for $n \to \infty$. Of special interest is the question whether the population described by (1.2.1) will settle down to some constant value $x_\infty(r)$ for $n \to \infty$, and here especially whether x_∞ is finite or zero. In order for the asymptotic value of x_n to be more significant, the result x_∞ should be independent of the starting value x_0 of the population. This is possible if x_∞ is an *attractor* of initial values $0 < x_0 < 1$. For $r < r^* = 3.5699...$ this is indeed the case and $x_\infty(r)$ is defined independently of x_0. But, as we saw above, the function $x_\infty(r)$ is not always unique. In order to capture even cyclic asymptotic behaviour, the following procedure yields an excellent representation of $x_\infty(r)$ in a single diagram. For a given value of r, one starts the iteration (1.2.1) with an arbitrary value $0 < x_0 < 1$. Then, one applies the mapping (1.2.1) for several hundred iterations, say, in order to eliminate any transient behaviour. Following this, one iterates the mapping for another, say, 100 iterations and plots the iterates thus obtained in a diagram versus the control parameter r. Fig. 1.8 shows $x_\infty(r)$ obtained using this procedure in the interval $0 < r < 4$.

More precisely, Fig. 1.8 was created in the following way. First the continuous interval $0 < r < 4$ of control parameters r was broken up into a discrete set of 1000 r values $r_j = 0.004 \times j, j = 1, ..., 1000$. For every one of the 1000 r values, the mapping (1.2.1) was started with $x_0 = 1/2$ and iterated 200 times to allow the iterates x_n to settle down. Following this step, the next 30 iterates of (1.2.1), $x_{201}, ..., x_{230}$, were marked by a dot in Fig. 1.8. Given the simplicity of (1.2.1) the resulting graph of $x_\infty(r)$ is bewildering. A *period doubling cascade* is followed by random, chaotic behaviour which starts at $r^* = 3.5699...$. The behaviour of the logistic map can be understood in detail. A thorough discussion of the logistic map is presented in Chapter 2. In this section we limit ourselves to a more "phenomenological" approach, discussing the most important features of $x_\infty(r)$ on an elementary level.

It is not difficult to understand the behaviour of $x_\infty(r)$ for $r < r^*$. The key to the structure of $x_\infty(r)$ is a careful analysis of the fixed points of the mapping function f_r in (1.2.1) as well as its iterates $f_r^{[2]}(x) = f_r(f_r(x))$, $f_r^{[3]}(x) = f_r(f_r(f_r(x)))$, etc. The mapping function f_r has two fixed points: $x_1^{(1)} = 0$ and $x_2^{(1)} = (r-1)/r$. A fixed point of a function is stable, i.e. *attracting*, if the modulus of the derivative of the mapping function evaluated at the fixed point is smaller than 1, and it is *repelling*

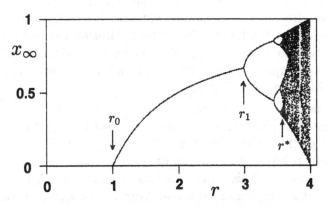

Fig. 1.8.   Asymptotic behaviour of the iterates of the logistic map.

if the modulus is larger than 1. The derivative of $f_r(x)$ is given by

$$\frac{d}{dx} f_r(x) = f_r'(x) = r(1 - 2x). \qquad (1.2.2)$$

For $0 < r < 1$ we have $|f_r'(x_1^{(1)})| = r < 1$, whereas $|f_r'(x_2^{(1)})| = 2 - r > 1$. Therefore, $x_1^{(1)}$ is attracting in $0 < r < 1$ whereas $x_2^{(1)}$ is repelling. This accounts for the behaviour of the graph of $x_\infty(r)$ in the interval $0 < r < 1$. At $r = r_0 = 1$, the stability properties of $x_1^{(1)}$ and $x_2^{(1)}$ exchange, and $x_2^{(1)}$ is attracting whereas $x_1^{(1)}$ is repelling. This accounts for the sharp corner in the graph of $x_\infty(r)$ at $r = r_0$. For $r > r_1 = 3$, neither of the two fixed points of f_r is stable. In order to understand the period doubling at $r = r_1$, it is necessary to consider the fixed points (and their stability properties) of the second iterate of f_r. The second iterate $f_r^{[2]}(x)$ is a quartic polynomial in x. Therefore, the fixed point equation $x_j^{(2)} = f_r^{[2]}(x_j^{(2)})$ has four solutions given by

$$x_1^{(2)} = 0, \quad x_2^{(2)} = (r-1)/r,$$

$$x_3^{(2)} = \frac{1}{2r} \left\{ r + 1 + \sqrt{(r+1)(r-3)} \right\},$$

$$x_4^{(2)} = \frac{1}{2r} \left\{ r + 1 - \sqrt{(r+1)(r-3)} \right\}. \qquad (1.2.3)$$

Since both $x_1^{(1)} = x_1^{(2)}$ and $x_2^{(1)} = x_2^{(2)}$ are unstable for $r > r_1$, the two solutions $x_3^{(2)}$ and $x_4^{(2)}$ take over for $r > r_1$ with consecutive iterates alternating between $x_3^{(2)}$ and $x_4^{(2)}$. This explains the doubling behaviour of x_∞ at $r = r_1$. Since there is only one branch of x_∞ for $r < r_1$, but two branches of x_∞ for $r > r_1$, the behaviour of x_∞ at $r = r_1$ is called a *bifurcation*. Because of its shape, which is due to the square roots in (1.2.3), it is called a *pitchfork bifurcation*. Bifurcation scenarios closely resembling the one shown in Fig. 1.8 can be observed in many physical systems. An example is the behaviour of a damped driven compass needle studied by Meissner and Schmidt (1986).

Two further bifurcations occur at $r = r_2$ where both $x_3^{(2)}$ and $x_4^{(2)}$ become unstable. The third iterate of f_r has to be consulted for an explanation of this behaviour. Since the third iterate is a polynomial of eighth degree, it has eight fixed points, four of which are identical to the fixed points of $f^{[2]}$. This leaves four fixed points which are visited by the iterates of x_n explaining the four branches of $x_\infty(r)$ in the interval $r_2 < r < r_3$.

A pattern emerges which explains the period doubling cascade seen in Fig. 1.8. At $r = r_k, k = 1, 2, ...$, the 2^{k-1} attracting fixed points of $f_r^{[k]}$ become unstable and $f_r^{[k+1]}$ has to be considered. Of the 2^{k+1} fixed points of this function, 2^k fixed points are identical to the 2^k unstable fixed points of $f_r^{[k]}$. This leaves 2^k attracting fixed points which are visited one by one by the iterates of $x_\infty(r)$. As a net result, the number of branches of $x_\infty(r)$ doubles whenever the control parameter r crosses one of the bifurcation values r_k.

As shown in Fig. 1.8, the bifurcation points r_k occur more and more frequently as $k \to \infty$. It was Feigenbaum's great discovery (Feigenbaum (1978, 1979)) that the sequence of bifurcation points r_k approaches a geometric sequence for $k \to \infty$ such that

$$\delta_k = \frac{r_k - r_{k-1}}{r_{k+1} - r_k} \tag{1.2.4}$$

converges to a limit δ for $k \to \infty$ called the *Feigenbaum number*. The Feigenbaum number is given by

$$\delta = \lim_{k\to\infty} \delta_k = 4.66920.... \tag{1.2.5}$$

It is *universal* for a large class of period doubling scenarios. Physical examples of this route to chaos include the driven pendulum (Baker and Gollub (1990)) and ion traps (Blümel (1995b)).

The bifurcation points r_k rapidly converge to $r_\infty = r^*$. Since the length of the cycles of x_n values in $r_k < r < r_{k+1}$ is equal to 2^k, the length of the cycle at r^* is infinite, i.e. it is not a cycle at all. At $r = r^*$ periodicity gives way to chaos. Thus, the approach to chaos in the logistic mapping is by means of a period doubling scenario whose end point (at finite $r = r^*$) is chaotic.

In addition to examination of individual trajectories, chaotic or not (Fig. 1.7), or examination of the long-time asymptotics which give rise to the bifurcation diagram of Fig. 1.8, the logistic map provides an opportunity to observe yet another hallmark of chaotic behaviour: extreme sensitivity to initial conditions. If we were to use (1.2.1) actually to model a population of, say, rabbits, following verification of the utility of the functional form itself, we would attempt to estimate both an initial population, x_0, and the growth rate, r. Since both are subject to experimental uncertainty, they will both be inexact. To what extent will this uncertainty affect our ability to predict the long-term solutions of the deterministic logistic dynamics? Fig. 1.9 indicates that there are regions in the (x_0, r) space where long-term predictions cannot be made.

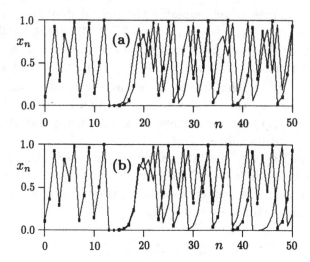

Fig. 1.9. Sensitivity of the logistic map to (a) initial conditions and (b) perturbation of the mapping itself.

In Fig. 1.9(a) we investigate the sensitivity of the logistic mapping (1.2.1) to the initial condition x_0 for $r = 4$. Just like in Fig. 1.7(c), we choose $x_0 = 0.1$ and obtain the first 50 iterates of x_0 marked with a full square in Fig. 1.9(a). We call these iterates the *reference trajectory*. Changing x_0 to $x_0 - 10^{-6}$, we obtain a trajectory which is very close to the reference trajectory for the first 15 iterates, but then departs significantly from the reference trajectory. Given the minute change in the initial condition ($\Delta x_0 = 10^{-6}$), the early departure of the perturbed trajectory from the reference trajectory is surprising. It is a direct expression of the fact that the logistic mapping, for $r = 4$, reacts exponentially sensitive to changes in the initial condition.

In some parameter regions the logistic mapping (1.2.1) is also exponentially sensitive to changes in the control parameter r. Changing r is equivalent to a perturbation of the mapping itself. Using the same reference trajectory as in Fig. 1.9(a), we change $r = 4$ to $r = 4 - 10^{-6}$. Fig. 1.9(b) shows that departures from the reference trajectory essentially occur as early as they do in Fig. 1.9(a). Thus, the logistic mapping is also sensitive in the control parameter r.

The fact that some, although clearly not all, regions of (x_0, r) parameter space indicate extraordinary sensitivity to initial conditions clearly precludes many types of prediction: taking the example of Lorenz (1963), prediction of the time and location of an individual tornado is not possible, even though meteorological models exist which successfully predict the onset of tornado season.

The origin of chaos in the logistic mapping (1.2.1) is the nonlinear term $\sim x^2$. Without this term (1.2.1) would read $x_{n+1} = rx_n$ and would have the simple analytical solution $x_n = r^n x_0$. In the presence of the nonlinearity in (1.2.1), explicit analytical solutions $x_n = P_{2^n}(r, x)$ exist for any finite n. The functions $P_{2^n}(r, x)$ are simple polynomials of degree 2^n in x with a control parameter r. There is, however, one difficulty with these analytical solutions: the length of the polynomials P_{2^n}, i.e. the number of terms to be written down when stating P_{2^n} explicitly, grows exponentially. This situation is not too bad for $r < r^*$ where the iterates x_n of (1.2.1) quickly converge to stable cycles. In chaotic regimes of (1.2.1), however, no such convergence to stable cycles occurs and high order polynomials $P_{2^n}(r, x)$ are required for the analytical prediction of x_n. Therefore, in the chaotic regime, the analytical solutions P_{2^n} are only of limited value for a prediction of the iterates x_n of x_0. The seriousness of this observation becomes clear with the following example. Suppose we are in the chaotic regime and we would like to write down the analytical prediction for x_{40} on a piece of paper. Even if we allocate only a space of 1 mm per term in P_{2^n}, the length of our formula would exceed the distance between the Earth and the Moon. Compare this with x_{200}, which was needed above to produce Fig. 1.8!

A note of caution, however, is in order now. While it is indeed true that for a generic r in the chaotic regime the length of an analytical formula for the prediction of x_n from x_0 grows exponentially in n, exceptions do occur at special values of r. For $r = 4$, e.g., a clever argument by Ulam and von Neumann (1947) shows that the iterates x_n of (1.2.1) can be written explicitly as

$$x_n = \sin^2[2^n \arcsin \sqrt{x_0}\,]. \qquad (1.2.6)$$

This, however, does not mean that we would not have to work "exponentially hard" actually to obtain x_n if an explicit numerical value is required. This is so because of the occurrence of the factor 2^n in (1.2.6) in conjunction with the trigonometric function $\sin^2$ which is π periodic. The periodicity of the $\sin^2$ function, which effectively truncates the leading decimal figures in its argument, requires that we have to keep exponentially many decimal digits in x_0 in order to produce the leading decimal figures of x_n. More about this mechanism that is at the heart of chaos can be found in Section 2.2.

In general one can prove that for a chaotic system no analytical "short cut" is possible. In other words: if we would like to predict the results of a chaotic system, the fastest (and in practice the *only*) way is simply to watch the system evolve. This has been expressed concisely by Ford (1989): a chaotic system is its own fastest computer.

This situation is serious. Although the logistic map (1.2.1) is a perfectly deterministic process, the complications of its dynamics seem to pose a serious limitation on our ability to predict the values of its iterates. Now, one could argue that this is not so bad. After all, (1.2.1) is only a "mathematical" construct which may have some peculiar properties that never occur in practice and therefore are not relevant for physical systems. But the contrary is true. As we saw in Section 1.1, the simplest physical systems can exhibit chaotic solutions. Moreover, we will see in the following chapters that physical systems can sometimes be mapped on simple mathematical systems such as the logistic equation (1.2.1). Thus, physical systems are generally as chaotic as certain formal mathematical mappings exemplified by the logistic equation. As a consequence, due to the presence of chaos, physics, supposed to be a "predictive" science, is now seen to have much less predictive power, at least as far as individual trajectories are concerned, than is generally acknowledged. Patterns, however, such as the Red Spot on Jupiter's surface, may be predicted and survive in the midst of chaos. For individual trajectories, however, the problem is a problem of principle and cannot be solved, e.g., with more powerful super-computers, or with advances in mathematical technique. The principal limitation of predictability, even in the face of perfectly deterministic dynamics, will be discussed briefly in the following section.

1.3 Determinism and predictability

Newtonian mechanics is *deterministic* as is the dynamics of the logistic equation of the previous section. This means that the time evolution of a classical system is unique once the initial conditions are specified. The reason for classical determinism is simple. Newton's equations of motion are formulated mathematically with the help of a set of ordinary differential equations. Under suitable conditions, which are practically always met in physical applications, it is possible to prove rigorously that the solutions of Newton's equations exist and are unique. The existence and uniqueness property of trajectories according to Newtonian mechanics was expressed in the post-Newtonian era as the principle of *determinism*, which eventually became one of the cornerstones of 18th and 19th century scientific philosophy. It was thought that the mechanical universe behaved like a great clock which, once set in motion, determines the entire future of the universe in an orderly and *predictable* manner (Peterson (1993)). This philosophy, created by Laplace in the 18th century, was (arguably) the most important driving force behind the industrial revolution of the 18th and 19th centuries. Unfortunately, we know today that the presence of chaos in many classical mechanical systems reveals that the hope instilled by the deterministic property of Newton's equations does not carry far.

Determinism, a formal mathematical property of Newton's equations, does not in general guarantee *predictability*, a much more useful physical concept. The "impotence" of determinism was realized by the eminent French mathematician Henri Poincaré in his epochal work on "New Methods in Celestial Mechanics" (Poincaré (1892, 1993)), where he pointed to the possibility of exponential instabilities in the dynamical equations of motion which are in fact the root of chaos. This aspect of Poincaré's work and its implications for physical predictability were discussed and reviewed in an excellent article by Brillouin (1960). Despite Poincaré's great insight, his ideas did not immediately catch on. In the early part of the 20th century only a few mathematicians occupied themselves seriously with questions of dynamical systems theory. It was not until the mid-1970s that the theory of chaos really took off. There are essentially two reasons for this delay. First of all, it is very difficult to imagine mentally the enormous complexity that chaos entails. Therefore, chaos became a "popular" subject only with the enormous progress in computer visualization techniques developed during the past two decades. Physics and mathematics would be very different today if Poincaré had had a PC! The second reason for the initial low impact of Poincaré's work was the advent of quantum mechanics. Quantum mechanics is such a successful theory that the focus of attention quickly shifted away from (nonlinear) dynamical systems theory to the solution of (linear) quantum problems in atomic, molecular, nuclear and solid state physics.

Quantum mechanics, of course, raises its own issues of determinism and predictability. The problem is not so much with the "machinery" of quantum mechanics, which produces correct results whenever applied to microscopic systems. The problem is with the interpretation of quantum mechanics. Born's *statistical interpretation* turned out to be the most successful interpretation so far, but it spelled the end of Laplacian determinism.

Some of the greatest physicists of all time, Einstein, de Broglie and Schrödinger amongst them, never did like this situation. "God does not play dice" was Einstein's famous remark (see, e.g., Pais (1991), p. 425). In order to save the idea of determinism, de Broglie searched for "hidden parameters" and Schrödinger studied the time evolution of wave packets.

But is determinism worth saving? Is it an important fundamental concept directly extracted from nature itself?

In 1955, Born wrote an article entitled "Is classical mechanics in fact deterministic?" (reprinted in Born (1969)). In this article Born tried to expose the emptiness of the concept of determinism. His argument is the following: If determinism is not a useful concept to begin with, an attempt to "save" determinism is meaningless from the outset. Investigating a system closely related to the box model discussed in Section 1.1, Born shows that determinism is not a concept extracted directly from observations. In his opinion determinism is an empty mathematical construct which has no correspondence in reality. In order to make his point as simple and direct as possible, Born considers the motion of a mass point on the x axis in the unit interval. The coordinate of the mass point is denoted by x, its velocity by v. There are no forces acting on the mass point in $0 < x < 1$, but the mass point is reflected elastically from the end points of the interval at $x = 0$ and $x = 1$. As a consequence, $|v|$ is constant in time. This system is obviously deterministic. Whenever the initial conditions $x_0 = x(t = 0)$ and $v_0 = v(t = 0)$ are known precisely, $x(t)$ can be computed trivially for any $t > 0$. The answer is

$$x(t) \; = \; v_0 t + x_0 \;\; \text{mod} \;\; 1. \tag{1.3.1}$$

But now Born argues that *precise* knowledge of x_0 or v_0 is not physical. No matter how good a measurement, there are always experimental uncertainties Δx_0 and Δv_0 in the initial position and the velocity that cannot be eliminated *in principle*. Even assuming no error in x_0, any uncertainty in v_0 turns out to be disasterous for the long-term predictability of $x(t)$ since there is always a time $t = t_c$ for which $\Delta x = \Delta v_0 t$ becomes larger than 1. From then on, the position of the mass point in the interval $[0, 1]$ is completely uncertain and unpredictable for $t > t_c$. The critical time t_c is given by $t_c = 1/\Delta v_0$. We point out that Born's example does not even show exponential instabilities. Neither is it a chaotic system in the technical sense. What the example does show is that the determinism contained in the equations of motion of the mass point is of practical relevance only over a finite interval of time, $t < t_c$, and irrelevant as far as the long-time properties of the system are concerned.

At this point Born argues that 20th century physics has celebrated its greatest triumphs by eliminating all concepts that are of no relevance for the observed phenomena. He points out that elimination of the concept of absolute space and time leads to the theory of relativity, and elimination of the concept of the possibility of simultaneous accurate measurement of conjugate dynamical variables leads to quantum mechanics. He calls this process of elimination the "heuristic principle of physics". Following the heuristic principle, he demands the elimination of the concept of determinism as a physical concept even in classical mechanics and wonders what it might lead up to.

In this context there are other concepts which should be discussed in some detail. There is, e.g., the concept of "specifying the initial conditions" of a dynamical system. This concept is problematic even from a purely mathematical point of view and has a lot to do with the properties of the number continuum. It was pointed out by the eminent mathematician Kaz that most of the real numbers in the unit interval $[0, 1]$ cannot be defined with a finite number of words (see, e.g., Schroeder (1991), p. 161). Specifying an initial condition with an infinite number of words or prescriptions again makes no physical sense. Accordingly, most numbers in $[0, 1]$ do not qualify as physical initial conditions which could reasonably be *specified*. What happens if a chaotic mechanical system starts with such an unspecifiable initial condition? We will see in Chapter 2 that the resulting motion of the system, although deterministic, is then *random*. Fig. 1.9 hints at the complexity which can follow. The existence of *deterministic randomness* is the ultimate reason for the apparent chaotic behaviour of the solutions of certain deterministic systems or processes, and resulted in the name *deterministic chaos* for this phenomenon. Therefore, surprisingly, determinism and randomness are not mutually exclusive concepts. Again, it was recent developments in nonlinear dynamics that attracted attention to this basic fact. It is clear that the presence of deterministic randomness poses severe restrictions on our ability to predict the time evolution of dynamical systems.

So what is the bottom line? Is prediction possible or is it not? Born himself notes that the problem of prediction is an exercise in double limits. In the case of Born's example, e.g., we are dealing with two variables, t and Δv_0, one of which tends to infinity, the other one to zero, respectively. The result (prediction possible or not) is unspecified without further information on how this limit is to be performed. If we resolve, e.g., that we want to predict only over a fixed time interval $0 < t < t^*$, then, obviously, reduction of the experimental error Δv_0 in the velocity v eventually results in $t_c < t^*$ and meaningful prediction is indeed possible. On the other hand, for fixed Δv_0, no prediction is possible for $t > t_c$.

The question of predictability within the deterministic structure of classical mechanics was clearly appreciated by many eminent researchers in nonlinear systems theory and theoretical physics (see, e.g., Brillouin (1960)). Borel (1914) adds an additional twist to the predictability discussion. He argues that the displacement of a lump of matter with mass on the order of $1\,\mathrm{g}$ by as little as $1\,\mathrm{cm}$ and as far away as, e.g., the star Sirius is enough to preclude any prediction of the motion of the molecules of a volume of a "classical" gas for any longer than a fraction of a second, even if the initial conditions of the gas molecules are known with mathematical precision. Borel's example shows that many physical systems are not only sensitive to initial conditions, but also to miniscule changes in system parameters. The sensitivity to system parameters is a fundamental additional handicap for accurate long-time predictions. In the face of Borel's example, Brillouin (1960) points out that the prediction of the motion of gas molecules is not only "very difficult", as pointed

out in some standard books on statistical mechanics, but strictly impossible given the fact that we cannot control the physical processes on the
distant stars, which nevertheless have a serious, non-negligible effect on
earthbound mechanical systems.

Although the above examples already provide strong arguments against
classical mechanics' presumed ability to predict, the box models discussed
in Section 1.1 add new aspects to the discussion. First of all we note that
for fixed $\varphi_0 < \pi/2$ the results of box R are strictly predictable in any
interval $0 < \varphi < \varphi_0$. This is so because the trajectories used to calculate
y^R all terminate within a finite time $t < t^* < \infty$, where t^* depends only on
φ_0. Therefore, pushing the experimental uncertainty in the initial angle
φ below a certain value determined by t^* always results in the possibility
of meaningful predictions.

The situation is quite different for box C. Suppose the initial angle φ
has an experimental uncertainty $\Delta\varphi$, i.e. the initial angle is only known
to lie between $\varphi - \Delta\varphi/2$ and $\varphi + \Delta\varphi/2$. A reasonable prediction for
the outcome y^C is only possible if for some sufficiently small $\Delta\varphi$ the
resulting values of the reaction function can be guaranteed to lie within a
y interval which is smaller than the unit interval $[0, 1]$. This is so because
the prediction "y^C is somewhere between 0 and 1" is obviously a pretty
safe bet and no meaningful prediction at all.

Armed with this definition of prediction one can now investigate whether the reaction function y^C can be predicted. The answer is negative. The
reaction function y^C cannot be predicted in general because, as shown in
Fig. 1.5, there are always φ intervals in which y^C displays structure on all
scales. In more concrete terms this means that for box C, and no matter
how small the error $\Delta\varphi$, there always exist φ intervals of length $\Delta\varphi$ in
which y^C can attain any value between 0 and 1. This means additionally
that the reaction function y^C is not experimentally resolvable.

The chaotic behaviour of box C shows that questions of measurement
theory and the concept of predictability are not just at the foundations
of quantum mechanics, but enter in an equally profound way already
on the classical level. This was recently emphasized by Sommerer and
Ott in an article by Naeye (1994). They argue that in addition to the
problem of predictability the problem of reproducability of measurements
in classically chaotic systems has to be discussed. The results of Fig. 1.9
indicate that the logistic map displays similar complexity. In fact, regions
which act sensitively to initial conditions, intertwined with regions where
prediction *is* possible, are generic in classical particle dynamics.

In conclusion, classical mechanics is not at all the safe ground of physics
for which it is sometimes portrayed. Because of the existence of chaos,
concepts long held to be part of the firm foundations of classical mechanics
are currently under serious discussion. As a first result, a long-cherished

notion of classical mechanics is gone forever: Determinism does not imply predictability.

1.4 Chaos and quantum mechanics

Following the turbulent developments in classical chaos theory the natural question to ask is whether chaos can occur in quantum mechanics as well. If there is chaos in quantum mechanics, how does one look for it and how does it manifest itself? In order to answer this question, we first have to realize that quantum mechanics comes in two layers. There is the statistical clicking of detectors, and there is Schrödinger's probability amplitude ψ whose absolute value squared gives the probability of occurrence of detector clicks. From all we know, the clicks occur in a purely random fashion. There simply is no dynamical theory according to which the occurrence of detector clicks can be predicted. This is the nondeterministic element of quantum mechanics so fiercely criticized by some of the most eminent physicists (see Section 1.3 above). The probability amplitude ψ is the deterministic element of quantum mechanics. Therefore it is on the level of the wave function ψ and its time evolution that we have to search for *quantum deterministic chaos* which might be the analogue of classical deterministic chaos.

The wave function ψ satisfies the partial differential equation

$$i\hbar \frac{\partial \psi}{\partial t} = \hat{H}\,\psi. \qquad (1.4.1)$$

This equation is Schrödinger's wave equation, where $\hbar$ is Planck's constant and $\hat{H}$ is the Hamiltonian of the system to be investigated. The Schrödinger equation is a deterministic wave equation. This means that once $\psi(t=0)$ is given, $\psi(t)$ can be calculated uniquely. From a conceptual point of view the situation is now completely analogous with classical mechanics, where chaos occurs in the deterministic equations of motion. If there is any deterministic quantum chaos, it must be found in the wave function ψ.

Let us first look for chaos in a bounded quantum system that does not depend explicitly on the time t. In order to solve the Schrödinger equation (1.4.1) one first solves for the spectrum of $\hat{H}$:

$$\hat{H}\,\varphi_n = E_n\,\varphi_n. \qquad (1.4.2)$$

Since according to assumption the system is bounded, the index n is discrete and can be assumed to run through the positive integers. The total wave function ψ is then given by

$$\psi = \sum_{n=1}^{\infty} A_n \exp(-iE_n t/\hbar)\,\varphi_n. \qquad (1.4.3)$$

The expansion amplitudes A_n in (1.4.3) have to be determined such that at time $t = 0$ the wave function fulfils the starting condition:

$$\psi(t = 0) = \sum_{n=1}^{\infty} A_n \varphi_n. \qquad (1.4.4)$$

We can always assume that the wave functions φ_n (the basis functions) are normalized:

$$\langle \varphi_n \mid \varphi_n \rangle = 1. \qquad (1.4.5)$$

Orthogonality follows automatically from the hermiticity of $\hat{H}$. We have

$$\langle \psi(t) \mid \psi(t) \rangle = \sum_{n=1}^{\infty} \mid A_n \mid^2 = 1. \qquad (1.4.6)$$

This shows that the expansion coefficients A_n are normalizable.

But how do we decide whether $\psi(t)$ contains, or does not contain, chaos? While classical chaos can be defined rigorously in mathematical terms (see Section 2.2 and Devaney (1992)), there is as yet no generally accepted definition of quantum chaos. The best we can do at this point is to develop and apply certain "tests" for quantum chaos which are constructed with the intention of revealing complicated behaviour of $\psi(t)$ reminiscent of classical chaos.

A popular test for the presence of deterministic chaos in $\psi(t)$ is to examine auto-correlation functions of $\psi(t)$, for instance,

$$\theta(t) = \langle \psi(0) \mid \psi(t) \rangle. \qquad (1.4.7)$$

If $\psi(t)$ shows regular behaviour, for instance periodicities, then $\theta(t)$ shows an oscillating behaviour. If $\theta(t)$ decays, this would be an indication that $\psi(t)$ develops complicated structures by looking less and less like its starting configuration. For the wave function (1.4.3) it is easy to calculate $\theta(t)$. The result is:

$$\theta(t) = \langle \psi(t = 0) \mid \psi(t) \rangle = \sum_{n=1}^{\infty} \mid A_n \mid^2 \exp(-iE_n t/\hbar). \qquad (1.4.8)$$

It is seen that $\theta(t)$ has a very simple form. It is the sum over a countable set of periodic functions with coefficients that vanish rapidly enough so that the sum over the coefficients is 1. Sums of this type have been studied intensively in the mathematical literature. Functions such as $\theta(t)$ are called *quasi-periodic* and it is known that they behave in a very orderly way which is reminiscent of simple periodic motion.

So, as a result, if there is any chaos in $\psi(t)$, the auto-correlation test does not show it. We would like to point out, however, that in the semi-classical limit, i.e. the limit where Planck's constant $\hbar$ is small compared with typical actions in the system, $\psi(t)$ can mimic a chaotic time evolution over some time interval $0 < t < t^*$, where t^* is known as the "break time". In this case, transiently, the correlation function $\theta(t)$ decays in the time interval $t < t^*$, but shows recurrent behaviour for $t > t^*$. An example of such a transient decay, indicative of "transiently quantum chaotic" behaviour, was presented, e.g., by Pomeau *et al.* (1986). Mathematically we are dealing here with a double limit $\hbar \to 0$, $t \to \infty$. For $\hbar$ sufficiently small, the break time t^* can be made arbitrarily large, resulting in an arbitrarily long "quantum chaotic" regime.

Another test for the presence of quantum chaos is to investigate the power spectrum of the auto-correlation function $\theta(t)$. It is immediately obvious from (1.4.8) that the power spectrum of a bounded system is a countable sum of δ functions. Since no chaos is present in (1.4.8), one may conjecture that for chaos to be present it is necessary that the power spectrum of $\theta(t)$ contains a continuous component. Other tests for quantum chaos are to compute the power spectrum of expectation values of dynamical variables, for instance the position $x(t) \equiv \langle\psi(t)|\hat{x}|\psi(t)\rangle$. Again, a purely discrete power spectrum indicates regular time evolution of ψ, whereas a continuous component indicates the presence of chaos. Obviously, a continuous component in the power spectrum is only a *necessary* condition for chaos. It is not sufficient since unbounded systems, for instance scattering systems, may show a continuous component without any sign of chaos in their dynamics. This state of affairs was known more than 20 years ago (Lebowitz and Penrose (1973), Hogg and Huberman (1982), Wunner (1989)).

In the absence of deterministic chaos in the time evolution of the wave functions of bounded systems, the focus of quantum chaos research shifted towards the identification of the *fingerprints* of classical chaos in the properties of ψ. The usual procedure is to start with a classically chaotic system, quantize it canonically, and then try to identify those characteristics of ψ in the semiclassical limit ($\hbar \to 0$) that give away the chaoticity of the underlying classically chaotic system.

Applied to the example of a bounded quantum system discussed above, one might choose to examine directly the properties of the stationary basis functions φ_n. This could, e.g., be done by investigating the coarse grained values of the wave functions in the position representation. Assuming that the quantum system is defined in the two-dimensional x, y space, one investigates the functions

$$\varphi_n(x, y) = |\langle x, y \mid \varphi_n \rangle| \qquad (1.4.9)$$

on a two-dimensional grid defined by

$$\varphi_n^{jk} = \varphi_n(x_j, y_k), \quad j = 1, 2, ..., N; \quad k = 1, 2, ..., M. \qquad (1.4.10)$$

The numbers φ_n^{jk} can then be investigated statistically. For instance, one may calculate the probability distribution $P_n(\varphi)$, i.e. the probability that one encounters a particular function value φ if one randomly picks a mesh point (j, k). Or one may look at correlations between φ_n^{jk} and its neighbours $\varphi_n^{j'k'}$. Additionally, classical phase-space information can be projected directly from quantum wave functions. Thus, classical and quantum dynamics can be compared on an equal footing. All these questions are still under active investigation.

But the most promising lead in quantum chaos research is based on the following thought. According to Heisenberg the operators of quantum mechanics can be represented by matrices. It is then natural to conjecture that the observables of a classically chaotic system should be represented by quasi-random matrices, matrices whose elements are quasi-random numbers. Beginning with the work of Wigner in the 1950s (see, e.g., Mehta (1991)), the theory of random matrices was already highly developed at the beginning of the 1960s (see, e.g., Porter (1965)). Many properties of random matrix ensembles are known by now (Mehta (1991)), and the matrices obtained from classically chaotic quantum systems can easily be checked for their "random" properties. Starting with the seminal work by Bohigas *et al.* in 1984, tests of this kind were indeed performed for many classically chaotic quantum systems. The results are promising. Except for some very special systems, the fingerprints of classically chaotic motion are clearly present in the statistical properties of matrix elements and spectra of quantum operators. Thus, random matrix theory turns out to be a powerful tool for the description and classification of atomic and molecular systems. The discussion of the fingerprints of classical chaos in atomic systems occupies a major part of this book.

2

Chaos: tools and concepts

In Chapter 1 we discussed some concepts of chaos, its manifestations and applications on an introductory level from a purely qualitative point of view. The concepts were introduced *ad hoc* and in a pictorial manner. We will now turn to a more detailed investigation of chaos in order to prepare the tools and concepts needed for the discussion of chaotic atomic and molecular systems.

For a long time researchers thought that every given nonlinear system required its own individual method of solution. If this were the case, there could not be any general theory of nonlinear systems. Rather, the science of nonlinear systems would resemble descriptive sciences such as 19th century biology or geology. The best one could offer would be a catalogue of nonlinear systems together with their individual properties and methods of solution (if any). Luckily, the situation is much more promising. Not so long ago it was shown by Feigenbaum (1978, 1979) that there is *universality* in chaos. Universality is the key property of chaos. Universality means that all nonlinear (chaotic) systems can be analysed using a common set of methods and tools. Thus, a given nonlinear system does not require special treatment. It is always amenable to a general analysis, whose elements are discussed in the following sections.

It was Poincaré who introduced a major revolution in the analysis of dynamical systems in classical mechanics. Instead of focussing on the time evolution of individual trajectories of a system, he emphasized the *global* point of view (Poincaré (1892, 1993)). Poincaré argued that it was much more important to know about the *qualitative* behaviour of the solutions of a given dynamical system in different parts of the classical phase space, than to know the detailed time evolution of a special solution. Poincaré developed this point of view by analysing the age old question of whether the solar system is stable. Clearly the answer to this question is a qualitative statement about a property of the solar system as a whole, rather than a statement about the details of the time evolution of individual planets. In the course of his investigation, Poincaré developed the method of *phase-space diagrams* or *phase-space portraits* for the qualitative assessment of the properties of solutions of ordinary differential equations. This method is still in use today. It is a powerful tool in the arsenal of the theory of differential equations (Braun (1975)) and dynamical systems (Tabor (1989)).

Another of Poincaré's new methods was the reduction of the continuous phase-space flow of a classical dynamical system to a discrete mapping. This is certainly one of the most useful techniques ever introduced into the theory of dynamical systems. Modern journals on nonlinear dynamics abound with graphical representations of Poincaré mappings. A quick glance into any one of these journals will attest to this fact. Because of the usefulness and the formal simplicity of mappings, this topic is introduced and discussed in Section 2.2.

The intricate structure of the set S of scattering singularities we encountered in connection with the reaction function y^C in Section 1.1 can be characterized using the concept of *fractals* introduced by Mandelbrot in 1975 (see also Mandelbrot (1977, 1983)). Fractals are discussed in Section 2.3.

A useful technique is the method of symbolic dynamics. We will encounter this topic in Section 2.4. Symbolic dynamics enables us to establish connections between dynamical systems. Often it is possible to map a given dynamical system onto an "older" one which has been studied before. If such a mapping is possible on the symbolic level, the two systems are dynamically equivalent.

A quantitative characterization of chaos is possible with the help of Lyapunov exponents. These exponents are numbers which essentially measure the degree of chaoticity of a given chaotic system. With reference to the discussion in Section 1.3, the Lyapunov exponents enable us to calculate quantitatively an estimate for the critical time t_c of a chaotic system, which limits our ability to predict the time evolution of the system for $t > t_c$.

The mathematical basis of chaos is the number continuum. The existence of deterministic randomness, e.g., a key feature of chaos, relies essentially on the properties of the number continuum. This is why we start our discussion of tools and concepts in chaos theory in the following section with a brief review of some elementary properties of the real numbers.

2.1 The number continuum

In what follows the set of natural numbers $\{1, 2, 3, ...\}$ is denoted by N, the set of integers $\{-2, -1, 0, 1, 2, ...\}$ by Z, the set of rational numbers by Q and the set of real numbers is denoted by R.

The number system
The natural numbers are *countable*. Any set of numbers that can be mapped one-to-one onto the natural numbers (or a subset thereof) is countable too. For example, Z is countable. To prove it, we define the following invertible mapping

$$f : \quad N \leftrightarrow Z$$

$$n \leftrightarrow \begin{cases} (n-1)/2, & \text{if } n \text{ is odd} \\ -n/2, & \text{if } n \text{ is even.} \end{cases} \qquad (2.1.1)$$

Cantor, the creator of set theory (Dauben (1979), Purkert and Ilgauds (1987)), proved the following result: the rational numbers are countable. His scheme is both simple and ingenious. First, we note that if the positive rational numbers can be counted, so can all the rational numbers by a scheme similar to (2.1.1). In order to count the positive rationals, Cantor used an invertible mapping from N onto Q called the *diagonal method*. This method is based on arranging the positive rational numbers in a square table as shown in Fig. 2.1. The rationals in the table are counted going diagonally through the table as indicated by the arrows in Fig. 2.1. This way, skipping all double entries in the table, every rational can be assigned a unique counting label. According to Fig. 2.1, the first few assignments are given by $1 \leftrightarrow 1$, $1/2 \leftrightarrow 2$, $2 \leftrightarrow 3$, $1/3 \leftrightarrow 4$, $3 \leftrightarrow 5$,

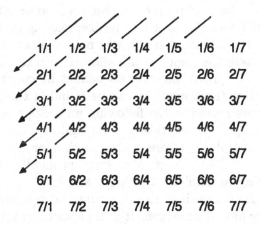

Fig. 2.1. Cantor's counting scheme for the rational numbers.

$1/4 \leftrightarrow 6$, $2/3 \leftrightarrow 7$, $3/2 \leftrightarrow 8$, $4 \leftrightarrow 9$, $1/5 \leftrightarrow 10$, $5 \leftrightarrow 11$, $1/6 \leftrightarrow 12$, $2/5 \leftrightarrow 13$ $3/4 \leftrightarrow 14$ $4/3 \leftrightarrow 15$.

There are members of the number continuum that are different from rational. The number $\sqrt{2}$, e.g., is not rational. The proof was known to the ancient Greeks: assume $\sqrt{2}$ is rational. Then, it can be written as $\sqrt{2} = n/m$ where n and m are integers and relatively prime. Squaring this relation we get $n^2 = 2m^2$. This means that n^2 contains a factor 2. But if n^2 contains a factor 2, so does n. Therefore, we write $n^2 = 4p$, where p is another integer. Thus, $m^2 = 2p$. Using the same argument as before, m contains a factor 2. But this contradicts the assumption that n and m are relatively prime. Thus we proved that $\sqrt{2}$ is not rational.

In fact, there are many more irrationals than rationals, so many that the irrationals cannot be counted. This means that a one-to-one mapping $N \leftrightarrow R$ does not exist.

The rationals are *dense* in the irrationals. This means that given any irrational number ν and any $\epsilon > 0$, there is always a rational number s that satisfies $|\nu - s| < \epsilon$. Although the rationals are dense in the *real number continuum* their measure is zero. This is easily demonstrated, e.g., by covering all rationals in $[0,1]$ by a set whose total length is arbitrarily small. The proof is simple. We use Cantor's counting scheme (see Fig. 2.1) to assign a sequence number n to every rational s in $[0,1]$. Then, we draw a line segment of length $2\epsilon^n$ from $s - \epsilon^n$ to $s + \epsilon^n$. This line segment covers s. The union of all these line segments then covers all rationals in $[0,1]$. The total length of this union set is given by

$$ l = \sum_{n=1}^{\infty} \epsilon^n = \frac{\epsilon}{1-\epsilon} \rightarrow 0, \quad \text{as } \epsilon \rightarrow 0. \qquad (2.1.2) $$

Thus we have proved that the measure of the rationals is zero.

Quite generally the measure of a countable set is zero. This does not imply that the measure of an uncountable set is automatically nonzero. This was proved by means of a counterexample, again by Cantor. He considered the set of all numbers which in base-3 notation can be written as $0.a_1a_2a_3...$, where the digits a_j are allowed to take only the values 0 and 2. It is easy to prove that this set cannot be counted (see Section 2.3). Geometrically, these numbers can be found in the interval $[0,1]$ in a set which is constructed in the following way. Start with the interval $[0,1]$ (see Fig. 2.2(a)). Now take out its middle third (corresponding to the numbers $0.1b_2b_3...$, where b_j can be 0, 1 or 2). We obtain the set shown in Fig. 2.2(b). Next, take out the middle third of what is left (corresponding to numbers $0.01b_3b_4...$ and $0.21b_3b_4...$). The result is shown in Fig. 2.2(c). Continue this procedure *ad infinitum*. What is left in the end clearly has no extension. It is of measure zero, but nevertheless uncountable. The set which is left by this "middle thirds" construction is

Cantor's *middle thirds set*. We denote it by the symbol C. It has recently attracted much attention in connection with chaotic scattering and decay processes (see Sections 1.1 above and 2.3 below, Chapter 8 and Chapter 9). Cantor's middle thirds set is also an example of a *fractal*, a concept very important in chaos theory (see Section 2.3 for more details).

The most important property that distinguishes real numbers from rational numbers is that the binary, ternary, ..., decimal, ... expansion of a rational number repeats itself periodically while the expansion of an irrational number is not periodic. This is also the most relevant property in connection with chaos.

There are a few special classes of irrational numbers. An *algebraic irrational* is a zero of a polynomial of finite degree with integer coefficients. Since $\sqrt{2}$ is a zero of $x^2 - 2 = 0$, it is certainly algebraic. Cantor proved that the set of algebraic irrationals is countable (Dunham (1990)).

Another example of an algebraic number is the *golden mean*

$$g = (\sqrt{5} - 1)/2. \tag{2.1.3}$$

Historically this number played an important role in ancient Greek geometry and in concepts of aesthetics and beauty. It is also a key number in chaos theory. In mathematics the significance of g lies in the fact that it is the "most irrational number". Its continued fraction expansion is:

$$g = \cfrac{1}{1 + \cfrac{1}{1 + \cfrac{1}{1 + \cdots}}}. \tag{2.1.4}$$

The proof is easy. From the special form of (2.1.4) we get immediately $g = 1/(1 + g)$, which yields $g^2 + g - 1 = 0$. The solutions are $g_{1,2} = -\frac{1}{2} \pm \frac{1}{2}\sqrt{5}$. Since g is positive, only the "+" sign is relevant. Thus, (2.1.3) and (2.1.4) are indeed the same.

All real numbers have a unique expansion of the form

$$a_0 + 1/(a_1 + 1/(a_2 + \cdots, \tag{2.1.5}$$

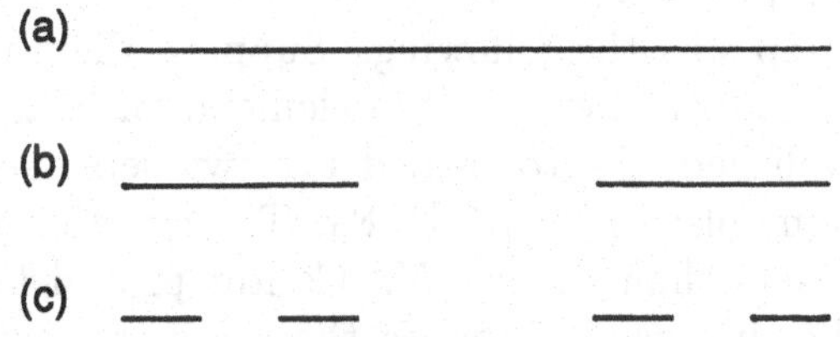

(a) ——————————————————————

(b) ————————   ————————

(c) —— ——   —— ——

Fig. 2.2. Sketch of the first three stages in the construction of Cantor's middle thirds set C.

where the a_i are natural numbers. Should the continued fraction (2.1.5) terminate, then the represented number is rational. In the continued fraction (2.1.4) representing the golden mean, all a_i are equal to 1. Thus this continued fraction has the slowest possible convergence. It is the least well approximated by rationals p/q, $p, q \in N$, for a given bound on q. This explains why g is called the "most irrational number".

Irrationals like π or e are not algebraic, i.e. they are not the zeros of any polynomial of finite degree with integer coefficients. The proof of this fact is very difficult, but was produced toward the end of the last century. Irrationals which are not algebraic are called *transcendental*. The transcendentals are not countable. The transcendentals are what remain of the real numbers after removing the rationals and the algebraic irrationals. Since the real numbers are not countable, but both rationals and algebraic irrationals are, the transcendentals are obviously not countable.

Technically it is the existence of transcendental numbers in the number continuum that makes chaos possible. We should note, however, that this does not prevent meaningful computer exploration of chaos despite the fact that computers only deal in rational approximations to algebraic and transcendental numbers (see, e.g., Hammel *et al.* (1987)). This fact is illustrated in Section 2.2, where we discuss some important examples of chaotic mappings.

The cardinality of sets
How big are sets? There are infinitely many natural numbers and there are infinitely many irrational numbers. The natural numbers can be counted, the irrational numbers cannot. Are there two different kinds of infinity? It was Cantor's genius to answer with a resounding "yes". He defined a sequence of *transfinite cardinals* to characterize the different types of infinity. Cantor first looked for the lowest type of infinity and found it in the set of natural numbers, N. This set, as everybody intuitively knows, is "infinite". This is so, because there is no end to the natural numbers. Given a number $n \in N$, one can always come up with $n+1$. which is also in N. Cantor assigned the transfinite cardinal number $\aleph_0$ to the set of integers. The *cardinality*, or "size", of a set S is denoted by $\mid S \mid$. Therefore, $\mid N \mid = \aleph_0$.

Cantor's next step was the following. Suppose there is a set S. If a one-to-one correspondence between the elements of S and the elements of N can be established, Cantor called the two sets "sets of the same cardinality". In symbols: $\mid N \mid = \mid S \mid = \aleph_0$. The question now is whether there exist larger sets than the set N. Cantor proved this in the affirmative in the following way. Suppose there are two symbols a and b. Consider the set of semi-infinite symbol sequences $S = \{abaabbbaab...,$ $aababaabbbaaba..., ababaababaab..., ...\}$ that are made up of the two symbols a and b. Is this set *countable*? If so, a one-to-one correspondence with the integers must be possible and the cardinality of S is $\aleph_0$. With the help of his "diagonal method" discussed above, Cantor showed that this is not so and that the elements of S are *not* countable. This means that no one-to-one correspondence with the elements of N can be established and that the number of elements in S necessarily must be of a higher order of infinity than $\aleph_0$. Cantor called this type of infinity $\aleph_1$. Here is the proof. Assume that the elements of S can be listed. For example:

element number 1: *aabaabbaababaa*...
element number 2: *bbabbabbaaaaba*...
element number 3: *aaabaaabaababa*...
element number 4: *bababababbbabab*...

...

Then, one can show easily that no matter how ingeniously this scheme is set up, there is always an element α of S which is not in the scheme. The construction of α is straightforward: Take as α the sequence $\alpha = babb$... which is constructed by simply flipping the symbol in the nth position of element number n in the above listing scheme and constructing the sequence α by listing the flipped symbols one by one. This way, α cannot be element number 1, since α and element number 1 differ already in the first position. It can also not be element number 2, since α and element number 2 differ in the second position, etc. In general, α cannot be element number n, since α and element number n differ in the nth position. Therefore, it cannot be any element in the listing scheme at all and the proof is complete. Thus, $\mid S \mid = \aleph_1 \neq \aleph_0$.

If we interpret a as "0" and b as "1", and if additionally we put "0." in front of every sequence, the set S is equivalent with the binary representation of all the real numbers in the unit interval. Therefore, as a corollary, we have established the fact that the real numbers in $[0, 1]$ are uncountable. Thus, a set is "uncountable" (or "overcountable") if its cardinality is at least $\aleph_1$, a higher order of infinity.

Now, is there something even more than $\aleph_1$? Yes, indeed, there is. Consider for instance the set of functions $f : x \to y$, $x \in [0, 1]$ and y either 0 or 1. The set of all these functions is denoted by F. These functions are only a modest subset of the set of all functions that can be defined in the unit interval. Nevertheless, applying again an idea very similar to the Cantor diagonal method, we can show that the cardinality of the set of those functions exceeds $\aleph_1$. We call it $\aleph_2$. The proof is the following: Assume that the set of functions f is of the same cardinality as the real numbers in $[0, 1]$. Then, by the very definition of cardinality, a one-to-one correspondence can be established between the functions f and the real numbers in $[0, 1]$. In other words, the functions f can be "counted" introducing a continuous "counting label" x. Given an $x \in [0, 1]$, there

corresponds a $f_x(y) \in F$ and vice versa. We will now show that no matter
how ingeniously this counting scheme is set up, there is always a function
g which does not belong to the scheme. Define g in the following way:

$$g(y) = \begin{cases} 0, & \text{if } f_y(y) = 1 \\ 1, & \text{if } f_y(y) = 0. \end{cases} \qquad (2.1.6)$$

Obviously g thus defined cannot be found in the counting scheme. It does
not help to say: OK, so just add g to the scheme. Why not? Simple. Add
g to the scheme, then from the resulting new scheme a function h can be
defined that is not in the scheme. Add this function, then a function i
can be defined which is not in the scheme, etc., etc. This establishes that
indeed $| F |= \aleph_2 \neq \aleph_1$.

It can be shown that the set of all subsets of a given infinite set J
is always of a higher cardinality than the cardinality of the original set
J. This observation has an immediate consequence: We can form the
set of all subsets of F and obtain a set G of cardinality $\aleph_3$. The set
of all subsets of G then has cardinality $\aleph_4$ etc. This shows that there
is an infinite hierarchy of infinities, an infinite number of $\aleph$s. Thus we
have established a countable progression of infinities, which helps us in
characterizing the sizes of sets we will encounter in the following sections.

2.2 Mappings

All the physical examples of chaos in atomic and molecular physics discussed in this book are analysed with the help of discrete mappings derived from continuous equations of motion and classical phase-space flows.
In fact, we have encountered mappings twice already in the Introduction.
The reaction functions y^R and y^C studied in Section 1.1 are mappings of
the initial inclination angle φ to the final position y on side d of the boxes
R and C, respectively. In Section 1.2 we studied the logistic mapping.
It is the purpose of this section to provide a more in-depth discussion of
technical concepts and terminology which arise in connection with mappings. Still, we present only introductory material. More complete ac-
counts of the mathematical theory of mappings and their connection with
chaos are available. We mention, e.g., the excellent books by Lichtenberg
and Lieberman (1983, 1992), Tabor (1989), Gutzwiller (1990), Devaney
(1992), Reichl (1992), Ott (1993), and the introductory article on chaos
by McCauley (1988).

A mapping M is a prescription that assigns elements of a set A to
elements of a set B. The set A is called the *domain* of the mapping, the
set B is called the *co-domain*. If A and B are sets of real or complex
numbers, then the mapping M can sometimes be expressed conveniently
with the help of a simple function f. The function f is then called a

mapping function. We have already encountered examples of mapping functions, the function $y^R(\varphi)$ defined in (1.1.1) in Section 1.1 and the function $f_r(x)$ of the logistic mapping defined in Section 1.2.

The term "mapping" has a second meaning. It is frequently used in the context of mapping an initial condition forward in time by using the mapping function iteratively. Thus, especially in the context of a dynamical system, the term "mapping" usually refers to the *iteration prescription*

$$x_{n+1} = f(x_n). \tag{2.2.1}$$

The logistic mapping defined in (1.2.1) is an example. The starting value x_0, also called the *seed*, or the *initial condition* of (2.2.1), defines the *orbit* of x_0, i.e. the sequence of iterates $x_n, n = 1, 2, \ldots$. Depending on the properties of the mapping function f, the iteration prescription (2.2.1) may or may not generate a chaotic orbit. In this context it is useful to distinguish between the following three types of mapping functions: (i) linear, (ii) nonlinear and (iii) chaotic.

An example of a linear mapping function is

$$f_{a,b}(x) = ax + b, \tag{2.2.2}$$

where a and b are real constants. Linear mappings, especially over finite-dimensional vector spaces, are of importance for analysing the stability properties of nonlinear mechanical systems. Linear mapping functions do not lead to chaos when used in (2.2.1). The situation is different with piece-wise linear functions. These functions are effectively nonlinear and may indeed lead to chaos.

For (2.2.2) the orbit of a seed x_0 can be calculated explicitly and analytically. For $a = 1$ the solution is $x_n = x_0 + nb$. For $a \neq 1$ the solution is

$$x_n = Ca^n + D, \tag{2.2.3}$$

where $D = b/(1 - a)$ and $C = x_0 - D$. For $|a| < 1$ the orbit of x_0 converges to $x_n \to D$ for $n \to \infty$. This result is independent of the seed x_0. Therefore, D is called an *attractor* of (2.2.1). For $|a| > 1$ the orbit of x_0 grows beyond all bounds. This growth, although exponential, is perfectly orderly and not technically classified as chaotic. We note, however, that in this case a simple modification of (2.2.1) can yield chaotic orbits. Define, e.g.,

$$x_{n+1} = f_{a,b}(x_n) \mod 1, \tag{2.2.4}$$

and (2.2.4) will exhibit strongly chaotic properties. The mapping (2.2.4) is then related to the *shift map* discussed below.

Nonlinear mapping functions, such as the function f_r of the logistic mapping discussed in Section 1.2, are the most important and the most useful type of mapping functions for the theory of chaos. Although usually quite innocuous in appearance ($f_r(x)$, e.g., is a simple quadratic function of x), they can produce astonishingly complex orbits when used in iteration prescriptions such as (2.2.1). We encountered examples of this complexity in Section 1.2 (see Figs. 1.7 – 1.9).

The third type of mapping function has appeared only recently in the chaos literature. Chaotic mapping functions occur in the context of chaotic scattering (see Chapter 9). These mapping functions cannot be expressed by any analytical prescription. An example of a chaotic mapping function was presented in Section 1.1, the reaction function $y^C(\varphi)$. Iterations of y^C arise in a natural way if we close box C at d with a totally reflecting wall (a wall of the same nature as the walls a, b, c already present) and monitor the positions y and angles of departure φ of M from d whenever it hits the wall d. This procedure is equivalent to reinjecting the mass point M into the box whenever it arrives at d. Denoting by (y_n, φ_n) the position and angle at impact number n with d, the sequence of impacts is described by a two-dimensional mapping function $\vec{f}(y, \varphi) = (f_1(y, \varphi), f_2(y, \varphi))$ according to

$$y_{n+1} = f_1(y_n, \varphi_n), \quad \varphi_{n+1} = f_2(y_n, \varphi_n). \qquad (2.2.5)$$

In a special case, we already determined the first component of the mapping function f_1. For $y = 1/2$, $f_1(y, \varphi)$ is identical to $y^C(\varphi)$. Imagine the degree of complexity the orbits of (2.2.5) will show if the mapping function y^C by itself is already as chaotic as shown in Fig. 1.5!

The operation of iteration is denoted by

$$x_n = \underbrace{f \circ f \circ \cdots \circ f}_{n}(x_0). \qquad (2.2.6)$$

The "o" symbol denotes *composition*, i.e. iterated application of the function f. The n-times iterated function f is abbreviated as $f^{[n]}$. The nth derivative of f is denoted by $f^{(n)}(x) \equiv \frac{d^n}{dx^n} f(x)$.

While the operations of taking the nth power of a function ($f^n(x)$) or differentiation of a function ($f^{(n)}(x)$) are well known and familiar in all physics contexts, the composition operation ($f^{[n]}(x)$) is not usually encountered in standard physics problems. This, however, does not mean that it is less significant. Iterations of simple maps, not necessarily related to or motivated by physical considerations, combined with computer graphics has led to an explosion in attempts to see how the process of iteration can generate (and store) images of remarkable complexity (see, e.g., Mandelbrot (1983), Peitgen and Richter (1986), Barnsley (1988), Peitgen *et al.* (1993)). Moreover, we will show later that, e.g., in the qualitative solution of differential equations arising from nonlinear mechanical systems, the composition operation occurs naturally. First introduced by Poincaré, this method consists in setting up a "target plane" in the phase space of the system. A system trajectory will pierce the target plane (also called the *surface of section*) in consecutive points $\vec{x}_n$, $n = 1, 2, \ldots$. The natural order of the points $\vec{x}_n$ defines a mapping $\vec{x}_n \to \vec{x}_{n+1}$, called the *return map* of the mechanical system. The iterates of the return map reveal valuable information about the qualitative behaviour of the solutions of the mechanical system. The return map is one of the most useful tools for visualizing chaos in dynamical systems. We will encounter this mapping frequently in this book. We discuss the return map and its properties in Section 3.2 in connection with a physical system, the double pendulum.

There are many special cases of orbits. The most important orbit is the *fixed point*. A fixed point of a mapping function f is a point that satisfies

$$f(x_0) = x_0. \tag{2.2.7}$$

For example, $x_0 = 1 - 1/r$ is a fixed point of the logistic map f_r. The fixed point is a special type of *periodic orbit*. Suppose that for a particular mapping function f we have

$$x_1 = f(x_0); \quad x_2 = f(x_1) = x_0. \tag{2.2.8}$$

Then, the orbit of the seed x_0 reads

$$x_0, \ x_1, \ x_0, \ x_1, \ x_0, \ x_1, \ldots, \tag{2.2.9}$$

a periodic orbit of period 2. Of course, x_1 can also act as the seed for this periodic orbit. A periodic orbit of period n is called an *n-cycle*. A fixed point, therefore, defines a *1-cycle*.

Fixed points and cycles are clearly manifest in Fig. 1.8. If, for instance, we start the logistic mapping with the seed $x_0=0$, we generate the orbit $0, 0, 0, \ldots$. Therefore, $x = 0$ is a fixed point of f_r for all r.

For $0 < r < 1$, Fig. 1.8 shows that $x_\infty = 0$. In other words, the iterates x_n of any initial condition $0 < x_0 < 1$ converge to zero with arbitrary precision. Therefore, $x = 0$ is an *attracting* fixed point of the logistic map for $0 < r < 1$.

For $r > 1$ the fixed point $x = 0$ loses its stability. This means that starting at $x_0 = \epsilon$ with ϵ arbitrarily small and positive, the iterates of ϵ will tend away from 0. Thus, for $r > 1$ we call $x = 0$ a repelling fixed point, or a *repeller*.

Generally, if x_0 is a fixed point of the iteration function f, x_0 is called *attracting* (or *stable*) if $\mid f^{(1)}(x_0) \mid < 1$; it is called *neutral* if $\mid f^{(1)}(x_0) \mid = 1$; and it is called *repelling* (or *unstable*) if $\mid f^{(1)}(x_0) \mid > 1$.

The concepts of attraction, neutrality and repulsion can immediately be generalized to periodic orbits. Suppose that x_0 is the seed of a periodic orbit of length n, i.e. $f^{[n]}(x_0) = x_0$. Then, clearly, x_0 is a fixed point of $f^{[n]}$ and we call it stable, neutral or repelling according to whether $\mid \frac{d}{dx} f^{[n]}(x_0) \mid < 1, = 1$, or > 1, respectively.

Applying the chain rule, we have

$$\frac{d}{dx} f^{[n]}(x_0) = f^{(1)}(x_{n-1}) \cdot f^{(1)}(x_{n-2}) \cdot \ldots \cdot f^{(1)}(x_0). \qquad (2.2.10)$$

Using this result, we have immediately

$$\frac{d}{dx} f^{[n]}(x_i) = \frac{d}{dx} f^{[n]}(x_j), \quad 0 \le i, j \le n - 1. \qquad (2.2.11)$$

This means that all members of an n cycle share exactly the same stability property, and it makes sense to call the whole orbit attracting, neutral, or repelling, respectively, depending on whether *one* of its elements has this property.

In order to gain some experience with the new concepts introduced above, we will now discuss the stability properties of the fixed points and periodic orbits of the logistic mapping. The following is also a more in-depth presentation of the period doubling scenario briefly discussed in Section 1.2.

Fig. 2.3 shows a qualitative sketch of Fig. 1.8. Solid lines indicate attracting fixed points and cycles, dashed lines indicate repelling fixed points and cycles.

From the discussion above, we know that $x_1^{[1]} = 0$ and $x_2^{[1]} = 1 - 1/r$ are fixed points of $f_r = f_r^{[1]}$. Calculating the derivative we get $f_r^{(1)}(x) = r - 2rx$. According to our criterion above, $f^{(1)}(0) = r$ indicates that $x = 0$ is attracting for $0 < r < 1$, neutral for $r = 1$ and repelling for $r > 1$.

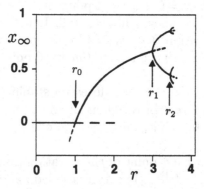

Fig. 2.3.   Qualitative sketch of the bifurcation diagram of the logistic mapping.

The second fixed point yields $f_r^{(1)}(x_1^{[2]}) = 2 - r$. Thus, $x_2^{[1]}$ is repelling for $0 < r < 1$, neutral for $r = 1$ and attracting for $r > 1$. Thus, at $r = 1$ the two fixed points exchange their stability properties. What happens at $r = 1$ can be called a *collision* of the two fixed points. They collide, exchange stability properties and scatter off in different directions.

The fixed point $x_2^{[1]}$ becomes neutral again at $r = 3$, and repelling for $r > 3$. Therefore, the first bifurcation of the logistic map happens at $r_1 = 3$.

The period-2 orbits can be calculated from

$$x = f_r^{[2]}(x) = r[rx(1-x)]\{1 - [rx(1-x)]\}. \qquad (2.2.12)$$

This equation has a factor x and therefore $x = 0$ is a solution. Taking out the factor x, we are left with

$$r^3 x^3 - 2r^3 x^2 + (r^3 + r^2)x + (1 - r^2) = 0. \qquad (2.2.13)$$

Since every fixed point is also a 2-cycle, the term $(x - 1) + 1/r$ must be a factor of (2.2.13). Dividing (2.2.13) by this factor, we obtain

$$r^2 x^2 - (r^2 + r)x + (r + 1) = 0. \qquad (2.2.14)$$

If the discriminant Δ of (2.2.14) is larger than zero, we obtain two real solutions corresponding to the 2-cycle. The discriminant is given by

$$\Delta = (r^2 + r)^2 - 4r^2(r + 1) = r^2(r+1)(r-3). \qquad (2.2.15)$$

For $\Delta = 0$ the 2-cycle degenerates and we obtain again $r = r_1 = 3$, the location of the first bifurcation. For $r > 3$ the period-2 fixed points are distinct and are given by

$$x_{1,2}^{[2]} = \frac{1}{2r}\left\{r + 1 \pm \sqrt{(r+1)(r-3)}\right\}. \qquad (2.2.16)$$

Thus, the value of the equilibrium population of the logistic mapping at the first bifurcation is $x_1^{[2]} = x_2^{[2]} = 2/3$. Indeed, $f_3(2/3) = 2/3$ is a period-1 fixed point or a degenerate period-2 cycle.

The stability of the 2-cycle can be obtained from

$$\frac{d}{dx} f_r^{[2]}(x_1^{[2]}) = -r^2 + 2r + 4. \qquad (2.2.17)$$

Beyond the first bifurcation, i.e. for $r > r_1$, this derivative is initially smaller than 1. Thus, the 2-cycle is attracting. The 2-cycle loses stability at $-r^2 + 2r + 4 = -1$, which yields $r_2 = 1 + \sqrt{6}$ and $x_{1,2}^{[2]} = \frac{1}{2(1+\sqrt{6})}\left\{2 + \sqrt{6} \pm \sqrt{2}\right\}$. This is exactly the location of the second bifurcation. By similar reasoning the locations r_j, $j = 3, 4, ...$, of further bifurcation points can be calculated. As mentioned in Section 1.2, the successive bifurcation points r_j form approximately a geometric series controlled by the Feigenbaum number δ defined in (1.2.5).

Having studied the properties of the logistic mapping, a mapping with practical applications, we will now encounter a more abstract mapping, the *shift map*. It is very useful for fundamental investigations in chaos theory. It is defined by

$$D(x) = \begin{cases} 2x, & \text{if } 0 \le x < 1/2 \\ 2x - 1, & \text{if } 1/2 \le x < 1. \end{cases} \tag{2.2.18}$$

A graphical representation of the mapping function D of the shift map is shown in Fig. 2.4. The definition (2.2.18) of the shift map is equivalent to the following alternative definition:

$$D(x) = 2x \mod 1. \tag{2.2.19}$$

The name "shift map" derives from the following observation: Let x_0 be a number in the unit interval $[0, 1]$. An example for x_0 in binary notation is

$$x_0 = 0.100101001.... \tag{2.2.20}$$

Calculating $D(x_0)$, we first have to multiply x_0 by 2. In binary notation this is merely a shift of the binary point one position to the right. We obtain

$$x_1' = 2x_0 = 1.00101001.... \tag{2.2.21}$$

In order to obtain the final result x_1, the modulo operation has to be applied to x_1'. But this is equivalent to discarding the integer part in (2.2.21). We obtain:

$$x_1 = 0.00101001.... \tag{2.2.22}$$

Thus, applying the shift map is equivalent to scanning through the binary digits of the seed x_0.

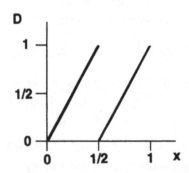

Fig. 2.4.   Graph of the mapping function of the shift map.

Since the binary digit sequence in x_0 can have any arbitrary complexity imaginable, we expect that the shift map for a *generic* seed x_0 produces the most complex orbits imaginable, in other words, the shift map is fully chaotic.

In order to prove this statement, imagine the following experiment. We choose the binary sequence of the seed x_0 according to the results of a coin toss experiment. We start our x_0 value by writing "$x_0 = 0.$". The binary digits after the binary point are determined in the following way: We toss a fair coin. If it comes up heads, we write "1". Otherwise we write "0". This way, i.e. by tossing a coin for every binary position after the binary point, we obtain a number x_0 whose binary expansion is completely random. Using this initial condition as the seed for an orbit generated by the shift map D we see that the shift map will produce numbers in the interval $[0, 1]$ that are as random as a coin toss (Ford (1983, 1989)).

The randomness observed in the shift map is one of the cornerstones of chaos. In general, i.e. for a generic initial condition, a chaotic system produces output that is indistinguishable from noise, even though this very output is produced by a perfectly orderly *deterministic* process, namely the repeated application of a single mapping function. It is one of the great discoveries of chaos research that determinism and randomness are not mutually exclusive concepts. Although, of course, a random process cannot be deterministic, a deterministic process can be totally random. This surprising fact is at the heart of *deterministic chaos*.

The calculation of fixed points and periodic obits for the shift map is straightforward. There are exactly two fixed points. Since for a fixed point of period 1 a single shift of the binary point to the right has to reproduce the seed, there are only two possibilities:

$$x_1^{[1]} = 0.0000... \qquad (2.2.23)$$

and

$$x_2^{[1]} = 0.1111... = \frac{1}{2} + \frac{1}{4} + \cdots = \sum_{k=1}^{\infty} \frac{1}{2^k} = 1. \qquad (2.2.24)$$

These fixed points are, of course, also 2-cycles. Besides these trivial 2-cycles, there is also one nontrivial 2-cycle with seed

$$x_1^{[2]} = 0.0101... = 1/3. \qquad (2.2.25)$$

The corresponding orbit is $O_1^{[2]} = \{1/3, 2/3\}$.

There are two 3-cycles with seeds

$$x_1^{[3]} = 0.001001... = 1/7 \qquad (2.2.26)$$

and

$$x_2^{[3]} = 0.011011... = 3/7. \qquad (2.2.27)$$

The corresponding orbits are: $O_1^{[3]} = \{1/7, 2/7, 4/7\}$ and $O_2^{[3]} = \{3/7, 6/7, 5/7\}$.

In order to save some writing, we introduce the "bar" notation for periodic orbits. For instance

$$x_1^{[3]} = 0.001001... = 0.\overline{001}. \qquad (2.2.28)$$

The bar over a group of numbers denotes infinite repetition of the whole group *ad infinitum*.

The shift map also has three 4-cycles with orbits $O_1^{[4]} = \{1/15, 2/15, 4/15, 8/15\}$, $O_2^{[4]} = \{1/5, 2/5, 4/5, 3/5\}$ and $O_3^{[4]} = \{7/15, 14/15, 13/15, 11/15\}$.

A natural response is to ask for the number N_n of periodic orbits for a given period n. Suppose that $2^n - 1$ is prime. Then, (i) we do not have to worry about sub-cycles and (ii) the seeds of the periodic orbits are given by $p_k/2^n - 1$, where p_k, $k = 1, 2, ..., N_k$, is prime. Thus, the number of periodic orbits of period n is exactly $\pi(2^n - 1)$, where $\pi(x)$ is the *prime number function*. It counts the number of primes smaller than or equal to x. ("1" is not counted as a prime number). This function was studied extensively in the mathematical literature. An explicit expression for $\pi(x)$, however, is not known. A good approximation is

$$\pi(x) \approx \frac{x}{\ln(x) - 1}. \qquad (2.2.29)$$

Using (2.2.29) we get

$$N_n = \pi(2^n - 1) \approx \frac{2^n - 1}{n \ln(2) - 1}. \qquad (2.2.30)$$

The approximation in (2.2.30) is acceptable for large n. This formula yields an important result: The number of periodic orbits of the shift map increases roughly exponentially with the length of the period. In other words, periodic orbits in the shift map proliferate exponentially.

If we check the periodic orbits, especially their representation in binary notation, it is obvious that the seeds for the periodic orbits of the shift map are the rational numbers in the unit interval $[0, 1]$. Since the rationals are *dense* in the irrationals, we obtain a second important result: the periodic orbits of the shift map are dense.

This result is significant since it applies not only to the shift map but to many other chaotic systems as well. It means that the periodic orbits of a chaotic system can be used as a skeleton for the expansion of physically interesting quantities. In fact, Gutzwiller (1971, 1990) used a periodic orbit expansion technique to arrive at the first known formula for the semiclassical quantization of chaotic systems: the trace formula. This quantization technique is discussed in Section 4.1.3.

The simplicity of the shift map allows us to prove another astonishing fact: the existence of a *dense orbit*. An orbit of a dynamical system is said to be dense if it comes arbitrarily close to any point in the domain of the mapping. Following Devaney (1992), we choose the seed

$$x_0 = (0\,1 \quad 00\,01\,10\,11 \quad 000\,001\,010\,...). \tag{2.2.31}$$

This sequence consists of all possible blocks of length one, two, ... containing all possible combinations of the symbols "0" and "1". The seed (2.2.31) is constructed in a very clever way. If an n-digit approximation to any given number in $[0, 1]$ is desired, one simply shifts the binary point in x_0 to the beginning of the n-block that among all other 2^n n-blocks most closely approximates the given number. Since the shift of the binary point is produced by a certain number of applications of the shift map, and since n can be any positive integer, the orbit of x_0 will indeed approach any number in $[0, 1]$ to any degree of accuracy. Thus, the shift map has at least one dense orbit.

Closely related to the notion of dense orbits is the concept of *transitivity* of a dynamical system. A dynamical system is transitive if for any two points x_1 and x_2 in the domain of the system there exists an orbit that visits both x_1 and x_2 to any prescribed accuracy. Obviously, the existence of a dense orbit ensures transitivity of the corresponding dynamical system. Thus, the shift map is transitive.

Especially for physical applications, the most important concept of dynamical systems theory is *sensitivity to initial conditions*. A dynamical system exhibits sensitive dependence on initial conditions, or simply *sensitivity*, if no matter how closely spaced two initial conditions, there is always a time when they will have evolved into image points which will be further apart than a prescribed value β. Therefore, a mapping f is sensitive if, for given $\beta > 0$ and for any $x \neq y$, $\epsilon > 0$ with $|x-y| < \epsilon$, there exists a positive integer n with $|f^{[n]}(x) - f^{[n]}(y)| > \beta$. Of course, certain obvious restrictions apply. For instance, the value of β cannot be larger than the "diameter" of the co-domain of the mapping. Also, it has to be noted that once the distance β is exceeded at some iterate number n^*, the distance between the iterates of x and y will not in general stay larger than β for $n > n^*$. It is no contradiction to the definition of sensitivity if $|f^{[m]}(x) - f^{[m]}(y)| < \beta$ for some $m > n^*$.

It is not a difficult matter to prove sensitivity of the shift map. Since x and y are not identical, the binary expansion of their difference has at least one binary digit "1" in some binary position after the binary point. Iteration of the shift map will bring this "1" closer and closer to the binary point until the difference between the iterates exceeds 1/2 no matter how close the two points were initially.

We are now ready for a definition of chaos. Summarizing a century of research in chaos, Devaney (1992) gives the following definition of chaos: A system is chaotic if
(C1) periodic orbits are dense,
(C2) the dynamical system is transitive, and
(C3) the system is sensitive.

According to the above discussion and on the basis of the results proved above, the shift map obviously satisfies all three criteria and is therefore a chaotic system. This is an important result. From now on the shift map can serve as a paradigm for the investigation of other mappings. Even more importantly, in some cases it is possible to establish a relationship between the shift map and a seemingly unrelated mapping. In this case we can make full use of our knowledge of the chaoticity of the shift map and assert that the related mapping is chaotic too. An example is a proof of the chaoticity of the logistic map for $r = 4$. The proof is due to Ulam and von Neumann (see, e.g., Campbell (1987), p. 233) and is constructed in the following way.

For $r = 4$, the shift map reads

$$x_{n+1} = 4x_n(1 - x_n).$$
(2.2.32)

Since $0 \le x_n \le 1$, we substitute

$$x_n = \sin^2(\theta_n),$$
(2.2.33)

where θ can be restricted to $0 \le \theta \le \pi$. Using (2.2.33) in (2.2.32), we obtain

$$\sin^2(\theta_{n+1}) = 4\sin^2(\theta_n)\cos^2(\theta_n) = \sin^2(2\theta_n).$$
(2.2.34)

Defining $\theta_n = y_n\pi$, the mapping for y_n is nothing but the shift map $y_{n+1} = 2y_n \bmod 1$. This proves that the logistic map at $r = 4$ is completely chaotic.

In order to allow for the largest possible class of chaotic systems, the *degree* of sensitivity is not specified in Devaney's definition of chaos. It turns out that many chaotic systems of practical importance are *exponentially* sensitive to initial conditions. In this case the sensitivity can be characterized quantitatively with the help of *Lyapunov exponents*.

Consider a one-dimensional mapping function $f(x)$ that defines the iteration $x_{n+1} = f(x_n)$. Suppose we start this iteration with two seeds x_0 and x_0' with $\Delta x_0 = |x_0' - x_0|$. Then, after n iterations, we have

$$\Delta x_n = |x_n' - x_n| = |f^{[n]}(x_0') - f^{[n]}(x_0)|.$$
(2.2.35)

If the mapping shows exponential sensitivity,

$$\Delta x_n \sim \exp(\lambda n)\Delta x_0. \tag{2.2.36}$$

The rate of expansion, λ, is called the Lyapunov exponent. If the initial conditions x_0 and x_0' are close together, Δx_n can be computed according to

$$\Delta x_n \approx \frac{d}{dx} f^{[n]}(x_0)\Delta x_0. \tag{2.2.37}$$

Comparing (2.2.37) with (2.2.36) motivates the definition of the Lyapunov exponent

$$\lambda = \lim_{n\to\infty} \frac{1}{n}\ln(\Delta x_n/\Delta x_0) = \lim_{n\to\infty} \frac{1}{n}\ln\left|\frac{d}{dx} f^{[n]}(x_0)\right|. \tag{2.2.38}$$

In general, the Lyapunov exponent λ depends on the starting value x_0. For seeds of dense orbits of the mapping f, however, the limit operation in (2.2.38) assures that the Lyapunov exponent is independent of the starting value, since the dense orbit will explore all the points of the domain of f.

Using the chain rule of differentiation in (2.2.38) we obtain an equivalent, alternative form of (2.2.38) according to

$$\lambda = \lim_{n\to\infty} \frac{1}{n}\sum_{k=0}^{n-1} \ln\left|f^{(1)}(x_k)\right|. \tag{2.2.39}$$

The calculation of Lyapunov exponents can easily be generalized to mappings of more than one variable.

Following Lichtenberg and Lieberman (1983) we consider the mapping

$$\vec{x}_{n+1} = \vec{M}(\vec{x}_n), \tag{2.2.40}$$

where $\vec{x}_{n+1}$ and $\vec{x}_n$ are N-dimensional vectors, respectively, and $\vec{M}$ is a vector function with N components. If two vectors $\vec{x}_0$ and $\vec{x}_0'$ are initially close with $\vec{w}_0 = \vec{x}_0' - \vec{x}_0$ and $|\vec{w}_0|$ is infinitesimally small, then the distances $\vec{w}_n = \vec{x}_n' - \vec{x}_n$ can be obtained from the mapping

$$\vec{w}_{n+1} = J(\vec{x}_n)\,\vec{w}_n, \tag{2.2.41}$$

where J is the Jacobian matrix of $\vec{M}$. We now define the mean rate of divergence for the two initially close vectors $\vec{x}_0$ and $\vec{x}_0'$:

$$\sigma(\vec{x}_0, \vec{w}_0) = \lim_{n\to\infty,\,|\vec{w}_0|\to 0} \left(\frac{1}{n}\right)\ln\left(\frac{|\vec{x}_n|}{|\vec{x}_0|}\right). \tag{2.2.42}$$

The definition of σ shows that σ, the mean rate of divergence, depends in general on the starting point $\vec{x}_0$ and also on the initial displacement $\vec{w}_0$. It can be proved that locally there exists at $\vec{x}_0$ a basis $\{\hat{e}_i\}, i = 1, ..., N$

such that $\sigma(\vec{x}_0, \hat{e}_i) = \sigma_i(\vec{x}_0)$, and that the N numbers σ_i exhaust all possibilities for the limit (2.2.42). In other words, although the vector $\vec{w}_0$ in $\sigma(\vec{x}_0, \vec{w}_0)$ can assume a continuum of directions, the result of the limit can be only one of the N numbers σ_i. In many physics applications the σ_i are real numbers and can be ordered according to their size:

$$\sigma_1 \geq \sigma_2 \geq \sigma_3 \geq \cdots \geq \sigma_N. \tag{2.2.43}$$

In analogy with the one-dimensional case discussed above, the $\sigma_i, i = 1, ..., N$ are called the *Lyapunov exponents*.

The fact that there are exactly N Lyapunov exponents is easily understood in the case of a periodic orbit with period p. Periodicity means that the orbit returns to its starting point after p applications of $\vec{M}$. In other words, $\vec{x}_p = \vec{x}_0$. If we denote by U the product of the Jacobians

$$U = \prod_{k=0}^{p-1} J(\vec{x}_k), \tag{2.2.44}$$

then

$$\vec{w}_p = U\vec{w}_0. \tag{2.2.45}$$

The propagator U has N (possibly complex) eigenvalues $\lambda_i, i = 1, ..., N$. They can be ordered according to size:

$$|\lambda_1| \geq |\lambda_2| \geq \cdots \geq |\lambda_N|. \tag{2.2.46}$$

Let $\hat{e}_i$ be the eigenvector corresponding to λ_i and $\vec{w}_0 = \hat{e}_i$. Then, we have

$$\vec{w}_p(\vec{x}_0) = U\vec{w}_0 = U\hat{e}_i = \lambda_i \hat{e}_i. \tag{2.2.47}$$

We conclude:

$$\sigma(\vec{x}_0, \hat{e}_i) = \frac{1}{p} \ln|\lambda_i| = \sigma_i. \tag{2.2.48}$$

If the eigenvalues of U are nondegenerate, we see now immediately why there are only N distinct values allowed for the limit in (2.2.42). If we expand $\vec{w} = c_1 \hat{e}_1 + c_2 \hat{e}_2 + \cdots + c_N \hat{e}_N$, then it is immediately clear that for a general $\vec{w}_0$ the first term in this expansion dominates in the time evolution. All $\vec{w}_0$ that are not orghogonal to $\hat{e}_1$ will have Lyapunov exponent σ_1. Only in the case where $\vec{w}$ is orthogonal to $\hat{e}_1$ (but not orthogonal to $\hat{e}_2$), will the Lyapunov exponent be given by σ_2. In general, only if $\vec{w}_0$ is orthogonal to all $\hat{e}_k$ with $k = 1, 2, ..., j-1$ (but not orthogonal to $\hat{e}_j$), will the Lyapunov exponent be given by σ_j. This argument shows that with probability 1 we will obtain the Lyapunov exponent σ_1 if we select a random initial displacement $\vec{w}_0$.

Of course the above arguments hold only in the case of a periodic orbit. If an orbit started at $\vec{x}_0$ is nonperiodic, then we are somewhat at a loss.

But for a chaotic mapping periodic orbits are dense. In other words, arbitrarily close to any nonperiodic orbit we find a periodic orbit. Thus, we can apply the above considerations. Therefore, in general, there are exactly N Lyapunov exponents that can be used to characterize the local rate of exponential divergence, and thus chaos itself.

To conclude this section we discuss the "baker's map" (Farmer *et al.* (1983)) as an example for an *area preserving* mapping in two dimensions. Area preservation is of utmost importance for Hamiltonian systems, since Liouville's theorem (Landau and Lifschitz (1970), Goldstein (1976)) guarantees the preservation of phase-space volume in the course of the time evolution of a Hamiltonian system. The baker's map is a transformation of the unit square onto itself. It is constructed in the following four steps illustrated in Fig. 2.5.

(1) Start with the unit square $[0,1] \times [0,1]$ (Fig. 2.5(a)).
(2) Compress the square to one-half its size in the y direction, expand it to double its size in the x direction (Fig. 2.5(b)).
(3) Cut the resulting rectangle in the y direction at $x = 1$ (see dashed line in Fig. 2.5(b)).
(4) Move the right hand piece of the rectangle on top of the left hand piece, as shown in Fig. 2.5(c).

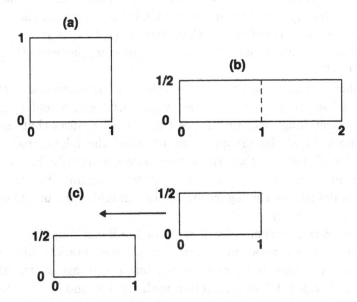

Fig. 2.5.   Construction of the baker's mapping.

The four steps of the mapping defined above can be represented concisely by the following formula:

$$(x_{n+1}, y_{n+1}) = \begin{cases} (2x_n, y_n/2) & \text{if } x_n \leq 1/2 \\ (2x_n - 1, (y_n + 1)/2) & \text{if } x_n > 1/2. \end{cases} \qquad (2.2.49)$$

According to its construction sketched in Fig. 2.5, the baker's map is obviously area preserving. It is called "the baker's map" because it is vaguely
reminiscent of how a baker kneads dough. The idea is to spread baking
ingredients homogeneously throughout the dough by repeatedly applying
(2.2.49). That this method is indeed quite efficient is demonstrated in the
sequence of four frames shown in Fig. 2.6. Fig. 2.6(a) shows an orderly
swirl of 50 "raisins" at the centre of the unit square. Fig. 2.6(b) shows
the positions of the raisins after one application of the mapping (2.2.49).
Fig. 2.6(c) shows the result after five applications. Although there are
still noticeable correlations in the positions of the raisins, one can already see how the raisins tend to a homogeneous distribution. A very
satisfactory result is achieved after only ten applications of the mapping,
as shown in Fig. 2.6(d). The "secret" of this quick equilibration is the
chaoticity of the baker's map. The baker's map is an example of how we
can apply our knowledge about the shift map. A careful examination of
(2.2.49) shows that the mapping in the x direction is identical to the shift
map defined in (2.2.18). Since we know that the shift map is chaotic, it
follows immediately that the mapping (2.2.49) is chaotic in the x direction. The complete proof of the chaoticity of (2.2.49) is more involved
and consists in proving that the baker's map complies with Devaney's
criteria C1 − C3.

The baker's map demonstrates the two basic mechanisms that lead
to chaos. Chaos requires a *stretching* mechanism and a *folding* mechanism. The stretching mechanism (see Fig. 2.5(b)) ensures that the Lyapunov exponents of the mapping are nonzero; the folding mechanism
(see Fig. 2.5(c)) ensures that the system stays bounded. For a simple
two-dimensional mapping such as the baker's mapping, the Lyapunov
exponent is defined as the logarithm of the stretching factor. Therefore,
we have $\sigma_{baker} = \ln(2) > 0$.

In the following section we will see how nonlinear mappings are also
responsible for the generation of fractal structures. Fractals play an important role in atomic and molecular scattering systems, as was already
discussed in Section 1.1 in connection with the box model C. They are
also useful for the description of atomic ionization processes.

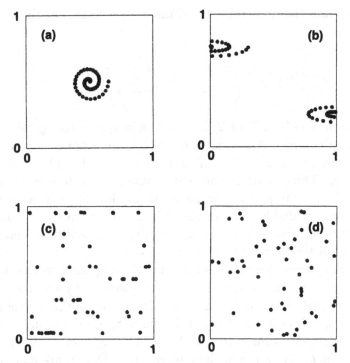

Fig. 2.6. Evolution of an initially orderly set of points after n applications of the baker's mapping. (a) Initial configuration, (b) $n = 1$, (c) $n = 5$ and (d) $n = 10$.

2.3 Fractals

In this section we study a type of sets so complicated in structure that they are best described as sets with noninteger dimension. These sets are known as *fractals*, a name coined by Mandelbrot (1975, 1977). It was Mandelbrot's great achievment to have recognized the outstanding role that fractals play in many natural phenomena (Mandelbrot (1983)). Fractal sets were first conceived by Cantor who used the fractal middle thirds set discussed in Section 2.1 to win an argument concerning the theory of Fourier series (Dauben (1979), Purkert and Ilgauds (1987)). As shown in Fig. 2.2 the Cantor set is constructed by successively removing middle thirds from the unit interval. The unit interval will occur frequently throughout this section. Therefore, we abbreviate the unit interval by the symbol A. We will now show that the successive removal of the middle thirds from A can also be achieved with the help of a mapping. The removal process itself can be interpreted as a model for the *ionization* or *dissociation* of an atom or molecule, and thus establishes a connection between fractals and atomic decay.

The mapping that generates the Cantor set is known as the *tent map*. It is defined by

$$T(x) = \begin{cases} 3x, & \text{for } x \le 1/2 \\ 3(1-x), & \text{for } x > 1/2. \end{cases} \tag{2.3.1}$$

Fig. 2.7 shows the graph of T in A. The shape of the graph explains the mapping's name. All points with $x < 0$ are mapped monotonically to $-\infty$. Points with $x > 1$ are first mapped to $3(1-x) < 0$ and then also to $-\infty$. Thus, none of the points outside A will ever be mapped into A. This is an important property. It implies that whenever a point of A is mapped outside A this point will never return to A. Thus, this property is called the *never-come-back* property. It facilitates appreciably the analysis of the tent map.

Let us now study the action of T when it is applied to the points in A. In order to make the connection with atomic physics, we interpret the unit interval A as an "atom" and the action of the mapping T as its dynamics, induced, e.g., by a periodically applied external field. The unit interval A, the atom at rest, is sketched in Fig. 2.8(a). After the first application of T, all points in the interval $A_0 = [0, 1/3]$ are mapped onto A. The same holds for the points in the interval $A_1 = [2/3, 1]$. The points in the middle thirds set $(1/3, 2/3)$ are mapped outside A. In subsequent applications of T they are mapped to $-\infty$, and are lost. Keeping with our atomic physics analogy we say that the points of the middle thirds set are *ionized* after the first application of T. In order to represent this situation graphically, we delete the middle thirds set from the unit interval. The result is shown in Fig. 2.8(b).

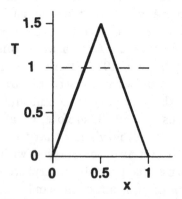

Fig. 2.7.   Graph of the tent map.

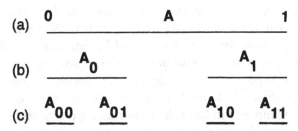

Fig. 2.8. The "ionization" process induced by the tent map T. (a) Initial condition, (b) remaining points after the first application of T, (c) remaining points after the second application of T.

Since A_0 is mapped onto A after the first application of T, the middle third of the set A_0 is ionized after the second application of T. The same holds for A_1. What remains are the sets A_{00}, A_{01}, A_{10} and A_{11}, as shown in Fig. 2.8(c).

It is by now obvious how this process continues. With every application of the tent map T, the middle third of all the disconnected intervals still present will be deleted. The total length of the surviving intervals after step number n is

$$P_B(n) = \left(\frac{2}{3}\right)^n = \exp(-\gamma n), \qquad (2.3.2)$$

where $\gamma = \ln(3/2)$. The notation P_B again derives from atomic physics. In atomic physics we are interested in the fraction of electron probability still bound by the atom after application of some ionizing process (such as, e.g., an external radiation field). Therefore, the subscript "B" in (2.3.2) stands for "bound". The constant γ is called the *decay constant* or *ionization rate*.

According to (2.3.2), there will be no probability left in the unit interval in the limit of $n \to \infty$. Therefore, we say that a point x_0 in A ionizes with probability 1 according to the ionization process defined by the tent map T. But is there really nothing left in A after the application of $n \to \infty$ mappings? The successive steps in the ionization process defined by repeated application of T (see Fig. 2.8) remind us strongly of the construction scheme of Cantor's middle thirds set C (Fig. 2.2) which was introduced and briefly discussed in Section 2.1. And indeed, there is a whole infinity of points left in A, even in the limit of an infinite number of applications of T. What kind of infinity? We can easily answer this question with the tools developed in Section 2.1. Let us introduce a notation for the set of points in A that never ionize. We call this set Λ^+. For the tent map T we have $\Lambda^+ = C$.

It is convenient to represent the real numbers in A in base-3 notation. This way, the real numbers in A_0 and A_1 can be written as $0.0a_2a_3...$ and $0.2b_1b_2b_3...$, respectively, where a_j and b_j can take any of the values 0, 1, or 2. The ionizing real numbers in the middle thirds set have the representation $0.1c_1c_2c_3...$, where the c_j, just like the a_j and the b_j, can take any of the values 0, 1, or 2. It is now obvious that, in the end, the points that never ionize will have the following representation: $0.d_1d_2d_3...$, where the numbers d_j can assume only the values 0 and 2.

We will now show that although the set Λ^+ of never ionizing points has measure zero, it is nevertheless an uncountable set of points whose cardinality is the same as the cardinality of A itself. The proof is not difficult. All we have to do is to take the numbers in Λ^+ in the ternary representation and map all $d_j = 2$ into $d_j = 1$. Since all possible combinations of 0's and 2's appear in the ternary expansion $0.d_1d_2d_3...$ of elements of Λ^+, the mapping $2 \to 1$ produces numbers $0.z_1z_2z_3...$ with all possible combinations of 0's and 1's. The numbers $0.z_1z_2z_3...$ can now be interpreted as the binary representation of real numbers, and since the z_j can assume both values 0 and 1 with no restriction, the set of all numbers $0.z_1z_2z_3...$ is clearly identical to A itself. Thus, the "$2 \to 1$ mapping" establishes a one-to-one correspondence between the elements of Λ^+ and the elements of A. This proves that Λ^+ and A are of the same cardinality. This is both counterintuitive and surprising since the unit interval A has length 1, whereas the Cantor set Λ^+, as argued above, has total length 0. The length of a set (or more generally its measure) is therefore not a reliable indication of its cardinality. Because of the one-to-one correspondence, we conclude $\mid C \mid = \mid \Lambda^+ \mid = \aleph_1$.

The set Λ^+ looks rather flimsy. It is often referred to as mere "dust" in the literature. But because of its size (its cardinality) it is capable of influencing profoundly the quantum mechanics of an ionizing atom, or an atomic or molecular scattering system. Thus, fractals play an important role in atomic and molecular physics.

Besides the tent map many other mappings can be used as models for ionizing systems. An example is the logistic mapping for $r > 4$. While for $r \leq 4$ the unit interval A is mapped onto itself, the logistic mapping shows ionizing behaviour for $r > 4$. The reason is clear from a quick glance at Fig. 2.9, which shows the mapping function f_5 of the logistic map: the mapping function f_5 resembles a smooth version of the tent map. Therefore, the unit interval A can ionize under the application of f_5 just like it did for the tent map. Therefore, we expect to see very similar ionization behaviour.

Indeed, the logistic map for $r > 4$ also leads to a "middle thirds" elimination process, although now the "middle thirds" are not exactly thirds any more. In fact, the length of the interval that leaves A after the first application of f_5 is $1/\sqrt{5} \approx 0.45$. But topologically there is no difference compared with the Cantor elimination process. Therefore, in analogy to the tent map, we expect to see exponential decay of A induced by f_5. This is indeed the case. An analytical approximation to the decay constant is obtained if we assume that the fraction of points that leave A in every step is the same as the fraction that leaves after the first application of f_5. With this assumption we obtain $P_B(n) = \exp(-\gamma n)$, where $\gamma = \ln[\sqrt{5}/(\sqrt{5} - 1)]$.

According to the above discussion, ionizing mappings, such as the tent map or the logistic map for $r > 4$, can lead to a fractal set Λ^+ of surviving points. We have already characterized the cardinality of these points and have made the surprising discovery that these fractal sets can be rather "large" as far as the number of their elements is concerned. At the same time fractal sets can be vanishingly small in "size". The Cantor set C is an example. According to Euclid (Heath (1956)), a *point* is that which has no part (i.e. no extension, no size). We assign dimension 0 to a point. According to the measure of the Cantor set we might be tempted to assign dimension 0 to the Cantor set C. Doubts arise when we consider the huge number of points in C; in fact, as we saw above, there are as "many" as there are on the one-dimensional real line itself. So, given this fact, do we have to assign dimension 1 to C? In order to answer this question, we have to say something about the concept of "dimension" itself.

So far we have encountered Cantor as a master of mathematical surprises. He was even called the first truly original mathematician since the time of the ancient Greeks (Dunham (1990)). Naturally, Cantor had

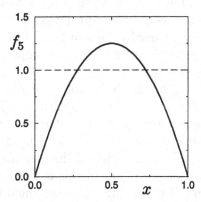

Fig. 2.9.   Mapping function f_5 of the logistic map.

something to say about dimensions too. Naively we all know what "dimension" is all about. A line segment has a certain number of points, a square has many, many more. Not so! says Cantor, and he goes on to show that a one-to-one mapping exists between the points of a line and the points of the unit square. This mapping is known as the "zipper mapping" and works like this. Take a point $P = (0.a_1a_2a_3..., 0.b_1b_2b_3...)$ of the unit square. To be specific, let us work in binary representation, where a_j and b_k can take any of the two values 0 or 1 without restriction. The two coordinates of P are now merged into a single number $Q = 0.a_1b_1a_2b_2a_3b_3...$, which can be interpreted as a point in the unit interval A. The mapping is one-to-one since Q can easily be "unzipped" to yield the two coordinates of the original point P. This mapping shows clearly that there are just as many points in the unit square as there are points in the unit interval. Therefore, "dimension" has nothing to do with the cardinality of point sets. Since the unit square can be mapped into A, so can the unit cube, and the unit cube in four dimensions, and so on for any dimension. This means, e.g., that a function of two variables, $f(x, y)$, can always be represented by another function $g(z)$ with only a single argument. It seems that the concept of "dimension" is utterly meaningless. Again, we have to say: not so! While it is true that the two arguments x and y of f can always be zipped into the single argument z of g, the mapping $x, y \rightarrow z$ scatters the points z quasi-randomly onto A. Thus, while the function $f(x, y)$ may have a simple analytical form, it is immediately obvious that the equivalent function $g(z)$ is usually very complicated. From this example we learn that, while the concept of "dimension" may not have any significance in principle, it will always be clear for the physicist what the "correct dimension" of an object is: it is the dimension in which the object looks simplest. But for a mathematician this is certainly not an acceptable definition. To begin with, it is circular since the word "dimension" occurs in its very "definition". Many scientists, including Cantor, worked on a more precise definition of dimension. A very successful attempt dates back to the work of Felix Hausdorff (1919). His advanced notion of fractal dimension is discussed in Section 8.2 in connection with a physical example of a fractal derived from an atomic decay problem.

If we want to get a grip on the problem of dimension, we have to find a property of the familiar integer-dimensional spaces that allows for a generalization. One such property is *self-similarity*. A line segment, e.g., is self-similar because a stretched (or compressed) version of the line segment looks just like the original line segment itself. The same holds for the square or for the cube. In this connection we introduce the *magnification factor* m. If we magnify a line segment by the factor m, the length of the resulting line segment is m times the length of the original line segment. If we magnify the length of the sides of a square by m, the area of the resulting square is m^2 times the area of the original square, and for the magnified cube we have m^3 times the original volume.

In general the contents of sets can be calculated by a mathematical function which is called the *measure*. We denote it by μ. Introducing the symbols L_a, S_a and K_a to denote a line segment of length a, a square of side length a and a cube with side length a, respectively, we have $\mu(L_a) = a$, $\mu(S_a) = a^2$ and $\mu(K_a) = a^3$. We are now ready to define the dimension of a set in a way that can be generalized. Let M denote a given set. By M' we denote the set that is obtained by magnifying the set M with the magnification factor m. We then define the dimension of the set M as d if the following scaling relation holds:

$$\mu(M') = m^d \mu(M). \tag{2.3.3}$$

Let us try this definition with our familiar sets L, S and K. We have

$$\mu(L_{ma}) = m \mu(L_a) = m^1 \mu(L_a). \tag{2.3.4}$$

Thus, this set is one-dimensional, as expected. For the square we obtain

$$\mu(S_{ma}) = m^2 \mu(S_a). \tag{2.3.5}$$

As expected, S_a turns out to be two-dimensional. The same trivial calculation shows that K_a is in fact three-dimensional.

Let us now turn to the Cantor set C and calculate its dimension using (2.3.3). Magnifying the Cantor set by a factor 3 gives us two identical copies of the original Cantor set. Denoting the magnified set by C', we obtain:

$$\mu(C') = 2\mu(C) = 3^d \mu(C). \tag{2.3.6}$$

From this relation, and assuming that $\mu(C) \neq 0$, we obtain immediately $3^d = 2$ or

$$d = \ln(2)/\ln(3) \approx 0.631. \tag{2.3.7}$$

This shows that the dimension of the Cantor set is in fact "fractal". It is certainly noninteger. For the following reasons it is very satisfactory that the dimension of the Cantor set turns out to be between 0 and 1. The Cantor set has measure zero. Therefore, it is hard to associate it with a dimension-1 object. On the other hand, it contains an uncountable number of points which is "too much" to associate the Cantor set with a dimension-0 object. Therefore, even intuitively, its dimension has to be somewhere in between 0 and 1.

There is a rule which can be used to calculate fractal dimensions. Suppose a fractal can be split into k congruent pieces, and magnifying one of the pieces by a factor m yields an identical copy of the original fractal. Then, the fractal dimension of this object is given by:

$$d = \frac{\ln(k)}{\ln(m)}. \tag{2.3.8}$$

Let us apply this formula to the Cantor set. The Cantor set consists of two pieces, each of which can be magnified by a factor $m = 3$ and yields a complete copy of the original set. Therefore, we have $k = 2$ and $m = 3$ and $d = \ln(2)/\ln(3)$ according to (2.3.8). This result agrees with what we computed using the dimension formula (2.3.3).

Another well-known fractal is the "Sierpinsky gasket" (Fig. 2.10). The Sierpinsky gasket consists of three congruent pieces. Magnified by 2 they are identical with the whole fractal. Therefore, the dimension of the Sierpinsky gasket is $d = \ln(3)/\ln(2) \approx 1.59$.

The Cantor set C, as well as the fractal shown in Fig. 2.10, are self-similar fractals with simple construction rules. Many fractals encountered in physical and mathematical applications are not at all that simple. In order to compute their dimensions, one has to use numerical methods. The following are two frequently employed numerical methods for computing fractal dimensions.

The box counting dimension
This method is probably the most widely used method for the determination of the fractal dimension of a set. The embedding space with dimension D is divided into D-dimensional boxes of side length ϵ. It is then determined how many of the boxes contain points of the fractal. This number is called $N(\epsilon)$. If $N(\epsilon)$ scales like ϵ^{-d} for $\epsilon \to 0$, the frac-

Fig. 2.10.   The Sierpinsky gasket, a classic fractal.

tal dimension of the set is determined and equal to d. The dimension obtained by this method is called the "box counting dimension".

The uncertainty dimension

The method of the uncertainty dimension was introduced in 1985 by Grebogi *et al.* Suppose we are given a fractal that can be covered by the interval $[0, 1]$. Also, suppose that we are provided with a mapping function f that generates the fractal and determines the orbits of seeds x_0 in $[0, 1]$. Choose $\epsilon > 0$ and $N = 1/\epsilon$ random points x_j in $[0, 1]$. Now, for every one of the N points determine its *lifetime* l_j. The lifetime is the number of times the seed x_j can be iterated forward with f without ionizing it. Then, for every x_j determine the lifetimes of $x_j + \epsilon$ and $x_j - \epsilon$. Call them $l_j^{(+)}$ and $l_j^{(-)}$, respectively. For $l_j = l_j^{(+)} = l_j^{(-)}$ the point x_j is called ϵ certain. Otherwise, the point x_j is called ϵ uncertain. This procedure measures the fraction of gaps in the fractal on a scale ϵ. Define $f(\epsilon)$ as the ratio of ϵ uncertain points and the total number of points N. Then, $f(\epsilon)$ is also called the uncertainty fraction.

For simple self-similar fractals $f(\epsilon)$ scales like a powerlaw: $f(\epsilon) \sim \epsilon^\nu$, where ν is called the uncertainty exponent. In this case the quantities needed in (2.3.8) for calculating the fractal dimension are easily available: $k = f(\epsilon) \cdot N$, $m = 1/\epsilon$. Therefore,

$$d = \frac{\ln[f(\epsilon)/\epsilon]}{\ln(1/\epsilon)} = 1 - \nu. \tag{2.3.9}$$

The generalization to D dimensions is straightforward. We obtain

$$d = D - \nu. \tag{2.3.10}$$

We conclude this section by pointing out an important relationship between the decay rate γ, the Lyapunov exponent λ and the fractal dimension d of a one-dimensional self-similar fractal. In the context of fractals, the Lyapunov exponent is the rate of stretching given by

$$\lambda = \ln(m). \tag{2.3.11}$$

If the fractal consists of k congruent sub-pieces that can be made equal to the whole fractal by stretching with the magnification factor m, then the dimension d of the set is given by (2.3.8). Also, from m and k we can calculate the decay rate. The fraction of probability remaining in every application of the fractal generating mapping is k/m. Since $P_B(n) = (k/m)^n = \exp[-n \ln(m/k)]$, we get the decay rate

$$\gamma = \ln(m/k). \tag{2.3.12}$$

Using (2.3.11) in (2.3.8) we have $d = \ln(k)/\lambda$, or $\ln(k) = d\lambda$. With these relations we have:

$$\gamma = \lambda - \ln(k) = \lambda(1-d). \qquad (2.3.13)$$

This is the relation between the decay rate γ, the Lyapunov exponent λ and the fractal dimension d we were looking for. The importance of (2.3.13) lies in the fact that the decay rate γ of a decaying system is also defined within quantum mechanics. The relation (2.3.13) allows us to predict (approximately) a quantum observable (γ) using purely classical input (λ and d).

2.4 Symbolic dynamics

Symbolic dynamics is one of the most powerful tools in the theory of chaotic systems. It is a qualitative method for characterizing the dynamics of a given nonlinear system. The power of symbolic dynamics shows whenever a "new" system N, whose properties are not yet known, can be mapped via symbolic dynamics onto an "old" dynamical system O that has already been thoroughly studied, and whose properties are understood. If such a mapping exists, the two systems N and O are dynamically equivalent. Obviously, symbolic dynamics has the potential to save a lot of work.

Symbolic dynamics can also be used for down-to-earth jobs such as, for instance, book keeping of system trajectories. This is illustrated in the following physical example proposed by Eckhardt in 1987. The system is shown in Fig. 2.11. A mass point, launched with impact parameter b, scatters elastically off three hard disks in the plane. Fig. 2.11(a) shows an example of a scattering trajectory that results for a specific choice of the impact parameter. As shown in Fig. 2.11, the particle bounces off disk A, flies to disk B, scatters, and leaves the three-disk set-up after bouncing three more times with disks A and C. A natural question to ask is whether there is a limit on the number of times a particle can scatter. The answer is no. Any number of scatterings is possible with any possible combination of disks. In order to list all possible trajectories in a convenient way, we use the labels of the three disks (A, B and C) as an *alphabet* of three *symbols*. With the help of the symbols, the time evolution of a scattering trajectory can be characterized by writing down a *word* as a string of the three basic symbols A, B and C in the same order as the particle bounces off the corresponding disks. Thus, the trajectory shown in Fig. 2.11(a) is characterized by the word "ABCAC". The word that characterizes a system trajectory is called the *itinerary* of

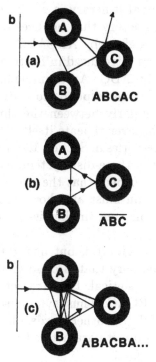

Fig. 2.11. The three-disk scattering system with three sample trajectories. (a) An exiting scattering trajectory, (b) a trapped periodic trajectory, (c) a trapped nonperiodic trajectory.

the trajectory. The set of all possible words is the *symbolic dynamics* of the system.

Since we have an alphabet, and we are allowed to form words, there should also be a grammar which tells us the rules of how to form words. The grammar of the three-disk system is remarkably simple. It has only one rule: Any combination of letters is allowed except for two identical consecutive letters. The restriction in the grammatical rule is immediately obvious since two identical letters in succession corresponds to two successive bounces off the same disk. This is obviously impossible in a force-free situation.

Since our grammatical rule does not impose any restrictions on the length of a symbol sequence, we are obviously allowed to write down infinitely long sequences. There are two kinds: periodic and aperiodic sequences. With the help of the mapping $A \to 0$, $B \to 1$ and $C \to 2$ the symbol sequences can be interpreted as real numbers in base-3 notation. On the basis of this analogy, we call finite (or infinite periodic) sequences "rational" and infinite aperiodic sequences "irrational". Thus, the itinerary of the trajectory shown in Fig. 2.11(a) is rational. An example of an infinite rational itinerary is shown in Fig. 2.11(b). It corresponds to a trapped trajectory that bounces periodically between the three disks. Its itinerary is given by ABCABC.... In analogy to the notation for rational numbers introduced in Section 2.1 we abbreviate this infinite symbol sequence by $\overline{ABC}$. An example of an irrational itinerary is shown in Fig. 2.11(c). It corresponds to a dynamically trapped trajectory that bounces irregularly between the three disks and produces a symbol sequence which never repeats itself. We already encountered dynamically trapped trajectories in Section 1.1 in connection with the dynamics of box C. Since the three-disk system, as well as box C, exhibit irrational trajectories, we suspect that the two systems are dynamically equivalent. This observation illustrates how symbolic dynamics can be used to establish connections and similarities between systems that look very different.

Symbolic dynamics is a powerful, but qualitative, tool. We can easily imagine a system trajectory that is close to the trajectory shown in Fig. 2.11(a), but not quite identical. Both trajectories are then characterized by the same word, ABCAC. Thus, symbol sequences do not specify system trajectories uniquely. In other words, there is no one-to-one correspondence between the impact parameters b and the words constructed from the alphabet { A, B, C }.

The application of symbolic dynamics to the three-disk scattering system helped us to focus right away on the essential questions: (i) Is it possible to have an infinite number of scatterings? (ii) Are there trapped trajectories, and, if yes, how many?

The trapped set of trajectories is especially interesting. It consists of two parts, the set L, which contains dynamically trapped trajectories that start outside the scattering system, and the set Λ of *invariant* trajectories that are completely decoupled from the outside and bounce forever between A, B and C, no matter whether the trajectory is run forwards or backwards in time. The set Λ can be further decomposed into rational and irrational members. The dynamically trapped trajectories of the variety L may "emulate" any trajectory in Λ for any length of time at any stage in their time evolution. Since the periodic orbits in Λ can have any length and their symbols, subject to the grammatical rule, can be chained in a great variety of patterns, it can now be appreciated that the scattering dynamics of the three-disk system is very complex indeed.

Since the set Λ of trapped orbits is not reachable from the outside, it "repells" scattering trajectories that eventually exit the system. Exiting scattering trajectories, which at some point in their time evolution may be "close" to a trapped trajectory, will eventually be repelled from the trapped trajectory and scatter off to infinity. This is the reason why the set Λ of unreachable trapped trajectories is also called a *repeller*. Repeller sets are very important for the qualitative characteristics of a dynamical system and even imprint their presence on the system's quantum mechanics. The repeller set of the three-disk scattering problem and its quantum implications were studied in a sequence of three classic papers by Gaspard and Rice (1989a–c). In this book symbolic dynamics is applied on an elementary level in the chapter on chaotic scattering (Chapter 9) and in the chapter on the one-dimensional helium atom (Chapter 10).

3
Chaos in classical mechanics

Poincaré (1892, reprinted (1993)) was the first to appreciate that exponential sensitivity in mechanical systems can lead to exceedingly complicated dynamical behaviour. Surprisingly, complicated systems are not necessary for chaos to emerge. In fact, chaos can be found in the simplest dynamical systems. Well known examples are the driven pendulum (Chirikov (1979), Baker and Gollub (1990)), the double pendulum (Shinbrot *et al.* (1992)), and the classical versions of the hydrogen atom in a strong magnetic (Friedrich and Wintgen (1989)) or microwave (Casati *et al.* (1987)) field.

In general it is not possible to understand the spectra and wave functions of highly excited atoms and molecules without reference to their classical dynamics. The correspondence principle, e.g., assumes knowledge of the classical Hamiltonian as a starting point. Since the Lagrangian and Hamiltonian formulations of classical mechanics provide the most natural bridge to quantum mechanics, we start this chapter with a brief review of elementary concepts in Lagrangian and Hamiltonian mechanics (see Section 3.1). The double pendulum, an example of a classically chaotic system, is investigated in Section 3.2. This is also the natural context in which to introduce the idea of Poincaré sections. With the help of Poincaré sections we can reduce the continuous motion of a mechanical system to a discrete mapping. This is essential for visualization and analysis of a chaotic system. A discussion of integrability and chaos in Section 3.3 concludes Chapter 3.

3.1 Lagrangian and Hamiltonian mechanics

The most basic entity in mechanics is the *mass point*. It is one of the earliest and most fruitful abstractions in physics. The mass point is an extrapolation from real slabs of matter to something that has no form and no extension. The only attributes of a mass point are its position, its velocity and its mass.

To locate a mass point in space we specify its Cartesian coordinates x, y and z. The velocity of the mass point is given by $v_x = dx/dt = \dot{x}$, $v_y = dy/dt = \dot{y}$ and $v_z = dz/dt = \dot{z}$. The "dot notation" dates back to Newton.

But not only Cartesian coordinates can serve to specify position. To locate a star in the sky, e.g., one uses angles. To locate a pearl on a bent wire one uses the arc length. Mechanical problems often become much simpler to solve when expressed in the right coordinate system. However, while a change of coordinate system is sometimes useful, it would be very cumbersome if the basic laws of classical mechanics depended on the specific coordinate system at hand. It was therefore a great breakthrough when Lagrange (1788) developed a formulation of classical mechanics that treats all coordinate systems on an equal footing. Even better, the formalism does not care about how many particles there are, nor how many coordinates are needed to specify their positions. Therefore, following Lagrange, we introduce the idea of generalized coordinates denoted by $q_\alpha, \alpha = 1, 2, ..., n$. Here, the q_α can represent anything from Cartesian coordinates to angles to arc lengths (they are also allowed to occur in mixtures), and n is the total number of coordinates needed to specify the system uniquely. The number n is called the number of degrees of freedom of the system, or simply the number of freedoms of the system. Unless otherwise specified, q means the whole set of $q_\alpha, \alpha = 1, 2, ..., n$.

The total time derivatives of the generalized coordinates dq_α/dt are called the generalized velocities. They are denoted by $\dot{q}_\alpha$. Again, $\dot{q}$ specifies the set $\dot{q}_\alpha$, $\alpha = 1, 2, ..., n$.

Now that we have introduced coordinates and velocities, the next question is how to predict the time evolution of a mechanical system. This is accomplished by solving a set of ordinary differential equations, the equations of motion, which can be derived from the principle of least action. It was discovered by Maupertuis and was further developed by Euler, Lagrange and Hamilton (d'Abro (1951)).

Lagrangian mechanics
In order to apply the principle of least action we first assign a function L, the Lagrangian function, to a mechanical system M

$$M \rightarrow L(q, \dot{q}; t). \qquad (3.1.1)$$

Then, the principle of least action states that the path a mechanical system takes from (q_1, t_1) to (q_2, t_2) is such as to minimize the *action integral*

$$S[q] = \int_{t_1}^{t_2} L(q, \dot{q}; t) \, dt. \tag{3.1.2}$$

There are cases where stationarity of S is sufficient. For our purposes, however, it is enough to assume that the actual motion of a system corresponds to a minimum of the action S.

Not only mechanics but all of physics can be derived from the principle of least action. There are appropriate Lagrangian functions for electrodynamics, quantum mechanics, hydrodynamics, etc., which all allow us to derive the basic equations of the respective discipline from the principle of least action. In this sense, the principle of least action is the most powerful "economy principle" known in physics since it is sufficient to know the principle of least action, and the rest can be derived. Nature as a whole seems to be organized according to this principle. The principle of least action can be found under various names in nearly every branch of science. For instance: the principle of least cost in economy or Fermat's principle of least time in optics.

The Lagrangian L and the action integral S are both functions in the sense that they are prescriptions of how to assign numbers to elements of their respective domains. But while L operates on numbers (the values of q, $\dot{q}$ and t at any given time t), the "function" S operates on functions, i.e. on test paths $q(t)$. The function $S[q]$ is thus often referred to as a functional.

Suppose that $q(t)$, $\dot{q}(t)$ is the actual path a mechanical system describes in $(q, \dot{q})$ space. Then, S is minimal. In other words, S increases if we deviate from the "right" path. Let us introduce functions $\eta(t)$, called *variational functions*. We require that $\eta(t_1) = 0$, $\eta(t_2) = 0$ and $|\eta(t)|$ small for $t_1 < t < t_2$. Let us calculate S for the new path $r(t) = q(t) + \eta(t)$, called a *variation* of the optimal path

$$S' = S[r] = \int_{t_1}^{t_2} L[r(t), \dot{r}(t); t] \, dt = \int_{t_1}^{t_2} L[q(t) + \eta(t), \dot{q}(t) + \dot{\eta}(t); t] \, dt$$

$$\approx S + \int_{t_1}^{t_2} \left\{ \frac{\partial L}{\partial q} \eta(t) + \frac{\partial L}{\partial \dot{q}} \dot{\eta}(t) \right\} dt. \tag{3.1.3}$$

The difference $S' - S$ is called variation of the action. It is denoted by δS. Therefore, with a partial integration, and taking into account that $\eta(t)$ vanishes at t_1 and t_2, we obtain:

$$\delta S = \int_{t_1}^{t_2} \left\{ -\frac{d}{dt} \frac{\partial L}{\partial \dot{q}} + \frac{\partial L}{\partial q} \right\} \eta(t) \, dt. \tag{3.1.4}$$

Since S is stationary, δS has to vanish. Since the functions $\eta(t)$ are arbitrary, the integrand in (3.1.4) vanishes only if

$$\frac{d}{dt}\frac{\partial L}{\partial \dot{q}} - \frac{\partial L}{\partial q} = 0. \tag{3.1.5}$$

The above equation is shorthand for

$$\frac{d}{dt}\frac{\partial L}{\partial \dot{q}_\alpha} - \frac{\partial L}{\partial q_\alpha} = 0, \qquad \alpha = 1, 2, ..., f. \tag{3.1.6}$$

It is useful to define

$$p_\alpha = \frac{\partial L}{\partial \dot{q}_\alpha}. \tag{3.1.7}$$

The quantities p_α are called generalized momenta. They can be used together with the coordinates q_α to define a system trajectory. The system trajectory evolves in the $2f$-dimensional space spanned by the f coordinates q and the f coordinates p. This space plays a central role in analytical mechanics. It is called the *phase space* of the system. A point in phase space uniquely defines the mechanical state of a system. In connection with Poincaré's method of surfaces of section, the phase space is also an important vehicle for the visualization of the qualitative behaviour of a given dynamical system. An example is presented in Section 3.2.

The set of equations (3.1.6) is a special case of the Euler equations of the calculus of variations (see, e.g., Arnold (1989)). They are referred to as the Euler-Lagrange equations in the literature. The Euler-Lagrange equations are ordinary second order differential equations for the generalized coordinates q_α.

The Lagrangian function L assigned to a mechanical system M is not unique. The same equations of motion are obtained if instead of working with L, we would have chosen $\bar{L} = L + dF/dt$, where F is any arbitrary function of the coordinates, the momenta and t. This fact is important for the theory of canonical transformations to be discussed below.

A last question remains: How do we choose the function L for a given mechanical system? A very important part of the answer is that L cannot, in principle, be derived by pure reasoning. L has to be chosen such that the equations of motion obtained from the principle of least action reflect the physical reality.

How to choose the appropriate Lagrangian that fits our every-day "mechanical universe" is answered by the following argument (Landau and Lifschitz (1970)). All our experience with mechanical systems was summarized by Newton in his famous laws. According to Newton,

$$m\ddot{x} = f(x), \tag{3.1.8}$$

where $f(x)$ is the force acting on the mass point m. If $f(x)$ can be derived from a potential, we can write

$$m\ddot{x} = -\frac{\partial V(x)}{\partial x}. \qquad (3.1.9)$$

Since L cannot depend on whether our particle goes left or right ($\dot{x} > 0$ or $\dot{x} < 0$), the simplest way of incorporating this fact into the Lagrangian is to demand that the Lagrangian depends only on the square of the velocity according to

$$L = L(x, \dot{x}^2; t). \qquad (3.1.10)$$

Using this particular form of L in the Euler-Lagrange equations of motion (3.1.6) we get

$$\frac{d}{dt}\frac{\partial L(x,\dot{x}^2;t)}{\partial \dot{x}} - \frac{\partial L}{\partial x} = \left[\frac{d}{dt}\frac{\partial L}{\partial(\dot{x}^2)}\right]2\dot{x} + \left[\frac{\partial L}{\partial(\dot{x}^2)}\right]2\ddot{x} - \frac{\partial L}{\partial x} = 0. \qquad (3.1.11)$$

Since there are no terms proportional to $\dot{x}$ in Newton's equations, we must have

$$\frac{d}{dt}\frac{\partial L}{\partial(\dot{x}^2)} = 0. \qquad (3.1.12)$$

But this immediately implies that $\partial L/\partial\dot{x}^2 = const$. Comparing the second term in (3.1.11) with Newton's law, the constant turns out to be half the mass of the particle. Thus we have

$$\frac{\partial L}{\partial \dot{x}^2} = m/2, \qquad (3.1.13)$$

or,

$$L(x,\dot{x};t) = \frac{1}{2}m\dot{x}^2 + A(x), \qquad (3.1.14)$$

where $A(x)$ is an as yet unknown function of x. But since according to Newton's law the third term in (3.1.11) must be $-\partial V/\partial x = f$, it follows immediately that (up to a global constant) $A(x) = -V(x)$. Thus, we have derived the form of the Lagrangian as

$$L = T - V, \qquad (3.1.15)$$

where $T = \frac{1}{2}m\dot{x}^2$ is the kinetic energy and $V(x)$ is the potential energy of the system.

The Lagrangian formalism is widely applied to solve mechanical problems. But besides Lagrange's formalism, there is a formalism first developed by Hamilton. Sometimes the Hamiltonian formalism presents certain advantages in solving mechanical problems. But the real power of a Hamiltonian formulation of classical mechanics shows on the conceptual level. Hamilton's method provides us with more insight into the structure of classical mechanics and serves as the starting point for many other related fields of physics, such as statistical mechanics, fluid dynamics and quantum mechanics.

Hamiltonian formulation
The Hamiltonian function, together with its associated canonical equations of motion, can be derived in the following way.

We start by calculating the total differential of the Lagrangian:

$$dL = \frac{\partial L}{\partial q} dq + \frac{\partial L}{\partial \dot{q}} d\dot{q} + \frac{\partial L}{\partial t} dt. \tag{3.1.16}$$

Now we use the Euler-Lagrange equations of motion (3.1.6) and the definition of the canonical momentum p in (3.1.7) to obtain

$$dL = \dot{p}\, dq + p\, d\dot{q} + \frac{\partial L}{\partial t} dt. \tag{3.1.17}$$

This can also be written as:

$$d\,[p\dot{q} - L] = \dot{q}\, dp - \dot{p}\, dq - \frac{\partial L}{\partial t} dt. \tag{3.1.18}$$

The quantity in the square brackets is defined as the Hamiltonian function H. Physically it is the total energy of the system. Explicitly,

$$H(p, q; t) = \sum_{\alpha} p_{\alpha} \dot{q}_{\alpha} - L. \tag{3.1.19}$$

Since H is a function of p, q and t, we have to express $\dot{q}$ as a function of p, q and t. This can be done with the help of (3.1.7). The total differential of (3.1.19) is given by

$$dH = \frac{\partial H}{\partial p} dp + \frac{\partial H}{\partial q} dq + \frac{\partial H}{\partial t} dt. \tag{3.1.20}$$

Comparing this expression with (3.1.18) we get immediately:

$$\dot{p}_{\alpha} = -\frac{\partial H}{\partial q_{\alpha}} \qquad \dot{q}_{\alpha} = \frac{\partial H}{\partial p_{\alpha}}. \tag{3.1.21}$$

These equations are known as the canonical equations of motion. We also have:

$$\frac{\partial H}{\partial t} = -\frac{\partial L}{\partial t}. \tag{3.1.22}$$

From (3.1.20) we obtain the total time derivative of the Hamiltonian function. It is given by

$$\frac{dH}{dt} = \frac{\partial H}{\partial p}\dot{p} + \frac{\partial H}{\partial q}\dot{q} + \frac{\partial H}{\partial t} = \frac{\partial H}{\partial t}. \tag{3.1.23}$$

Not explicitly time dependent systems are called autonomous. For autonomous systems $\partial H/\partial t = 0$ and we have $H = E = const$, i.e. the total energy of the system is conserved. Clearly the system of equations (3.1.21) is more symmetric than the set (3.1.6) of second order differential equations obtained from the Lagrangian formalism.

The transformation from the Lagrangian representation to the Hamiltonian formalism is a special case of a *Legendre transformation*. A Legendre transformation, quite generally, is the following: Suppose we have a function $g(x,y)$ with x and y as its two independent variables. The total differential of this function is given by

$$dg = \frac{\partial g}{\partial x}\,dx + \frac{\partial g}{\partial y}\,dy = u(x,y)\,dx + v(x,y)\,dy. \qquad (3.1.24)$$

Now, suppose that instead of x we would like to introduce u as an independent variable and we look for a function $h(u,y)$ such that

$$dh = \frac{\partial h}{\partial u}\,du + \frac{\partial h}{\partial y}\,dy. \qquad (3.1.25)$$

The function $h = g - ux$ does the job since

$$dh = dg - d(ux) = \frac{\partial g}{\partial x}\,dx + \frac{\partial g}{\partial y}\,dy - x\,du - u\,dx = -x\,du + v\,dy.$$
$$(3.1.26)$$

Comparing the above procedure with our derivation of the Hamiltonian function, we see that the transformation from the Lagrangian function to the Hamiltonian function is a Legendre transformation.

Legendre transformations are very useful in physics in general. In thermodynamics, e.g., they can be used for switching between thermodynamic potentials. Another application of Legendre transformations is in the theory of canonical transformations to be discussed next.

Canonical transformations
One of the main advantages of the Hamiltonian formalism is that it treats coordinates and momenta on an equal footing. In order to simplify the formulation and solution of mechanical problems it is often useful to define new coordinates and momenta which are mixtures of the old coordinates and momenta. But in order not to destroy the basic structure of the theory, only transformations which preserve the canonical structure (3.1.21) are allowed. These transformations are also called *canonical transformations* because they respect the canonical structure of the equations of motion (3.1.21). A transformation from the old momenta and coordinates p and q, to new momenta P and new coordinates Q according to

$$P = P(p,q;t), \quad Q = Q(p,q;t) \qquad (3.1.27)$$

is canonical if the equations of motion for the new coordinates and momenta are given by

$$\dot{P} = -\frac{\partial \tilde{H}(P,Q;t)}{\partial Q}, \quad \dot{Q} = \frac{\partial \tilde{H}(P,Q;t)}{\partial P}, \tag{3.1.28}$$

where $\tilde{H}$ is the new Hamiltonian. But since the structure of (3.1.28) is canonical, it can also be derived from the principle of least action:

$$\delta \int_{t_1}^{t_2} \tilde{L}\, dt = \delta \int_{t_1}^{t_2} \left(\sum_\alpha P_\alpha \dot{Q}_\alpha - \tilde{H}(P,Q;t) \right) dt = 0. \tag{3.1.29}$$

The actual motion of the system does not depend on whether it is described with the help of the old or the new set of coordinates. Therefore, the original principle of least action still holds:

$$\delta \int_{t_1}^{t_2} L\, dt = \delta \int_{t_1}^{t_2} \left(\sum_\alpha p_\alpha \dot{q}_\alpha - H(p,q;t) \right) dt = 0. \tag{3.1.30}$$

We saw above that the Lagrangian function L is only determined up to the total time derivative of an arbitrary function F. Since both integrals, (3.1.29) as well as (3.1.30), describe the same motion, the integrands can differ only by the quantity dF/dt. We obtain

$$\sum_\alpha p_\alpha \dot{q}_\alpha - H = \sum_\alpha P_\alpha \dot{Q}_\alpha - \tilde{H} + \frac{d}{dt} F. \tag{3.1.31}$$

We see now that the indeterminacy of the Lagrangian is very important for the existence of canonical transformations. The function F in (3.1.31) is called the generating function of the canonical transformation.

There are four possible forms of the generating function F for which Goldstein (1976) introduced the following notation:

$$F_1(q,Q;t)\, ; \quad F_2(q,P;t)\, ; \quad F_3(p,Q;t)\, ; \quad F_4(p,P;t). \tag{3.1.32}$$

Which form of F to choose depends on the problem at hand. Choosing, e.g., F_1 in (3.1.31), yields

$$\sum_\alpha p_\alpha \dot{q}_\alpha - H = \sum_\alpha P_\alpha \dot{Q}_\alpha - \tilde{H} + \sum_\alpha \left(\frac{\partial F_1}{\partial q_\alpha} \dot{q}_\alpha + \frac{\partial F_1}{\partial Q_\alpha} \dot{Q}_\alpha \right) + \frac{\partial F_1}{\partial t}. \tag{3.1.33}$$

Comparing terms, we get

$$p_\alpha = \frac{\partial F_1}{\partial q_\alpha}\, ; \quad P_\alpha = -\frac{\partial F_1}{\partial Q_\alpha}\, ; \quad \tilde{H} = H + \frac{\partial F_1}{\partial t}. \tag{3.1.34}$$

This means that the canonical transformation is determined once F_1, the generating function, is known.

The generating function F_2 can be obtained by a Legendre transformation from F_1. Defining F_1 as the Legendre transform of F_2 we have

$$F_1(q, Q; t) = F_2(q, P; t) - \sum_{\alpha} P_\alpha Q_\alpha. \qquad (3.1.35)$$

Using (3.1.35) in (3.1.33) we get

$$p_\alpha = \frac{\partial F_2}{\partial q_\alpha} \; ; \quad Q_\alpha = \frac{\partial F_2}{\partial P_\alpha} \; ; \quad \tilde{H} = H + \frac{\partial F_2}{\partial t}. \qquad (3.1.36)$$

Similar relations can be derived for F_3 and F_4.

We can use canonical transformations to solve mechanical problems. Suppose that we can find a canonical transformation such that the original Hamiltonian $H(p, q; t)$ is transformed into the new Hamiltonian $\tilde{H}(P; t)$, i.e. the new Hamiltonian $\tilde{H}$ does not depend at all on the position coordinates. Then, according to (3.1.6), we have:

$$\dot{P}_\alpha = -\frac{\partial \tilde{H}}{\partial Q_\alpha} = 0. \qquad (3.1.37)$$

But this means that the new momenta P are *constants of the motion*. Therefore, the equations (3.1.37) can be integrated immediately:

$$P_\alpha = C_\alpha, \qquad (3.1.38)$$

where the C_α are arbitrary constants (consistent with the boundary conditions of the mechanical problem). For the Q-equations we obtain:

$$\dot{Q}_\alpha = \frac{\partial \tilde{H}}{\partial P_\alpha} = \omega_\alpha, \qquad (3.1.39)$$

where the ω_α are functions of the P_α. But since all the P_α are constant, so are the ω_α and we can immediately integrate the equations for Q_α:

$$Q_\alpha = \omega_\alpha t. \qquad (3.1.40)$$

The existence of constants of the motion is exceptional in the set of all mechanical systems. A full set of f constants of the motion is an even rarer case. It guarantees that explicit solutions of the equations of motion can be found. In this case the Hamiltonian system is said to be *integrable*. According to (3.1.38) and (3.1.40), the solutions of an integrable mechanical system are very simple. Thus integrability is equivalent with absence of chaos. Of the many mechanical systems of physical significance, only very few are known to be integrable. Many more show chaos and are therefore not integrable. Thus it seems that the occurrence of chaos in mechanical systems is the rule rather than the exception, and we should be prepared to encounter complete integrability only for very few, very special dynamical systems. This point is illustrated in the following section where we discuss the double pendulum. Although a simple two degree of freedom mechanical system, it displays complexity and chaos in its dynamics. Therefore, not even for a system as simple as the double pendulum does an integrating canonical transformation exist. A more detailed discussion of integrability and chaos is presented in Section 3.3.

Integrable systems are mostly characterized by obvious symmetries, such as rotational invariance, which allows for the reduction of the number of degrees of freedom, and thereby integration of the problem. Not so obvious "hidden" dynamical symmetries may sometimes exist. They can also be used for the integration of a given mechanical problem. But they are hard to find. An example is the Toda system discussed in Section 3.3. Since a given mechanical problem is most likely not integrable, we should not expect to be able to "solve" a given mechanical problem by means of a suitable canonical transformation. However, canonical transformations are very useful to simplify mechanics problems, or to eliminate some degrees of freedom by exploiting geometric or dynamical symmetries. Applied in this way, canonical transformations are an indispensible tool in the theory of dynamical systems. We shall encounter an instructive example of a simplifying canonical transformation in Section 6.4 in connection with chaotic surface state electrons.

3.2 The double pendulum

The purpose of this section is to illustrate the methods of Lagrangian and Hamiltonian mechanics with the help of a simple mechanical system: the double pendulum. It is shown that although the equations of motion for this system look very simple, the double pendulum is a chaotic system.

Consider a double pendulum in a uniform gravitational field with gravitational acceleration g free to move in the $x - y$ plane as shown in Fig. 3.1. The two mass points m_1 and m_2 are connected by two rigid, massless rods of length l. For the description of the locations (x_1, y_1), (x_2, y_2) of m_1 and m_2, we choose the generalized coordinates ϑ and φ. We consider only the case $m_1 = m_2 = m$. Introducing the unit of time, $t_0 = \sqrt{l/g}$, the unit of energy, $E_0 = mgl$, and a notation for the difference of the two angles, $\Delta = \vartheta - \varphi$, the Lagrangian of the pendulum is given by

$$L = T - U, \tag{3.2.1}$$

where

$$T = \frac{1}{2}\left[2\dot{\vartheta}^2 + \dot{\varphi}^2 + 2\dot{\vartheta}\dot{\varphi}\cos(\Delta)\right], \tag{3.2.2}$$

and

$$U = -2\cos(\vartheta) - \cos(\varphi). \tag{3.2.3}$$

The total energy of the system is

$$E = T + U. \tag{3.2.4}$$

The Euler-Lagrange equations of motion are

$$2\ddot{\vartheta} + \ddot{\varphi}\cos(\Delta) + \dot{\varphi}^2\sin(\Delta) + 2\sin(\vartheta) = 0,$$

$$\ddot{\varphi} + \ddot{\vartheta}\cos(\Delta) - \dot{\vartheta}^2\sin(\Delta) + \sin(\varphi) = 0. \tag{3.2.5}$$

Defining $z_1 = \vartheta$, $z_2 = \varphi$, $z_3 = \dot{\vartheta}$, and $z_4 = \dot{\varphi}$, the system (3.2.5) can be written as a system of four first order differential equations

$$\dot{z}_1 = z_3, \quad \dot{z}_2 = z_4,$$

$$\dot{z}_3 = \frac{\sin(z_2)\cos(\Delta) - 2\sin(z_1) - [z_3^2\cos(\Delta) + z_4^2]\sin(\Delta)}{2 - \cos^2(\Delta)},$$

$$\dot{z}_4 = \frac{2\sin(z_1)\cos(\Delta) - 2\sin(z_2) + [z_4^2\cos(\Delta) + 2z_3^2]\sin(\Delta)}{2 - \cos^2(\Delta)}. \tag{3.2.6}$$

Thus, the pendulum phase space is four-dimensional. For the special case considered here, the equations of motion (3.2.6) agree with those derived by Shinbrot *et al.* (1992).

For a given set of initial conditions $z_k(t = 0)$, $k = 1, ..., 4$, the solution of (3.2.6) defines a system trajectory. Since the energy is conserved, this trajectory winds in a three-dimensional sub-space of the four-dimensional phase space of the pendulum. It is not easy to imagine the motion in such a high-dimensional space, and we need to devise a visualization method that gives some clues as to the qualitative nature of the system trajectories. One particularly useful method, the method of the *surface of*

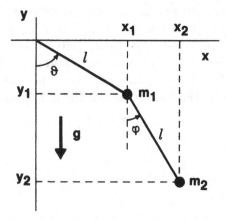

Fig. 3.1. Sketch of a double pendulum.

section, was suggested by Poincaré and is discussed below. But in order to get a first impression for how complicated the phase-space trajectories of a double pendulum can be, it is enough for the moment to simply solve (3.2.6) numerically for one set of initial conditions, and to plot the projection of the resulting system trajectory in the ϑ-φ plane of the pendulum phase space. The result is shown in Fig. 3.2 for $E = -1$, $z_1(0) = 1$, $z_2(0) = 1/2$, $z_3(0) = 0$ and $z_4(0) = \sqrt{2}\{E + 2\cos[z_1(0)] + \cos[z_2(0)]\}^{1/2}$.

The trajectory in Fig. 3.2 is obviously very complicated. This demonstrates that the double pendulum, even in its simplified version studied here, is capable of exhibiting very complicated motion.

We will now formulate the dynamics of the double pendulum within the Hamiltonian approach. The generalized momenta of the double pendulum are given by

$$p_\vartheta = \frac{\partial L}{\partial \dot{\vartheta}} = 2\dot{\vartheta} + \dot{\varphi}\cos(\Delta), \quad p_\varphi = \frac{\partial L}{\partial \dot{\varphi}} = \dot{\varphi} + \dot{\vartheta}\cos(\Delta). \quad (3.2.7)$$

The Hamiltonian is defined as

$$H(\vartheta, \varphi, p_\vartheta, p_\varphi) = p_\vartheta \dot{\vartheta} + p_\varphi \dot{\varphi} - L =$$

$$\dot{\vartheta}^2 + \frac{1}{2}\dot{\varphi}^2 + \dot{\vartheta}\dot{\varphi}\cos(\Delta) - 2\cos(\vartheta) - \cos(\varphi). \quad (3.2.8)$$

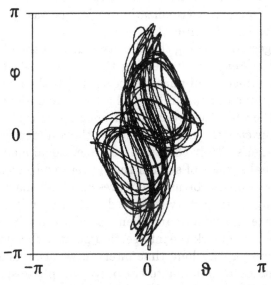

Fig. 3.2. Projection of a phase-space trajectory of the double pendulum on the ϑ-φ plane.

According to (3.2.4) above, the Hamiltonian equals the total energy E and is a conserved quantity. But since H is a function of the angles and the canonical momenta, the $\dot{\vartheta}$ and $\dot{\varphi}$ terms in (3.2.8) have to be expressed as functions of ϑ, φ, p_ϑ and p_φ. This is easily done if we first solve (3.2.7) for $\dot{\vartheta}$ and $\dot{\varphi}$, and then integrate the canonical equations $\dot{\vartheta} = \partial H/\partial p_\vartheta$ and $\dot{\varphi} = \partial H/\partial p_\varphi$ (interpreted as partial differential equations in ϑ, φ, p_ϑ, p_φ) with respect to p_ϑ and p_φ. The result is

$$H = \frac{1}{2-\cos^2(\Delta)}\left\{\frac{1}{2}p_\vartheta^2 - p_\vartheta p_\varphi \cos(\Delta) + p_\varphi^2\right\} - 2\cos(\vartheta) - \cos(\varphi).$$

$$(3.2.9)$$

The resulting canonical equations of motion are

$$\dot{\vartheta} = \frac{1}{2-\cos^2(\Delta)}[p_\vartheta - p_\varphi \cos(\Delta)], \quad \dot{\varphi} = \frac{1}{2-\cos^2(\Delta)}[2p_\varphi - p_\vartheta \cos(\Delta)],$$

$$\dot{p}_\vartheta = \frac{\sin(2\Delta)[p_\vartheta^2/2 - p_\vartheta p_\varphi \cos(\Delta) + p_\varphi^2]}{[2-\cos^2(\Delta)]^2} - \frac{p_\vartheta p_\varphi \sin(\Delta)}{2-\cos^2(\Delta)} - 2\sin(\vartheta),$$

$$\dot{p}_\varphi = -\frac{\sin(2\Delta)[p_\vartheta^2/2 - p_\vartheta p_\varphi \cos(\Delta) + p_\varphi^2]}{[2-\cos^2(\Delta)]^2} + \frac{p_\vartheta p_\varphi \sin(\Delta)}{2-\cos^2(\Delta)} - \sin(\varphi).$$

$$(3.2.10)$$

As they stand, the equations of motion (3.2.10) do not provide any additional insight or advantage over the equivalent set of equations (3.2.6) derived on the basis of the Lagrangian method. The Hamiltonian formulation of the problem, however, can be combined with Poincaré's method of the surface of section, a visualization technique vastly superior to the projection method used in Fig. 3.2.

Poincaré suggested reducing the dimensionality of the phase-space flow of a dynamical system by intersecting the energy surface with another suitably chosen sub-space of phase space such that the intersection of the two manifolds is one dimension lower than the dimensionality of the energy surface. The resulting lower-dimensional space is called the *surface of section*. In our case the energy surface is three-dimensional. Intersecting it with another suitably chosen three-dimensional manifold generates a two-dimensional surface of section that is very convenient to visualize. The surface of section can be used to represent phase-space trajectories in the following way. A trajectory is started with some initial conditions on the energy surface. Every now and then, the trajectory intersects the surface of section and we mark the intersection point with a dot. If we follow the trajectory over a very long time interval, we accumulate a considerable set of points corresponding to the particular phase-space trajectory. From the visual appearance of this point set, we can conclude whether the phase-space trajectory is regular or chaotic. A regular phase-space trajectory produces a point set arranged in a one-dimensional geometric pattern on the two-dimensional surface of section whereas a chaotic trajectory generates a point set that appears to fill in two-dimensional areas of the two-dimensional surface of section. There is one subtlety to be discussed at this point. Sometimes the energy surface is topologically complicated. We will encounter an example of this difficulty below. In this case not every intersection of the phase-space trajectory with the surface of section should be marked since otherwise marking all intersections results in a confusing overlay of phase-space portraits corresponding to disjoint portions of the energy surface. With this precaution in mind a well chosen Poincaré section provides us with an excellent idea of the location of phase-space regions in which the motion is simple, and the phase-space regions in which the motion is chaotic.

We construct the surface of section for the double pendulum in the following way.

(i) First we notice that given the total energy E of the double pendulum and three of the four dynamical variables, the fourth can be expressed as a function of the given three. We choose p_ϑ to be this "dependent" variable.

(ii) We now choose the three-dimensional sub-space to be intersected with the energy surface. We define this space by $\vartheta = 0$. It is a three-dimensional hyper-plane of the four-dimensional phase space. The resulting two-dimensional surface of section is the (φ, p_φ) plane.

(iii) We choose a definite energy E and start a trajectory with some initial conditions on the corresponding energy surface. The trajectory intersects the (φ, p_φ) plane in a countable set of points. But according to the above discussion, not all of these points should be marked. This is seen in the following way. For $H = E$ and $\vartheta = 0$ we solve (3.2.9) for p_ϑ. The result is

$$p_\vartheta - p_\varphi \cos(\varphi) = \pm\sqrt{2} \times$$

$$\left\{ [E + 2 + \cos(\varphi)] [2 - \cos^2(\varphi)] + \frac{1}{2} p_\varphi^2 \cos^2(\varphi) - p_\varphi^2 \right\}^{1/2}. \quad (3.2.11)$$

This expression shows that for given φ and p_φ there are two branches for p_ϑ that result in two distinct Poincaré sections. Therefore, naively marking a point in the (φ, p_φ) plane whenever the phase-space trajectory satisfies $\vartheta(t) = 0$ would result in an overlay of two different sections. In order to separate properly the two sections, we mark a point only if $\sigma [p_\vartheta - p_\varphi \cos(\varphi)] > 0$, where σ is the "branch switch". Once selected it has to be kept fixed during the production of the corresponding Poincaré section. The switch σ can take the values ± 1.

(iv) For given energy E not all phase-space points are dynamically reachable. The accessible region on the surface of section can be computed
from (3.2.11). For given φ and ϑ, p_φ in (3.2.11) is a function of p_ϑ
only, and can be maximized (minimized) with respect to this vari-
able. We obtain

$$p_\varphi^{(min,max)} = \mp\sqrt{2E + 4\cos(\vartheta) + 2\cos(\varphi)}. \qquad (3.2.12)$$

Thus, the accessible region in the (φ, p_φ) plane is $p_\varphi^{(min)} \leq p_\varphi \leq$
$p_\varphi^{(max)}$

We are now ready to plot a Poincaré surface of section for the double
pendulum. We choose $E = -2$ and 19 different initial conditions defined
by $\vartheta^{(j)}(t = 0) = 0$, $\varphi^{(j)}(t = 0) = 0$, $p_\varphi^{(j)}(t = 0) = j/7$, $j = -9, ..., +9$.
The initial condition for p_ϑ can be computed from (3.2.11). For $\sigma = 1$,
the resulting Poincaré section is shown in Fig. 3.3(a). For every one of

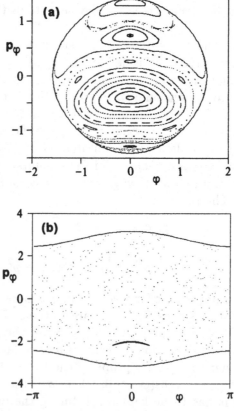

Fig. 3.3.   Poincaré section for the double pendulum. (a) $E = -2$, (b) $E = 2$.
The full lines indicate the dynamically accessible regions.

the 19 initial conditions, the corresponding phase-space trajectory was integrated numerically in the time interval $0 < t < 5000$ and $\vartheta = 0$-points were marked in the φ-p_φ plane according to (iii). The dynamically accessible region, computed according to (3.2.12), is shown as the full line framing the section points. Fig. 3.3(a) shows that at $E = -2$ the motion of the double pendulum is mostly regular. However, nearly all the regularity is destroyed if the energy is raised to $E = 2$, as shown in Fig. 3.3(b). In this case we chose $\vartheta^{(j)}(t = 0) = 0$, $\varphi^{(j)}(t = 0) = 0$, $p_\varphi^{(j)}(t = 0) = j/7$, $j = -18, ..., +18$ in steps of 2. Fig. 3.3(b) shows that only a small regular island persists in a "chaotic sea". This proves visually that the double pendulum is indeed a chaotic system.

Besides the one little regular island at $\varphi \approx 0$ and $p_\varphi \approx -2$ there are undoubtedly more regular islands in the phase space of the double pendulum at $E = 2$. We missed them by our rather coarse choice of initial conditions. As indicated in Fig. 3.3(b), their total area in phase space is probably very small. Nevertheless, Fig. 3.3(b) illustrates an important feature of the phase space of most physical systems: the phase space contains an intricate mixture of regular and chaotic regions. The system is said to exhibit a *mixed phase space*.

As illustrated by Figs. 3.3(a) and (b), Poincaré sections are a very powerful tool for the visual inspection and classification of the dynamics of a given Hamiltonian. The double pendulum illustrates that for autonomous systems with two degrees of freedom a Poincaré section can immediately suggest whether a given Hamiltonian allows for the existence of chaos or not. Moreover, it tells us the locations of chaotic and regular regions in phase space.

3.3 Integrability versus chaos

Constants of the motion were already discussed briefly at the end of Section 3.1. Here we attempt to eliminate some commonly encountered misconceptions surrounding the topic of the very existence of constants of the motion. Also, we present some examples for the occurrence of nontrivial constants of the motion in mechanical systems of current interest.

Suppose we investigate an autonomous Hamiltonian system with f degrees of freedom. Then, the solution of the $2f$ first order canonical equations (3.1.21) contains $2f$ integration constants C_k, $k = 1, ..., 2f$ according to

$$p_\alpha = p_\alpha(t; C_1, ..., C_{2f}), \quad q_\alpha = q_\alpha(t; C_1, ..., C_{2f}). \tag{3.3.1}$$

Following Landau and Lifschitz (1970) we solve (3.3.1) for the $2f$ constants C_k as functions of p_α and q_α. This defines $2f$ functions $C_k(p_\alpha(t),$

$q_\alpha(t))$ such that

$$\frac{d}{dt} C_k(p_\alpha(t), q_\alpha(t)) = 0. \qquad (3.3.2)$$

Thus, the functions C_k are constants of the motion just as defined and discussed in Section 3.1.

The above scheme suggests that all Hamiltonian systems are integrable, since all Hamiltonian systems seem to possess a maximal set C_k, $k = 1, ..., 2f$ of constants of the motion. Indeed, this is the impression conveyed by traditional textbooks on classical mechanics (see, e.g., Landau and Lifschitz (1970), Symon (1971), Goldstein (1976)). But if all Hamiltonian systems have a maximal set of constants of the motion, how can we reconcile this fact with the occurrence of chaos in most Hamiltonian systems?

The answer to this question is that, although the constants C_k undoubtedly exist, their analytical properties may be so complicated that they do not impose any restrictions on the motion of the system. This is immediately clear since the process of finding the constants C_k involves the inversion of a system of $2f$ nonlinear coupled equations. Theorems in mathematics assure us of the *local* existence of $2f$ explicit functions C_k, but *globally*, i.e. for all values of p, q and t, they may only be defined with the help of infinitely many branches. Therefore, we can divide the constants of the motion into two classes, "useful" and "useless". The "useful" constants of the motion possess a simple analytical structure, a finite number of branches, and are valid for all time t. Such constants actually restrict the motion of the system to a sub-manifold of phase space. Thus, the presence of a useful constant of the motion results in a simplification of the mechanical system at hand. The analytical properties of the useless constants are so unbelievably intricate and complex that they do not result in a reduction of the dimensionality of phase space. Their presence is no obstacle for chaos.

The "C_k puzzle" and its solution was well known to Poincaré. In fact he uses this example in his book on "Science and Hypothesis" (1902). It is also discussed in great detail by Brillouin (1960).

According to Noether's theorem (Arnold (1989)) symmetries of a mechanical system are always accompanied by constants of the motion. According to Section 3.1, system symmetries can be "obvious" (e.g. geometric) or "hidden". Examples for obvious symmetries that lead to constants of the motion are invariance with respect to time translations, spatial translations and rotations. Invariance with respect to time leads to the conservation of energy, spatial and rotational symmetries lead to the conservation of linear and angular momentum, respectively (see, e.g., Landau and Lifschitz (1970)). "Hidden" symmetries cannot be associated with any obvious geometric invariances and are very hard to identify. Hidden symmetries lead to unexpected constants of the motion.

A famous example for the presence of a hidden symmetry is the Toda Hamiltonian (see, e.g., Lichtenberg and Lieberman (1983)). In its simplest form it describes the dynamics of three particles moving on a ring subjected to repulsive two-body forces according to

$$H^{(Toda)} = \frac{1}{2}(p_1^2 + p_2^2 + p_3^3) +$$

$$\exp[-|\phi_1 - \phi_3|] + \exp[-|\phi_2 - \phi_1|] + \exp[-|\phi_3 - \phi_2|] - 3. \qquad (3.3.3)$$

The Toda Hamiltonian (3.3.3) exhibits a simple symmetry that results in a trivial constant of the motion $P = p_1 + p_2 + p_3$. Using this constant, the Hamiltonian (3.3.3) can be brought into the form

$$H = \frac{1}{2}(p_x^2 + p_y^2) + \frac{1}{24}[\exp(2y + 2\sqrt{3}x) + \exp(2y - 2\sqrt{3}x) + \exp(-4y)] - 1/8.$$
$$(3.3.4)$$

This Hamiltonian does not exhibit any obvious symmetries and we would suspect, given its complicated appearance, that its dynamics is chaotic. It came therefore as a complete surprise when Ford *et al.* showed in 1973 that its Poincaré sections look regular, indicating the existence of a hidden simple constant of the motion. Based on these findings, Hénon (1974) was able to isolate the constant of the motion and to state it explicitly. It is given by

$$C = 8p_x(p_x^2 - 3p_y^2) + (p_x + \sqrt{3}p_y)\exp[(2y - 2\sqrt{3}x)]$$

$$-2p_x\exp(-4y) + (p_x - \sqrt{3}p_y)\exp[(2y + 2\sqrt{3}x)]. \qquad (3.3.5)$$

This constant of the motion is far from simple, but the crucial fact is that it can be stated analytically in closed form. Thus, its analytical properties are simple enough to act *isolating*, i.e. it restricts the dynamics sufficiently enough so that phase-space trajectories are no longer allowed to roam uninhibited throughout the three-dimensional energy surface of (3.3.4). The constant (3.3.5) restricts them to lie in a sub-space of the energy surface giving rise to completely regular Poincaré sections. According to our classification above, the constant (3.3.5) is certainly in the category of useful constants of the motion. In conjunction with the other constants of the motion it is powerful enough to rule out the occurrence of chaos in this system.

We end this section with an example for an unexpected constant of the motion (Blümel *et al.* (1989a)) that was recently discovered to play a role in the dynamics of a Hamiltonian that approximately describes two charged particles in a Paul trap (Paul *et al.* (1958), Paul (1990)).

The Paul trap is a device of considerable importance for high-resolution laser spectroscopy. Moreover, the Paul trap has been proposed as the centrepiece of a new generation of ultra-stable atomic clocks (see, e.g., Itano *et al.* (1983), Wineland *et al.* (1984)). Related trap designs have very recently come into prominence from their use in first observation of gaseous Bose-Einstein condensation (Anderson *et al.* (1995), Collins (1995)).

In appropriate units, an approximate Hamiltonian of two ions in a Paul trap is given by

$$H = \frac{1}{2}(p_1^2 + p_2^2) + V(q_1, q_2),$$
(3.3.6)

where

$$V = \frac{1}{2}q_1^2 + \frac{1}{2}\lambda^2 q_2^2 + \frac{\nu^2}{2q_1^2} + \frac{1}{(q_1^2 + q_2^2)^{1/2}}.$$
(3.3.7)

According to (3.3.7) the Hamiltonian (3.3.6) contains two control parameters, λ and ν. It is generally chaotic, but for $\lambda = 1/2$ and $\lambda = 2$ analytical constants of the motion exist that result in regular motion. It can be shown by direct differentiation with respect to time that for $\lambda = 2$ and $\lambda = 1/2$ the two functions

$$F(q_1, q_2, \dot{q}_1, \dot{q}_2) = q_2\dot{q}_1^2 - \dot{q}_2 q_1\dot{q}_1 + \frac{q_2}{(q_1^2 + q_2^2)^{1/2}} - q_1^2 q_2 + \frac{\nu^2 q_2}{q_1^2}$$
(3.3.8)

and

$$G(q_1, q_2, \dot{q}_1, \dot{q}_2) = \left\{ \frac{\nu^2}{q_1} + q_1\dot{q}_2^2 - \dot{q}_1 q_2\dot{q}_2 + \frac{q_1}{(q_1^2 + q_2^2)^{1/2}} - \frac{1}{4}q_1 q_2^2 \right\}^2$$

$$+ \frac{\nu^2}{q_1^2}(q_1\dot{q}_1 + q_2\dot{q}_2)^2 + \nu^2(q_1^2 + q_2^2),$$
(3.3.9)

respectively, are constants of the motion. While the existence of the function F can be understood on the basis of the separability of the equations of motion of (3.3.6) in semiparabolic coordinates (Farrelly and Howard (1994) and references therein), the function G (for $\nu^2 > 0$) is not connected with any simple symmetry of H.

An analytical constant of the motion is a precious asset. It can be used to (partially) integrate the equations of motion or to rule out chaos in situations where the dynamics appears to be "complicated" but is nevertheless regular and integrable (Blümel (1993a)).

4

Chaos in quantum mechanics

Einstein (1917) appreciated early on that within the "old" pre-1925/26 quantum mechanics absence of integrability is a serious obstacle for the quantization of classical systems. Therefore, in retrospect not surprisingly, the quantization problem was not adequately solved until the advent of the "new" quantum mechanics by Heisenberg, Born, Jordan and Schrödinger. The new quantum mechanics did not rely at all on the notion of classical paths, and this way, unwittingly, sidestepped the chaos problem. Within the framework of the new theory, any classical system can be quantized, including classically chaotic systems. But while the quantization of integrable systems is straightforward, the quantization of classically chaotic systems, even today, presents a formidable technical challenge. This is especially true for quantization in the semiclassical regime, where the quantum numbers involved are large. In fact, efficient semiclassical quantization rules for chaotic systems were not known until Gutzwiller (1971, 1990) intoduced periodic orbit expansions. Gutzwiller's method is discussed in Section 4.1.3 below. It is important to emphasize here that the existence of chaos in certain classical systems in no way introduces conceptual problems into the framework of modern quantum theory, although, let it be emphasized again, chaos came back with a vengeance from the "old" days of quantum mechanics. Even given all the modern day computer power accessible to the "practitioner" of quantum mechanics, chaos is the ultimate reason for the slow progress in the numerical computation of even moderately excited states in such important, but chaotic problems as the helium atom.

With chaos in classical dynamical systems well established, the question arises whether quantum systems are able to display exponential sensitivity and chaos. The answer is that most quantum systems do not. Not even if their classical counterparts are chaotic. We can say that chaos is "suppressed" on the quantum level. An example of this suppression effect is discussed in Chapter 5. There is, however, a class of systems that shows exponential sensitivity and chaos even on the quantum level. Such systems were first discussed by Eidson and Fox (1986) in the context of laser physics. These two authors were able to demonstrate theoretically that the quantum dynamics of a two-level atom coupled to a classical radiation field can show genuine chaos if the atom is allowed to act back on the field. Systems of this kind can be described as semi-quantum (Cooper *et al.* (1994)) since the quantum system is coupled to at least one classical degree of freedom. It seems that such systems arise quite naturally in many fields of physics. The possibility of "true" quantum chaos, i.e. exponential sensitivity and complexity on the quantum level, is also not ruled out.

Given the abovementioned bewildering cornucopia of quantum systems that in one way or another all invoke the notion of chaos, we have to ask the question: what exactly is "quantum chaos"? We think that "quantum chaos" comes in three varieties: (I) quantized chaos, (II) semi-quantum chaos and (III) quantum chaos. We refer to these three categories as type I, II and III quantum chaos. The division of quantum chaos into these three types arises naturally if quantum systems are characterized according to whether they do or do not show exponential sensitivity and chaos. The three different types of quantum systems are discussed in Sections 4.1, 4.2 and 4.3, respectively. A short preview of the three different types of quantum chaos follows.

Quantized chaos (Wintgen (1993)), or type I quantum chaos, is the most important type of quantum chaos research. It is discussed in Section 4.1. Type I quantum chaos is concerned with the quantization of classically chaotic systems in the semiclassical (but nonclassical) regime. Given the fact that already the helium atom is a classically chaotic system (Richter and Wintgen (1990a,b)) this type of research is the most important of the three types for atomic and molecular physics. Given its importance, it is not surprising that this type of quantum chaos research has many names. As discussed above, we refer to it as "type I", or "quantized chaos", while Berry (1989) calls it "quantum chaology". We use all three names synonymously. Quantum chaology looks for the signatures of classical chaos on the quantum level. Its most powerful tools are random matrix theory (Section 4.1.1), level dynamics (Section 4.1.2) and periodic orbit expansions (Section 4.1.3). The primary focus of attention of type I research is bounded autonomous systems with a discrete spectrum. The quantum dynamics of these systems does not show any exponential instabilities characteristic of classical chaos (Lebowitz and Penrose (1973), Hogg and Huberman (1982)). Quantum chaology is so far the most successful branch of quantum chaos research. The most important result is the universal behaviour of classically chaotic quantum systems which is, e.g., reflected in the fluctuation statistics of their energy levels. The results of quantum chaology are also "universal" in another sense: they are not particular to quantum mechanics, but are applicable to all branches of physics concerned with wave phenomena, e.g. acoustics (Weaver (1989), Bohigas *et al.* (1991)), electrodynamics (Stöckmann and Stein (1990)) and hydrodynamics (Blümel *et al.* (1992)). The results of quantum chaology have shaped the intuition of a new generation of physicists and have been applied successfully in atomic and molecular physics (see, e.g., Casati *et al.* (1987), Friedrich and Wintgen (1989)).

Type II quantum chaos is discussed in Section 4.2. It arises naturally in molecular physics in the form of the dynamic Born-Oppenheimer approximation (Blümel and Esser (1994)). In the dynamic Born-Oppenheimer approximation chaos may occur in both the classical and the quantum subsystem, although neither the classical nor the quantum systems by themselves are chaotic. Type II quantum chaos was also identified in a nuclear physics context (Bulgac (1991)).

Besides "quantized chaos" and "semi-quantum chaos" there is, at least in principle, a third possibility: type III quantum chaos, i.e. fully quantized systems that nevertheless exhibit exponential sensitivity and chaos in analogy to classically chaotic systems. An example of such a system is presented in Section 4.3. Although an increasing number of publications are concerned with identifying genuine quantum chaos (Pomeau *et al.* (1986), Luck *et al.* (1988), Chirikov (1991), Berry (1992), Weigert (1993), Blümel (1994a)), these systems are still under active investigation.

4.1 Quantized chaos

None of the classically chaotic quantum systems so far investigated in the atomic and molecular physics literature exhibits type III quantum chaos. On the other hand, atomic and molecular physics systems provide excellent examples for quantized chaos, the topic of this section. The attractive feature of the term "quantized chaos" is that it does not imply anything about what happens to the classical chaos when it is quantized. Usually, especially in bounded time independent quantum systems, classical chaos does not survive the quantization process. The quantized system does not exhibit any instabilities, or sensitivity to initial conditions, e.g. sensitivity to small variations in the wave function at time $t = 0$.

Over the past decade quantized chaos has become quite an "industry". It has been realized that except for the field-free hydrogen atom and related two-body atomic systems, all atoms and molecules, starting with the helium atom, can exhibit chaotic behaviour when treated as classical systems. Although the quantum dynamics of these systems do not show any signs of chaos, such as exponential sensitivity to initial conditions, their wave functions and spectra are strongly influenced by the underlying classical chaos. Wave functions and spectra of classically chaotic systems are best analyzed within the framework of random matrix theory. This technique is introduced in Section 4.1.1.

If a quantum system depends on an external parameter, such as the strength of an externally applied field, its energy levels are a function of the parameter. It is possible to interpret the parameter as a fictitious "time" and consider the "dynamics" of the energy levels with respect to changes of the parameter. The resulting level dynamics may contain interesting information about the chaoticity of the system. This method is discussed briefly in Section 4.1.2. It is generalized to the dynamics of resonances in the complex energy plane in Chapter 10.

Quantum calculations for a classically chaotic system are extremely hard to perform. If more than just the ground state and a few excited states are required, semiclassical methods may be employed. But it was not before the work of Gutzwiller about two decades ago that a semi-classical quantization scheme became available that is powerful enough to deal with chaos. Gutzwiller's central result is the *trace formula* which is derived in Section 4.1.3.

4.1.1 Random matrix theory

Why was it necessary to develop quantum mechanics in the first place? It was necessary because experiments in light scattering and emission from atoms revealed that, apart from capture- and line-broadening effects, the atom is not capable of radiating at any arbitrary frequency. The emission occurs only at a discrete set of sharply defined frequencies, called *spectral lines*. The spectral lines were subsequently interpreted by Bohr (1913a,b) as the transitions of an atom between a set of discrete stationary atomic states. The states define *energy levels*, and we can interpret the existence of spectral lines as transitions between energy levels. The sum total of all atomic energy levels make up the *atomic spectrum*. But discrete spectra are not only known in atomic physics; they also occur in nuclear physics, since in the low-energy regime atomic nuclei, too, can only emit or absorb at discrete frequencies. But while atoms and molecules under not too exotic conditions radiate in a frequency regime that spans the range from microwave to optical frequencies, the transition frequencies of atomic nuclei are usually in the X-ray or γ-ray regime.

But even in fields different from atomic and nuclear physics the concept of discrete spectra occurs. Consider, for instance, a microwave resonator. A closed ideal microwave resonator can only support a countable number of discrete electromagnetic field configurations called *modes*. There is one and only one frequency associated with every single one of the microwave modes. This does not exclude the occurrence of degeneracy, i.e. a situation in which two different modes have the same frequency. Even rooms, such as concert halls, exhibit a definite discrete spectrum of resonance frequencies. Sets of discrete frequencies may also occur as the eigenmodes of vibration of elastic membranes, or blocks of any solid material.

Since the phenomenon of spectra is so ubiquitous in physics it is important to have a tool to characterize energy or frequency spectra. Such a tool is provided by the theory of (energy) level statistics. Following closely the standard literature (Haake (1991), Mehta (1991)) we will now introduce some elements of this field.

Assume that we have a set of energy levels E_j, $j = 1, 2, ..., M$, where M can be finite or infinite. The first and most elementary thing we can do with a set of energy levels is to count them. We introduce the counting function

$$\mathcal{N}(E) = \sum_{j=1}^{M} \theta(E - E_j), \tag{4.1.1}$$

which counts the number of energy levels with energy less than E. Here, $\theta(x)$ is Heaviside's step function. It is defined as

$$\theta(x) = \begin{cases} 0, & \text{for } x < 0 \\ 1/2, & \text{for } x = 0 \\ 1, & \text{for } x > 0. \end{cases} \tag{4.1.2}$$

The function $\mathcal{N}(E)$ jumps by one unit whenever an energy level E_j is encountered. An example is shown in Fig. 4.1: it is now obvious why the

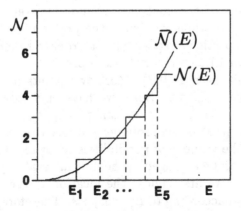

Fig. 4.1.   Sketch of a staircase function.

counting function $\mathcal{N}$ is also called the staircase function. With the help of the staircase function we define the level density $\rho(E)$ of the spectrum:

$$\rho(E) = \frac{d\mathcal{N}(E)}{dE}. \tag{4.1.3}$$

Since the staircase function is a sum of Heaviside functions and the derivative of a Heaviside function is a δ function, the density $\rho(E)$ is given by

$$\rho(E) = \sum_{j=1}^{M} \delta(E - E_j). \tag{4.1.4}$$

The staircase function $\mathcal{N}(E)$ is interpolated by a mean staircase $\bar{\mathcal{N}}(E)$, a smooth fit to the actual staircase function $\mathcal{N}(E)$. The mean staircase is shown as the full line interpolating the steps in Fig. 4.1. The mean staircase allows us to define the mean level density $\bar{\rho}(E)$ as

$$\bar{\rho}(E) = \frac{d\bar{\mathcal{N}}(E)}{dE}. \tag{4.1.5}$$

Next we define the energy level spacings

$$s_j = E_{j+1} - E_j. \tag{4.1.6}$$

With the help of the mean level density $\bar{\rho}$ we define the mean level spacing according to

$$\bar{s}(E) = 1/\bar{\rho}(E). \tag{4.1.7}$$

Note that the mean spacing is energy dependent. We eliminate the energy dependence by defining the normalized level spacings

$$x_j = s_j/\bar{s}(E). \tag{4.1.8}$$

With this definition the spacings x_j are normalized to mean spacing 1. The property of interest is therefore no longer the mean level spacing, but the fluctuations of the spacings around the average. In order to characterize the fluctuations we define the probability density $P(x)$ for the occurrence of a particular spacing x. In order to become familiar with this new concept we discuss a simple example.

Consider the unit interval $[0, 1]$. Mark two points in this interval, E_1 and E_2, as shown in Fig. 4.2. Since we have only two points, there is obviously only one spacing. Let us call this spacing s. It is given by $s = E_2 - E_1$. Now, what is the probability of occurrence of this spacing? This is obviously the number of possibilities according to which the two points E_1 and E_2 can be arranged in the unit interval to yield exactly the spacing s. Suppose we first choose the location of the point E_1. Then the point E_2 is fixed according to $E_2 = E_1 + s$. Therefore, E_1 can assume any value in the interval $[0, 1 - s]$, and the probability of occurrence of

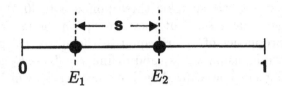

Fig. 4.2. Two randomly placed points in the unit interval.

the spacing s must be proportional to $1 - s$. Therefore, the probability of encountering the spacing s for two randomly placed points in the unit interval is

$$P_2(s) = 2(1-s), \qquad (4.1.9)$$

where the prefactor in (4.1.9) was chosen such that $\int P_2(s)ds = 1$. The mean spacing $\bar{s}$ is determined according to

$$\bar{s} = \int_0^1 s\, P_2(s)\, ds = 1/3. \qquad (4.1.10)$$

Then, according to (4.1.8), the normalized spacing x is given by

$$x = s/\bar{s} = 3s. \qquad (4.1.11)$$

Since we must have $P_2(s)ds = P_2(x)dx$, we get

$$P_2(x) = P_2(s)\, ds/dx = P_2(s)/3 = \frac{2}{3}(1 - x/3). \qquad (4.1.12)$$

The graph of this function is shown in Fig. 4.3.

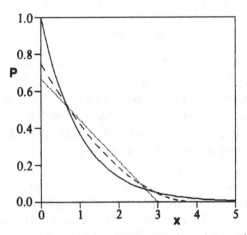

Fig. 4.3. Spacing probabilities of two (dotted line) and three (dashed line) points randomly placed in the unit interval. The limiting distribution P_∞ for infinitely many points in the unit interval (full line) is also shown.

Let us now suppose we sprinkle three points onto $[0, 1]$. Now there are two spacings, s_1 and s_2. But they are equivalent in the sense that the statistical properties of s_1 are equal to the statistical properties of s_2. Therefore, we focus on $s \equiv s_1$ and define $s = E_1 - E_2$ as before. A particular spacing s can be realized by first choosing E_1, which determines $E_2 = E_1 + s$, and then selecting an arbitrary point E_3 out of the rest of the interval. The probability $P_3(s)$ is therefore proportional to

$$P_3(s) \sim \int_0^{1-s} dE_1 \int_{E_2}^1 dE_3 = \int_0^{1-s} dE_1 \int_{E_1+s}^1 dE_3 = (1-s)^2 / 2.$$
(4.1.13)

Again, we have to normalize, and we obtain

$$P_3(s) = 3(1-s)^2.$$
(4.1.14)

The mean spacing is given by:

$$\bar{s} = \int_0^1 s P_3(s) ds = 1/4.$$
(4.1.15)

This makes sense, since three points divide the unit interval into four pieces, which, on average, should all be of equal length. Indroducing, as before, the normalized spacing x, we get:

$$P_3(x) = \frac{3}{4}(1 - x/4)^2.$$
(4.1.16)

This function is also shown in Fig. 4.3.

Continuing this procedure, we might first try the four-point case and then go to the limit of infinitely many points sprinkled onto the interval. In this case we get the limiting distribution (see Fig. 4.3)

$$P_\infty(x) = \exp(-x).$$
(4.1.17)

This distribution is called a Poissonian. Fig. 4.3 shows how the probability distributions $P_n(x)$, $n = 2, 3, \ldots$ tend to $P_\infty(x)$ in the limit $n \to \infty$. The Poissonian distribution is peaked and finite at $x = 0$. This means that the x_j values in the unit interval have a tendency to cluster. Thus, if the spacing statistics of a sequence of energy levels is Poissonian, the spectrum is characterized by level clustering.

Intuitively one might think that the spectrum of a chaotic system is completely uncorrelated, quasi-random. According to the above, the probability distribution of the spacings of energy levels is then expected to be Poissonian and the spectrum should show level clustering. Surprisingly, this is not the case. The investigation of many classically chaotic quantum systems shows that their energy levels are actually strongly correlated, showing level repulsion rather than level clustering.

It turns out that the energy level statistics of complicated atoms and nuclei (Rosenzweig and Porter (1960)), vibrating membranes and vibrating solids (Weaver (1989), Bohigas *et al.* (1991)), oddly shaped microwave cavities (Stöckmann and Stein (1990)) as well as the quantized versions of classically chaotic quantum systems (see, e.g., Gutzwiller (1990)) show approximately the same type of level statistics, which is close to a single universal function $P_W(x)$ proposed by Wigner in the 1950s. We will now proceed to derive this universal function from first principles and then compare it with experimental results.

The Hamiltonian of a conservative system is Hermitian, since the total probability is conserved. We will also assume that the Hamiltonian is time reversal invariant. In this case the Hamiltonian matrix is real and symmetric. In order to achieve maximum transparency with a minimum of formal derivations our investigation of the probability distribution $P_W(x)$ is restricted to random 2×2 Hamiltonian matrices. Pedagogically this is the most effective approach taken, e.g., by Bohr and Mottelson (1969) and Haake (1991). Assuming total randomness for the matrix elements of the Hamiltonian corresponds to a minimum knowledge (maximum entropy) assumption consistent with the notion of "chaos". However, we have to demand that the 2×2 real symmetric matrices considered are invariant under orthogonal transformations, which reflects our freedom in the choice of a set of basis functions to express the Hamiltonian as a Hamiltonian matrix. This additional condition is the main reason for the energy levels not to show a Poissonian nearest neighbour spacing statistics.

The Hamiltonian matrix is given by:

$$H = \begin{pmatrix} H_{11} & H_{12} \\ H_{12} & H_{22} \end{pmatrix}. \tag{4.1.18}$$

When one studies random objects, the first question is always: what is the probability of occurrence of individual objects in the ensemble? In our context the question is: what is the probability of occurrence of a specific Hamiltonian matrix H? So, clearly, we need a probability density $p(H)$ which assigns a probability to a given matrix H. Apparently the matrix H is specified by stating the three matrix elements H_{11}, H_{12} and H_{22}. Accordingly, the density p depends on three arguments,

$$p(H) = p(H_{11}, H_{12}, H_{22}). \tag{4.1.19}$$

According to our "minimum knowledge" assumption, there should not be any correlations in the probability density p. This means that p factorizes into the product of three densities according to:

$$p(H) = p_{11}(H_{11}) \, p_{12}(H_{12}) \, p_{22}(H_{22}). \tag{4.1.20}$$

Also, we have to satisfy the normalization condition

$$\int_{-\infty}^{\infty} p(H)\, dH_{11}\, dH_{12}\, dH_{22} \;=\; 1. \tag{4.1.21}$$

Invariance under orthogonal transformations implies

$$p(H') \;=\; p(H), \tag{4.1.22}$$

where

$$H' \;=\; O^T H O \tag{4.1.23}$$

and

$$O \;=\; \begin{pmatrix} \cos(\theta) & \sin(\theta) \\ -\sin(\theta) & \cos(\theta) \end{pmatrix} \tag{4.1.24}$$

is a general orthogonal 2×2 matrix. In order to use (4.1.22) to derive the explicit form of the functions p_{11}, p_{12} and p_{22} in (4.1.20) it is enough to consider infinitesimal transformations O in which $\theta \to 0$. Such an infinitesimal transformation is given by:

$$O \;=\; \begin{pmatrix} 1 & \theta \\ -\theta & 1 \end{pmatrix}. \tag{4.1.25}$$

From (4.1.23) we get

$$H'_{11} \;=\; H_{11} - 2\theta H_{12}, \qquad H'_{22} \;=\; H_{22} + 2\theta H_{12},$$

$$H'_{12} \;=\; H_{12} + \theta(H_{11} - H_{22}). \tag{4.1.26}$$

According to (4.1.22) we get:

$$p(H) \;=\; p(H)\left\{ 1 - \theta\left[2H_{12}\frac{d\ln p_{11}}{dH_{11}} - 2H_{12}\frac{d\ln p_{22}}{dH_{22}} \right.\right.$$

$$\left.\left. - (H_{11} - H_{22})\frac{d\ln p_{12}}{dH_{12}} \right]\right\}. \tag{4.1.27}$$

Since the infinitesimal angle θ is arbitrary, its coefficient in (4.1.27) must vanish. This condition yields three decoupled differential equations for the three independent variables H_{11}, H_{12} and H_{22}. Introducing the integration constants A, B and C, we solve the resulting differential equations and obtain

$$p(H) \;=\; C\exp[-A(H_{11}^2 + H_{22}^2 + 2H_{12}^2) - B(H_{11} + H_{22})]. \tag{4.1.28}$$

By properly choosing the zero of energy, we can always arrange for $H_{11} + H_{22} = 0$. The constant A can be set to 1 by properly choosing the unit of energy. But we may just as well carry it along. The term in the exponent of (4.1.28) can be expressed as the trace of the square of the Hamiltonian matrix H, and thus we finally obtain:

$$p(H) = C \exp(-A \operatorname{Tr} H^2). \tag{4.1.29}$$

It can be shown that the result (4.1.29) obtained in the context of 2×2 matrices in fact holds for arbitrary $M \times M$ matrices.

Next is the derivation of the explicit form of the Wigner distribution $P_W(x)$. The starting point is the Hamiltonian (4.1.18) whose matrix elements are assumed to be distributed according to (4.1.29). Its eigenvalues are given by

$$E_{\pm} = \frac{1}{2}(H_{11} + H_{22}) \pm \frac{1}{2}[(H_{11} - H_{22})^2 + 4H_{12}^2]^{1/2}. \tag{4.1.30}$$

With the help of the eigenvalues $E_{\pm}$, the diagonal matrix

$$D = \begin{pmatrix} E_+ & 0 \\ 0 & E_- \end{pmatrix} \tag{4.1.31}$$

and the orthogonal transformation

$$\Omega = \begin{pmatrix} \cos(\varphi) & -\sin(\varphi) \\ \sin(\varphi) & \cos(\varphi) \end{pmatrix}, \tag{4.1.32}$$

we can write the matrix H as:

$$H = \Omega D \Omega^T. \tag{4.1.33}$$

This yields immediately the following transformation between the original variables H_{11}, H_{12}, H_{22} and E_+, E_-, φ, respectively:

$$H_{11} = E_+ \cos^2(\varphi) + E_- \sin^2(\varphi), \quad H_{22} = E_+ \sin^2(\varphi) + E_- \cos^2(\varphi),$$

$$H_{12} = (E_+ - E_-) \cos(\varphi) \sin(\varphi). \tag{4.1.34}$$

The determinant of the Jacobian of this transformation is given by:

$$\det(J) = \det \frac{\partial(H_{11}, H_{22}, H_{12})}{\partial(E_+, E_-, \varphi)} = E_+ - E_-. \tag{4.1.35}$$

With the help of (4.1.35) we can now easily transform the probability distribution $p(H)$ into $p(E_+, E_-, \varphi)$. With

$$p(H_{11}, H_{12}, H_{22}) dH_{11} dH_{12} dH_{22} = p(E_+, E_-, \varphi) dE_+ dE_- d\varphi \tag{4.1.36}$$

we get immediately:

$$p(E_+, E_-, \varphi) = p(H) \det(J). \tag{4.1.37}$$

For the evaluation of $p(H)$ we need the trace of H^2 in the new variables. It is given by:

$$\operatorname{Tr} H^2 = E_+^2 + E_-^2. \tag{4.1.38}$$

Since neither $\det(J)$ nor $\operatorname{Tr} H^2$ depend on φ we obtain the distribution $p(E_+, E_-, \varphi)$ in the form:

$$p(E_+, E_-) \;=\; C \,|\, E_+ - E_- \,|\, \exp\left[-A(E_+^2 + E_-^2)\right]. \qquad (4.1.39)$$

In order to express (4.1.39) in terms of the spacing

$$s \;=\; E_+ \,-\, E_-, \qquad (4.1.40)$$

we introduce the auxiliary variable $z = (E_+ + E_-)/2$. This results in

$$p(s, z) \;=\; C \, s \, \exp[-A(s^2/2 + 2z^2)]. \qquad (4.1.41)$$

Integrating (4.1.41) over the auxiliary variable z we obtain the properly normalized spacing distribution

$$P(s) \;=\; 2\,A\,s\,\exp(-As^2), \quad s > 0. \qquad (4.1.42)$$

The mean spacing $\bar{s}$ is given by

$$\bar{s} \;=\; \int_0^\infty s P(s)\,ds \;=\; \frac{1}{2}\sqrt{\pi/A}. \qquad (4.1.43)$$

The mean spacing is normalized to 1 by choosing $A = \pi/4$. This choice yields the final result, the Wigner distribution $P_W(x)$, in the form

$$P_W(x) \;=\; \frac{\pi}{2}\,x\,\exp(-\frac{\pi}{4}\,x^2). \qquad (4.1.44)$$

The Wigner distribution is shown as the dotted line in Fig. 4.4.

Wigner was the first to suggest the application of random matrix theory to complicated quantum mechanical systems, such as atoms and nuclei.

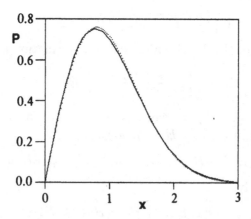

Fig. 4.4. Wigner distribution $P_W(x)$ (dotted line) and the limiting distribution $P_L(x)$ for the normalized eigenvalue spacings of $M \times M$ matrices with $M \to \infty$ (full line). The function $P_L(x)$ was plotted using the data listed in Table 4.3 of Haake (1991).

Thus, he is generally credited with creating the field of random matrix theory.

A surprising result of advanced random matrix theory is that the normalized level spacing distribution $P_L(x)$ for $M \times M$ matrices in the limit of $M \to \infty$ is very close to $P_W(x)$, which was obtained on the basis of $M = 2$. The distribution $P_L(x)$ is shown as the full line in Fig. 4.4. Fig. 4.4 shows that the difference between P_W and P_L is indeed very small.

How does the Wignerian law compare with the level spacing statistics obtained in actual experiments? We expect Wignerian statistics to hold for the spectra of complicated quantum systems with many degrees of freedom. In this case the Hamiltonian operator describing such a system is also a very complicated object, with complicated, quasi-random, matrix elements. Complicated quantum systems with many degrees of freedom are, e.g., atomic nuclei and atoms. In this case, it was shown by many authors that the agreement between the Wigner prediction and the experimentally obtained level spacings statistics is very good. In the introduction to his book on random matrix theory Mehta (1991) summarizes the available experimental data on the level spacing of atoms and nuclei. Very good agreement between the experimental level spacings statistics and the Wigner prediction is obtained.

In both nuclear and atomic physics there are complications with the notion of "levels". Beyond a certain excitation energy the energy levels acquire an appreciable width, i.e. the levels no longer correspond to bounded states. In atomic physics there is an additional complication with the existence of Coulomb accumulation points in the spectrum. Therefore, one has to be careful when comparing the spacings of experimental "levels" with the theoretical predictions. But if only bounded states or sharp resonances away from accumulation points are included in the level spacing statistics, there should be no theoretical objections.

Should we call the agreement between the experimental nearest neighbour spacing statistics in many-body systems and the Wignerian random matrix prediction a manifestation of quantum chaos? Certainly not. Chaos is about complicated things happening in *simple* systems. So although there may be chaos in a heavy atomic nucleus, or the electron system surrounding it, these systems are by no means simple. The real challenge is to set up a simple, but chaotic, system, and to show that it, too, exhibits Wignerian statistics. If this works out, we can say that at least as far as one aspect is concerned – nearest neighbour spacing statistics – the simple, but chaotic, system behaves just like the complicated many degrees of freedom systems. We have already discussed an example of a suitable "simple" system: the hydrogen atom in a strong magnetic field. For this system Wintgen and Friedrich (1986) computed the nearest neighbour spacing statistics. The result is shown in Fig. 4.5. The nearest neighbour spacing statistics is in good agreement with the Wigner prediction. Apparently, the underlying chaos in this system is able to generate enough complexity and "randomness" in the matrix elements of the system Hamiltonian that certain statistical features of its quantum spectrum are consistent with the predictions of random matrix theory.

Since a quantum system is an example of a wave system, we expect the same results to apply to all systems that involve waves. We may think about the abovementioned sound waves in a concert hall, for instance. If we measure the frequency spectrum of a concert hall, then the prediction is that its nearest neighbour statistics is Wignerian. Complications arise in this case due to the strong damping and absorption of sound that is usually encountered in a concert hall. For the elastic vibration modes of an aluminium block, however, the Wignerian law was actually verified by Weaver (1989), although its physical origins are much debated (Bohigas *et al.* (1992)). Absorption is less of a problem for microwave resonances in a microwave cavity. Level repulsion is expected provided that the microwave cavity shows classical chaos. Indeed it was verified experimentally that the nearest neighbour statistics of the resonance modes in a stadium shaped microwave cavity (a classically chaotic system) is Wignerian (Stöckmann and Stein (1990)).

Based on the above, it is probably not a good idea to restrict our investigations of chaos to quantum systems. We should allow other wave systems in our investigations to make full use of the *universality* which chaos

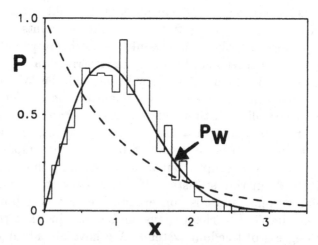

Fig. 4.5. Nearest neighbour spacing statistics of the hydrogen atom in a strong magnetic field. (From Wintgen and Friedrich (1986).)

brings about in these systems. Therefore, "microwave chaos", "acoustic chaos", and, of course, also "quantum chaos" are sub-fields of the universal field of *wave chaos*. This is indeed the way many researchers nowadays interpret these new directions of research in wave mechanics. So, if we want to decide, for instance, whether a microwave cavity is chaotic, we investigate the geometric optics limit of the microwave system. On the basis of the classical rays we can immediately decide whether a given microwave cavity is chaotic or not. A rectangular cavity, for instance, is not chaotic at all, since rays bouncing in a rectangular cavity do not behave in a chaotic way. This was illustrated in Section 1.1 above with the help of box R. The Lyapunov exponent of rays bouncing in a rectangular box is zero. But if we shape the microwave cavity according to box C, or shape it like a football stadium, then most classical rays bounce chaotically in the cavity and we say that the cavity is chaotic. Wigner nearest neighbour statistics characterizes the frequency spectrum in chaotic cavities.

In summary, random matrix theory is a very useful tool in the investigation of the spectra of classically chaotic wave systems.

To conclude this section we point out that the Wigner statistics is valid only if the system has integer spin and is invariant under an antiunitary symmetry such as time reversal symmetry. If all the antiunitary symmetries are broken, the nearest neighbour statistics is expected to be

$$P_U(x) = \frac{32x^2}{\pi^2} \exp\left(-4x^2/\pi\right). \qquad (4.1.45)$$

The subscript "U" in (4.1.45) stands for "unitary" since the system Hamiltonian is invariant under general unitary transformations in case all antiunitary symmetries are broken. The effects of time reversal symmetry breaking on the level statistics were recently investigated experimentally by So *et al.* (1995) and Stoffregen *et al.* (1995) by measuring the resonance spectrum of quasi-two-dimensional microwave cavities in the presence of time reversal breaking elements such as ferrites and directional couplers. These experiments are of considerable relevance for quantum mechanics since quasi-two-dimensional microwave cavities are generally considered to be excellent models for two-dimensional quantum billiards.

In case the system has half-integer spin and an antiunitary symmetry is active, we expect all levels of the system to be Kramers degenerate. In this case the nearest neighbour distribution of energy level spacings between Kramers degenerate pairs is given by

$$P_S(x) = \frac{2^{18}x^4}{3^6\pi^3} \exp\left(-64x^2/9\pi\right). \qquad (4.1.46)$$

Here the subscript "S" stands for "symplectic", a Greek word which literally means "entangled". If the system has half-integer spin and all antiunitary symmetries are broken, Kramers degeneracy is lifted and we return to the case P_U.

The three different cases P_W, P_U and P_S are characterized by the Dyson parameter β. This parameter indicates the degree of level repulsion for $x \to 0$. According to (4.1.44) – (4.1.46) the Dyson parameter takes the values $\beta = 1$, 2 and 4. According to (4.1.28) the matrix elements of the Hamiltonian are Gaussian distributed. Since for $\beta = 1$, 2 and 4 the Hamiltonian is invariant under orthogonal, unitary and symplectic matrix transformations, respectively, the three different random matrix ensembles are also referred to as the Gaussian orthogonal ensemble (GOE), the Gaussian unitary ensemble (GUE) and the Gaussian symplectic ensemble (GSE). Apart from determining its level statistics, the symmetry class of a given Hamiltonian H also influences such important physical characteristics as the localization length of the system. The localization length determines how fast system wave functions decay in a given basis set. This fundamental concept, including its modification by system symmetries, is discussed in Chapter 5.

4.1.2 Level dynamics

Another useful tool for studying the quantum mechanics of classically chaotic systems is *level dynamics*. In this approach, the real energy eigenvalues E_n of a Hermitian Hamiltonian

$$\hat{H}(\epsilon) \;=\; \hat{H}_0 + \epsilon \hat{V} \tag{4.1.47}$$

are studied as a function of the perturbation strength ϵ. It is assumed that neither $\hat{H}_0$ nor $\hat{V}$ depend on the control parameter ϵ. Within the framework of level dynamics the energy eigenvalues E_n are identified with fictitious particles. This analogy stems from the fact that the behaviour of E_n as a function of ϵ is described by a set of first order differential equations in the perturbation strength ϵ which acts as a fictitious time. This approach dates back to the work of Dyson (1962a–c), who studied the spectra of random unitary matrices. He was able to identify the N eigenphases of an $N \times N$ random unitary matrix with a set of N particles on the unit circle. The subject was further developed by Pechukas (1983) and Yukawa (1985).

In order to derive the dynamical equations of motion of level dynamics we start with the eigenvalue equation

$$\hat{H}(\epsilon) \, |n(\epsilon)\rangle = E_n(\epsilon) \, |n(\epsilon)\rangle. \tag{4.1.48}$$

We define the matrix elements

$$V_{nm}(\epsilon) = \langle n(\epsilon)|\hat{V}|m(\epsilon)\rangle. \tag{4.1.49}$$

As indicated in (4.1.48) and (4.1.49), the energies E_n, the matrix elements V_{nm}, as well as the states $|n\rangle$, are functions of ϵ. Since $\hat{H}$ is Hermitian, the states $|n\rangle$ can be assumed to be orthonormal and complete according to

$$\langle n|m\rangle = \delta_{nm}, \qquad \sum_n |n\rangle\langle n| = 1. \tag{4.1.50}$$

In the following we also assume that $\hat{H}$ is time reversal invariant. Therefore, states and matrix elements can be assumed to be real (Messiah (1979)). Differentiating the orthogonality relation in (4.1.50) with respect to ϵ we obtain

$$\langle \dot{n}|m\rangle = -\langle n|\dot{m}\rangle. \tag{4.1.51}$$

We introduced a dot notation to indicate differentiation with respect to ϵ. A simple corollary of (4.1.51) is the observation that the variation of a given state $|n\rangle$ with ϵ is orthogonal to the state itself:

$$\langle \dot{n}|n\rangle = 0. \tag{4.1.52}$$

Using (4.1.52) we derive immediately

$$\dot{E}_n(\epsilon) = \frac{d}{d\epsilon}\langle n|\hat{H}|n\rangle = V_{nn}(\epsilon). \tag{4.1.53}$$

This set of equations is known as the *Hellman-Feynman theorem* (see, e.g., Hirschfelder *et al.* (1954)). It is a set of simple first order differential equations for the energy levels E_n. But (4.1.53) is not a closed system of differential equations since we do not know the behaviour of the perturbation matrix elements V_{nn} as a function of ϵ. In an attempt to close the system (4.1.53) we compute

$$\dot{V}_{nn} = 2\sum_m \langle \dot{n}|m\rangle V_{mn}. \tag{4.1.54}$$

We used the completeness relation in (4.1.50). Because of (4.1.52) the sum in (4.1.54) ranges only over $m \neq n$. Still, we need an explicit expression for $\langle \dot{n}|m\rangle$ for $m \neq n$. Differentiating the matrix element $\langle n|\hat{H}|m\rangle$ with respect to ϵ and using (4.1.51), we obtain

$$\langle \dot{n}|m\rangle = \frac{V_{nm}}{E_n - E_m}, \quad \text{for } n \neq m. \tag{4.1.55}$$

Therefore,

$$\dot{V}_{nn} = 2\sum_{m \neq n} \frac{V_{nm}^2}{E_n - E_m}. \tag{4.1.56}$$

Differential equations for the off-diagonal matrix elements can be derived along similar lines. Together with (4.1.53) and (4.1.56) we obtain the following set of coupled equations as the final result for the equations of motion of level dynamics

$$\dot{E}_n(\epsilon) = V_{nn}(\epsilon), \qquad (4.1.57a)$$

$$\dot{V}_{nn} = 2 \sum_{m \neq n} \frac{V_{nm}^2}{E_n - E_m}, \qquad (4.1.57b)$$

$$\dot{V}_{nm} = \frac{V_{nm}(V_{nn} - V_{mm})}{E_m - E_n} + \sum_{l \neq n,m} V_{nl} V_{lm} \left\{ \frac{1}{E_n - E_l} + \frac{1}{E_m - E_l} \right\}.$$
$$(4.1.57c)$$

A similar set of equations was first derived by Pechukas in 1983. A set equivalent with (4.1.57) was subsequently derived by Yukawa (1985). The set (4.1.57) can be solved once the initial conditions at $\epsilon = 0$ are known. Diagonalizing $\hat{H}_0$, we obtain $|n(\epsilon = 0)\rangle$ and $E_n(\epsilon = 0)$. With this information, $V_{nm}(\epsilon = 0)$ can be calculated and the initial conditions at $\epsilon = 0$ are known.

The outstanding feature of the set (4.1.57) is that it is completely self-contained. Its structure does not depend on the specific form of the Hamiltonian (4.1.47). In fact, it describes the dynamics of energy levels for all Hamiltonian systems that can be split into two parts according to (4.1.47). The characteristics of a particular Hamiltonian enter only in the initial conditions. Thus we obtain the important result that if the system "forgets" the initial conditions after a "while" $\tilde{\epsilon}$, we can use the methods of equilibrium statistical mechanics to compute the statistical properties of the energy levels of the system. Pechukas (1983) used this approach to predict Wignerian statistics for the level fluctuations of chaotic systems. As discussed in the previous section, there is now ample evidence for the correctness of this prediction.

The equations of motion (4.1.57) can be derived as the canonical equations of a Hamiltonian. Defining the "angular momenta"

$$L_{nm} = (E_n - E_m) V_{nm}, \qquad (4.1.58)$$

we obtain

$$\dot{E}_n = V_{nn}, \qquad (4.1.59a)$$

$$\dot{V}_{nn} = 2 \sum_{m \neq n} \frac{L_{nm}^2}{(E_n - E_m)^3}, \qquad (4.1.59b)$$

$$\dot{L}_{nm} = \sum_{l \neq n,m} L_{nl} L_{lm} \left\{ \frac{1}{(E_n - E_l)^2} - \frac{1}{(E_m - E_l)^2} \right\}. \qquad (4.1.59c)$$

In addition to the "angular momentum" variables (4.1.58) we introduce the "position" and "momentum" variables x_n and p_n according to

$$x_j \equiv E_j; \quad p_j \equiv V_{jj}. \qquad (4.1.60)$$

With the help of the new dynamical variables we define the Hamiltonian

$$\mathcal{H} = \frac{1}{2} \sum_n p_n^2 + \frac{1}{2} \sum_{m \neq n} \frac{L_{nm}^2}{(x_m - x_n)^2}. \qquad (4.1.61)$$

The canonical equations of motion for $\mathcal{H}$ are given by

$$\dot{x}_n = \frac{\partial \mathcal{H}}{\partial p_n}, \quad \dot{p}_n = -\frac{\partial \mathcal{H}}{\partial x_n}, \quad \dot{L}_{nm} = \{\mathcal{H}, L_{nm}\}, \qquad (4.1.62)$$

where we used the angular momentum brackets (Haake (1991))

$$\{L_{mn}, L_{ij}\} = \frac{1}{2} \{\delta_{mj} L_{ni} + \delta_{ni} L_{mj} - \delta_{nj} L_{mi} - \delta_{mi} L_{nj}\}. \qquad (4.1.63)$$

Performing the differentiations in (4.1.62) it is seen that the canonical equations (4.1.62) are indeed equivalent with (4.1.57).

The analytical investigation of the statistical properties of the energy levels E_n on the basis of (4.1.62) requires assumptions on the ergodicity of the motion determined by (4.1.62). If H_0 is integrable, but $\hat{H}(\epsilon)$ is not for $\epsilon' \sim 1$, say, then, as "time" ϵ evolves from 0 to ϵ', the system switches from a regular to a chaotic regime. The transition region between the two regimes is best studied by integrating (4.1.62) numerically. A numerical investigation along these lines was reported by Yang and Burgdörfer (1991).

The methods of level dynamics can be extended to *resonance dynamics* in case the energy levels E_n acquire a width in the presence of a continuum. Resonance dynamics of the one-dimensional helium atom is discussed in Section 10.5.2.

4.1.3 The trace formula

Before 1925/26 quantum mechanics was based on *ad hoc* quantization procedures such as the Bohr-Sommerfeld quantization method. The disadvantage of this method is that it is coordinate dependent, and usually works only in Cartesian coordinates. In 1917 Einstein suggested a coordinate independent quantization procedure, an important improvement over the Bohr-Sommerfeld quantization scheme. Einstein's quantization method was subsequently extended and improved by Brillouin (1926) and Keller (1958). Therefore, this quantization scheme is referred to as EBK quantization in the literature. But even Einstein's method fails in cases where the classical system to be quantized does not possess the complete set of constants of the motion. Einstein was fully aware of the drawbacks of his method and refers to the three-body system as an example where his quantization procedure fails. With the advent of Heisenberg's and Schrödinger's quantum mechanics Einstein's problem became seemingly irrelevant since the "new" quantum mechanics works in all cases, including completely chaotic systems where both Bohr-Sommerfeld quantization and EBK quantization fail. Following the 1925/26 "quantum revolution", research in quantum mechanics was mainly focussed on finding exactly solvable quantum systems, postponing the problem of quantizing classically chaotic systems, or referring them to numerical solution. But it soon became apparent that the solvable quantum problems are mostly connected to cases where the classical mechanics is separable, an even more restrictive condition than integrability. Chaotic problems cannot be reduced in this way, thus making analytical solutions impossible to obtain. After having appreciated this problem, there developed a need for semiclassical procedures that work in the chaotic regime. Such procedures were not available until Gutzwiller developed his formalism of periodic orbit quantization. His central result is the trace formula presented below. With possible refinements such as cycle expansions (Cvitanović and Eckhardt (1989), Artuso *et al.* (1990a,b)) to improve its convergence properties, the trace formula is still the starting point and centrepiece of all semiclassical work on classically chaotic quantum systems.

In order to exhibit the basic ideas of Gutzwiller's method, we follow an elegant derivation of the trace formula given by Miller (1975). We restrict ourselves to a two degree of freedom bounded autonomous system with Hamiltonian $\hat{H}$. The spectrum of $\hat{H}$ is discrete and determined by

$$\hat{H} \mid n\rangle = E_n \mid n\rangle, \qquad (4.1.64)$$

where $\mid n\rangle$ are the eigenstates and E_n are the discrete energy levels of the system. The density of states is defined by (Brenig (1975), Gutzwiller (1990))

$$\rho(E) = Tr\left[\delta(E - \hat{H})\right]. \qquad (4.1.65)$$

In the representation (4.1.64) the density (4.1.65) can be evaluated explicitly and reads

$$\rho(E) = \sum_n \langle n \mid \delta(E - \hat{H}) \mid n\rangle = \sum_n \delta(E - E_n). \qquad (4.1.66)$$

Representing the δ function by its Fourier integral, (4.1.65) can be written as

$$\rho(E) \;=\; \frac{1}{2\pi\hbar}\int_{-\infty}^{\infty} dt\,\exp(iEt/\hbar)\,Tr[\exp(-i\hat{H}t/\hbar)] \;=$$

$$\frac{1}{2\pi\hbar}\int_{-\infty}^{\infty} dt\,\exp(iEt/\hbar)\int d^2\vec{q}\,K(\vec{q},\vec{q};t), \qquad (4.1.67)$$

where $K(\vec{q},\vec{q};t)$ denotes the diagonal matrix element of the propagator $\hat{K} = \exp(-i\hat{H}t/\hbar)$ in the position representation. The complete set of matrix elements of $\hat{K}$ is defined by

$$K(\vec{q}_2,\vec{q}_1;t) \;=\; \langle \vec{q}_2 \mid \exp(-i\hat{H}t/\hbar) \mid \vec{q}_1\rangle. \qquad (4.1.68)$$

In the semiclassical approximation the matrix elements (4.1.68) take the form

$$K^{(sc)}(\vec{q}_2,\vec{q}_1;t) \;\sim\; \exp[i\Phi(\vec{q}_2,\vec{q}_1)/\hbar)], \qquad (4.1.69)$$

where $\Phi(\vec{q}_2,\vec{q}_1)$ is the classical action along a trajectory connecting $\vec{q}_1$ and $\vec{q}_2$ in time t. The spatial integral in (4.1.67) can now be evaluated in stationary phase approximation. The stationary phase condition requires

$$\frac{\partial\Phi(\vec{q},\vec{q})}{\partial\vec{q}} \;=\; \frac{\partial\Phi(\vec{q}_2,\vec{q}_1)}{\partial\vec{q}_1} + \frac{\partial\Phi(\vec{q}_2,\vec{q}_1)}{\partial\vec{q}_2}\bigg|_{\vec{q}_1=\vec{q}_2=\vec{q}} \;=\; 0. \qquad (4.1.70)$$

But since

$$\frac{\partial\Phi(\vec{q}_2,\vec{q}_1)}{\partial\vec{q}_2} \;=\; \vec{p}_2, \qquad \frac{\partial\Phi(\vec{q}_2,\vec{q}_1)}{\partial\vec{q}_1} \;=\; -\vec{p}_1, \qquad (4.1.71)$$

where $\vec{p}_1$ and $\vec{p}_2$ are the canonically conjugate momenta, the stationary phase condition (4.1.70) is now seen to be a *periodic orbit condition*, since besides the condition $\vec{q}_1 = \vec{q}_2 = \vec{q}$ that derived from the trace operation, we see from (4.1.71) that (4.1.70) demands that $\vec{p}_1 = \vec{p}_2$ at the coinciding end points of the classical trajectory. Thus, in the semiclassical approximation, only periodic orbits contribute in the evaluation of the level density ρ. It is somewhat tedious but straightforward actually to evaluate the integral in (4.1.67). The necessary algebra is done, e.g., in Gutzwiller (1971) and Reichl (1992). The result is

$$\rho(E) \;=\; \rho_0(E) \,-\, \frac{1}{2\pi\hbar}\,\Im m\,\sum_p T_p \sum_{n=1}^{\infty} \frac{\exp[in(\Phi_p(E)/\hbar - \mu_p\pi/2)]}{i\,\sinh[n\lambda_p(E)/2]}. \qquad (4.1.72)$$

This is the trace formula, first derived in a physics context by Gutzwiller in 1971. Similar periodic orbit formulae were also derived in the context of abstract mathematical dynamical systems (Selberg (1956), McKean (1972)). The sum in (4.1.72) extends over all possible periodic orbits of the shortest nonvanishing length indicated by the summation index p. These orbits are also called *primitive periodic orbits*. The summation over the index n takes repeated traversals of the primitive periodic orbits into account. The quantity T_p in (4.1.72) is the traversal time of the primitive periodic orbit p, $\Phi_p(E)$ is the action of the periodic orbit at energy E and $\lambda_p(E)$ is the Lyapunov exponent of a Poincaré mapping with a Poincaré section constructed from a phase-space plane intersecting the periodic orbit p orthogonally at an arbitrary position along the orbit. The constants μ_p are characteristic for a given periodic orbit. Just like the Lyapunov exponents λ_p, the numbers μ_p depend on the dynamical behaviour of trajectories started in the immediate vicinity of the periodic orbit number p. The vicinity is best explored with a ray bundle with the orbit p as its guiding centre. The numbers μ_p are then computed from the turning points, focal points and caustics of the ray bundle (Miller (1975), Gutzwiller (1990)). The density ρ_0 is the average level density defined by

$$\rho_0(E) \;=\; \frac{1}{\hbar^2} \int d^2\vec{q}\, d^2\vec{p}\; \delta[E - \hat{H}(\vec{p},\vec{q})]. \qquad (4.1.73)$$

There are two ways in which the trace formula (4.1.72) can be used to deal with quantum problems. The trace formula can be applied "forwards" or "backwards". By "forward application" of the trace formula we mean that (4.1.72) is used to calculate the level density of a given quantum system on the basis of purely classical input represented by the periods, actions, Lyapunov exponents and characteristic numbers of the associated classical periodic orbits. The trace formula is applied "backwards" if the level density $\rho(E)$ is given, obtained, for instance, experimentally or numerically, and periodic orbit information is extracted from the level density by a generalized Fourier transformation that uses the particular mathematical structure of (4.1.72). The backward application of (4.1.72) in the case of the diamagnetic hydrogen problem, e.g., was pioneered by Wintgen (1988). He exploited the fact that for scaling systems the backward application of the trace formula is particularly simple and yields particularly illuminating results. The main point of the idea is illustrated in the following example. Suppose, as is the case for many atomic physics systems (see, e.g., Chapter 10), that the action $\Phi(E)$ in (4.1.72) scales according to

$$\Phi(E) \;=\; E^\alpha\; \Phi(E = E_0), \qquad (4.1.74)$$

where E_0 is some constant reference energy. Then, the traversal time of a periodic orbit is given by

$$T(E) \;=\; \frac{\partial \Phi(E)}{\partial E} \;=\; \alpha\, E^{\alpha-1}\, \Phi(E = E_0). \qquad (4.1.75)$$

We define the fluctuating part of the level density according to

$$\tilde{\rho}(E) = \rho(E) - \rho_0(E). \tag{4.1.76}$$

With the help of the variable

$$x = E^\alpha \tag{4.1.77}$$

we define the Fourier transform of $\tilde{\rho}$ with respect to x according to

$$R(\Phi) = \int_{-\infty}^{\infty} dx \, \tilde{\rho} \, \exp(-ix\Phi). \tag{4.1.78}$$

Assuming now that ρ is given by (4.1.72) and that the Lyapunov exponents $\lambda_p(E)$ in (4.1.72) are slowly varying functions of E (in some important atomic physics applications they are independent of E), the Fourier transform (4.1.78) yields the result

$$R(\Phi) \sim \sum_{pn} A_{pn} \, \delta[\Phi - n\Phi_p(E = E_0)], \tag{4.1.79}$$

where A_{pn} are finite complex amplitudes that can be computed from (4.1.72). According to (4.1.79) the Fourier transform $R(\Phi)$ is sharply peaked at scaled actions that correspond to multiples of the scaled actions of the primitive periodic orbits at the reference energy E_0. Using this method Wintgen (1988) was able to extract periodic orbit information from experimentally and numerically determined level densities for the hydrogen atom in a strong magnetic field. Similar methods revealed the presence of simple periodic orbits in the level densities of the one-dimensional (Blümel and Reinhardt (1992)) and the three-dimensional (Ezra *et al.* (1991), Wintgen *et al.* (1993)) helium atom.

The forward application of (4.1.72) is much more difficult to perform, but has been accomplished in many cases. Gutzwiller (1971) was the first to apply the trace formula to a real quantum mechanical problem. It consists of electrons with an asymmetric mass tensor moving in a Coulomb potential. This problem is important in semiconductor physics. Gutzwiller was able to compute good approximations to the first few quantum eigenstates of this system.

For classically chaotic quantum systems the forward application of the trace formula is difficult because of the following reasons. (i) In a chaotic system the number of periodic orbits proliferates exponentially; (ii) the orbits have to be computed numerically; and (iii) there are convergence problems with (4.1.72) that have to be circumvented with appropriate resummation prescriptions such as, e.g., cycle expansions (Cvitanović and Eckhardt (1989), Artuso *et al.* (1990a,b)). Nevertheless, sometimes valuable information on the structure of atomic states can be obtained by retaining only the shortest orbits in the expansion (4.1.72). This was demonstrated recently, e.g., by Ezra *et al.* (1991) and Wintgen *et al.* (1993). These authors were able to show that the Rydberg series of certain intra-shell resonances in the helium atom are well reproduced on the basis of only a single classical periodic orbit.

4.2 Semi-quantum chaos

Suppose a certain quantum system is governed by the Hamiltonian $\hat{H}$. What is its spectrum? It is an elementary fact that this question makes sense only after the boundary conditions imposed on the system wave functions have been specified. Thus, the boundary conditions are of crucial importance for the spectrum of $\hat{H}$. Confining a quantum system with the help of rigid stationary "walls" leads to boundary conditions on the wave functions that result in a discrete spectrum for $\hat{H}$. In this case the dynamics of the quantum system is simple. As discussed in Section 1.4 the time evolution of such a system is regular and does not show any chaos or exponential instabilities in its wave functions. But what if we make the boundaries time dependent? Or, even better, allow the quantum system to modify its boundaries in a dynamical way? It was proved recently (Blümel and Esser (1994), Cooper *et al.* (1994)), that in this case the quantum system may exhibit genuine chaos and complexity in the sense of classical deterministic chaos discussed in Chapters 2 and 3. In particular the quantum wave functions show exponential sensitivity to initial wave function configurations which can be characterized by a positive Lyapunov exponent. Truly quantum chaotic systems of this nature were introduced as type II systems in the introduction to this chapter. Type II systems, i.e. wave systems that dynamically modify their boundaries, are not exceptional. We may think, e.g., of sound waves in a balloon or electromagnetic waves in an ideally conducting cavity with movable walls. In both cases the pressure generated by the waves acts on the boundary and modifies its shape or position. Yet another source of type II systems is provided by atomic and molecular physics in the framework of the Born-Oppenheimer approximation (Born and Oppenheimer (1927)). Exponentially sensitive wave functions may arise if the Born-Oppenheimer approximation is used dynamically. By this we mean that the nuclear coordinates are not considered as static parameters in the electronic wave functions, but as dynamical variables that move in response to the electronic forces. We call this scheme the *dynamic Born-Oppenheimer approximation*. The dynamic Born-Oppenheimer approximation can be used whenever a given system naturally divides into two interacting subsystems, one of which is treated quantum mechanically whereas the other is treated in the classical approximation. The classical subsystem can then be interpreted as a moving boundary acted upon and driven by the forces due to the quantum subsystem. Thus, the quantum system dynamically modifies its boundaries, which in turn strongly influence the quantum system. This "give and take" between the quantum system and its boundary may result in chaotic dynamics for both the classical and the quantum subsystem. We illustrate this mechanism with the help of a schematic model molecule which is analysed in the dynamic Born-Oppenheimer approximation.

The model molecule is shown in Fig. 4.6. The "skeleton" of the molecule is represented by a square well potential with movable infinitely high walls at positions $\pm q(t)/2$. The skeleton represents the nuclei and the core electrons of the molecule. The skeleton has an effective mass M and moves in a potential $V(q)$. Furthermore we assume that the model molecule has a single valence electron of mass m confined in the space between the movable walls (see Fig. 4.6). The Hamiltonian of the electron is given by

$$\hat{K}(q) = -\frac{\hbar^2}{2m}\frac{\partial^2}{\partial x^2}, \quad -\frac{q}{2} \le x \le \frac{q}{2}. \tag{4.2.1}$$

The remaining part of the molecular Hamiltonian consists of the kinetic energy of the nuclei and their interaction potential $V(q)$. Thus, the complete model Hamiltonian is given by

$$\hat{H} = \hat{K}(q) + \frac{\hat{P}^2}{2M} + V(q), \tag{4.2.2}$$

where $\hat{P} = -i\hbar\partial/\partial q$ is the effective momentum of the walls. In order to make our model as simple as possible, we choose the harmonic potential

$$V(q) = V_0\,(q - Q)^2 \tag{4.2.3}$$

for the skeleton potential. Its simplicity notwithstanding, the model reflects all the essential features of a diatomic molecule in an illustrative way.

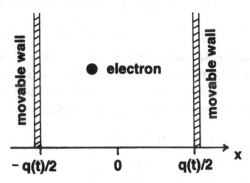

Fig. 4.6.   Sketch of the model molecule.

We solve our model by expanding the electronic wave function in the set of eigenstates of $\hat{K}(q)$ for a given q, i.e.

$$\psi^{\pm}(x,t) \;=\; \sum_n A_n^{\pm}(t)\,\phi_n^{\pm}(x,q), \qquad (4.2.4)$$

where $\phi_n^{+}(x,q) = (2/q)^{1/2}\cos[\pi(2n-1)x/q]$ and $\phi_n^{-}(x,q) = (2/q)^{1/2}\sin[2\pi n x/q]$, $n = 1, 2, ...$, are the eigenstates of even and odd parity of (4.2.1), respectively. A further simplification of the model is introduced by restricting the electronic states to the first two modes of positive parity only. The amplitudes of the two basis states are denoted by A_1 and A_2. Inserting (4.2.4) into the time dependent Schrödinger equation and paying special attention to the nonvanishing $\dot{q}$ terms we find

$$i\dot{A}_n \;=\; \sum_{k=1}^{2} D_{nk}\,A_k, \qquad n = 1, 2, \qquad (4.2.5)$$

where the Hermitian 2×2 matrix D_{nk} is given by

$$D_{nk} \;=\; \begin{pmatrix} \epsilon_1/\hbar q^2 & -i\mu_{12}\dot{q}/q \\ i\mu_{12}\dot{q}/q & \epsilon_2/\hbar q^2 \end{pmatrix}. \qquad (4.2.6)$$

Here, $\epsilon_n = [\hbar\pi(2n-1)]^2/2m$, $n = 1, 2$, $\mu_{12} = 3/4$. Note the q dependence in (4.2.6) which is crucial for the emergence of chaos. Expressing the kinetic energy $K(q) = \langle\psi\mid\hat{K}(q)\mid\psi\rangle$ in terms of the coefficients A_1 and A_2 we get

$$K(A_1, A_2, q) \;=\; \left[\epsilon_1\mid A_1\mid^2 + \epsilon_2\mid A_2\mid^2\right]/q^2. \qquad (4.2.7)$$

In the usual Born-Oppenheimer picture, the sum of $K(A_1, A_2, q)$ and $V(q)$ is the "adiabatic potential" for the molecular coordinate q. For the computation of the adiabatic potential, q is treated as a fixed parameter. In the dynamic Born-Oppenheimer approximation discussed above, we interpret q as a classical dynamical variable, with the result that the molecular vibrations are described by the Hamiltonian function

$$H_v \;=\; P^2/2M \;+\; K(A_1, A_2, q) \;+\; V(q). \qquad (4.2.8)$$

In order to eliminate an unimportant global phase in our two-state approximation, it is convenient to represent the quantum evolution of the electronic wave function by Bloch variable. The Bloch variables x, y and z are defined via the density matrix $\rho_{mn} = A_m A_n^*$ according to

$$x = \rho_{12} + \rho_{21}, \quad y = i[\rho_{21} - \rho_{12}], \quad z = \rho_{22} - \rho_{11}. \qquad (4.2.9)$$

With (4.2.5) and (4.2.6) we obtain

$$\dot{x} = -\omega_0 y/q^2 - 2\mu_{12}Pz/Mq, \quad \dot{y} = \omega_0 x/q^2,$$

$$\dot{z} = 2\mu_{12}Px/Mq, \qquad (4.2.10)$$

where $\omega_0 = (\epsilon_2 - \epsilon_1)/\hbar$. In order to close the system of equations (4.2.10) we add the equations of motion for P and q. They are obtained by expressing $K(A_1, A_2, q)$ with the help of the Bloch variable z. We obtain

$$K(z, q) = (\epsilon_+ + z\epsilon_-)/q^2, \qquad (4.2.11)$$

where $\epsilon_\pm = (\epsilon_2 \pm \epsilon_1)/2$. The force (the "quantum pressure") exerted by the quantum subsystem on the classical subsystem is given by

$$-\delta K(q)/\delta q, \qquad (4.2.12)$$

where, in addition to the explicit $1/q^2$ dependence in (4.2.11), one has to take the implicit q dependence of z into account. With $\delta z/\delta q = \dot{z}/\dot{q} = M\dot{z}/P$ and (4.2.10) we find $\delta z/\delta q = 2\mu_{12}x/q$. Hence the equations of motion for the classical subsystem are

$$\dot{q} = P/M, \quad \dot{P} = -\delta V/\delta q + 2[\epsilon_+ + \epsilon_-(z - \mu_{12}x)]/q^3. \qquad (4.2.13)$$

The equations (4.2.10) and (4.2.13) are a closed set of coupled equations for the two interacting subsystems. The integrals of motion for the system are the radius of the Bloch sphere

$$r^2 = x^2 + y^2 + z^2 = 1 \qquad (4.2.14)$$

and the total energy

$$E = P^2/2M + V(q) + K(z, q). \qquad (4.2.15)$$

It is useful to introduce dimensionless variables. We define $\eta = q/Q$, $p = PQ/\hbar$, $\tau = \pi^2 \hbar t/2mQ^2$, $\alpha = m/M$ and $\nu_0 = MV_0Q^4/\hbar^2$. In these units the total energy E is measured in units of $\epsilon = \pi^2\hbar^2/2mQ^2$.

The salient features of the dynamics of our model molecule are best exhibited with the help of Poincaré sections that have already proved useful in the analysis of the double pendulum presented in Section 3.2. Fig. 4.7 shows the η-p projection of an $x = 0$ surface of section of a trajectory for $\alpha = 0.1$, $\nu_0 = 10$ and $E = 4$ started at $\theta = 0.95\pi$, $x = \sin(\theta)$, $y = 0$, $z = \cos(\theta)$ and $\eta = 1.42$. The resulting η-p Poincaré section clearly shows chaotic features. This indicates that the classical dynamics of the skeleton of the model molecule is chaotic. But the most striking feature of the model molecule is its fully chaotic quantum dynamics. This is proved by Fig. 4.8, which shows the chaotic quantum flow of the molecule on the "southern hemisphere" of the Bloch sphere. Fig. 4.8 was produced in the following way. First we defined the Poincaré section by $p = 0$, $dp/dt > 0$. Then, we ran 40 trajectories in (x, y, z, η, p) space for $\alpha = 0.1$, $\nu_0 = 10$ and $E = 6$ starting at the 40 different initial conditions

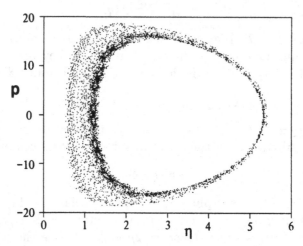

Fig. 4.7. $x = 0$ Poincaré section for the model molecule in the dynamic Born-Oppenheimer approximation. (From Blümel and Esser (1994).)

$x_j = \sin(\theta_j)$, $y_j = 0$, $z_j = \cos(\theta_j)$, $\eta_j = 2.4$, $\theta_j = j\pi/40$, $j = 1, 2, ..., 40$. In order to represent the resulting section points in the plane, we projected them onto the $x - y$ plane (the equatorial plane) of the Bloch sphere. In order to obtain a unique representation we divided the Bloch sphere into a "northern" and a "southern" hemisphere according to $z > 0$ and $z < 0$, respectively. Only the projections of points with $z < 0$ ("southern hemisphere") are shown in Fig. 4.8. We see that the southern hemisphere of the Bloch sphere is mostly chaotic. But this means that in the "chaotic sea" of Fig. 4.8 the quantum dynamics of the model molecule is *genuinely chaotic*.

One possible origin of chaos is the nonlinear resistance of the quantum particle against localization inside the box of width q. Therefore, the occurrence of quantum chaos in this system may be a direct consequence of Heisenberg's uncertainty principle. Therefore, the quantum pressure of the electron on the walls is called the "Heisenberg pressure".

In order to prove that the quantum dynamics on the Bloch sphere of the model molecule is genuinely chaotic, Blümel and Esser (1994) calculated the Euclidean distance $d(\tau) = [(x(\tau) - x'(\tau))^2 + (y(\tau) - y'(\tau))^2 + (z(\tau) - z'(\tau))^2]^{1/2}$ between two initially close trajectories. The result is shown in Fig. 4.9. It proves that for $\tau < 300$ the distance $d(\tau)$ grows exponentially. This corresponds to a positive Lyapunov exponent for the quantum subsystem. At $\tau \approx 300$ the exponential growth of $d(\tau)$ breaks.

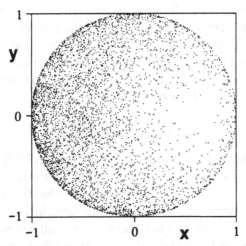

Fig. 4.8. Quantum chaos on the southern hemisphere of the Bloch sphere of the model molecule. (From Blümel and Esser (1994).)

This is natural since d cannot be larger than the diameter of the Bloch sphere, which equals 2.

In summary, we have shown in this section that even in one of the simplest conceivable models for a diatomic molecule, the coupling of classical and quantum degrees of freedom can lead to genuine chaos in both the quantum and the classical subsystems. This proves the existence of type II quantum chaos. This result is of general importance since type II quantum chaos may occur in any system that divides in a natural way into a

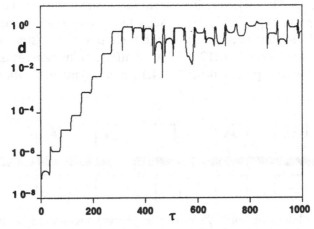

Fig. 4.9. Euclidean distance $d(\tau)$ of two initially close trajectories on the Bloch sphere. (From Blümel and Esser (1994).)

quantum subsystem and a classical subsystem. According to the above discussion, the occurrence of this type of chaos is not restricted to the quantum domain. It can occur in all wave systems that are dynamically coupled to their boundaries. Interestingly, all examples for type II quantum chaos discussed so far in the literature are *autonomous* systems, i.e. no coupling to external forces or fields is necessary.

4.3 Quantum chaos

In the preceding section we studied a system that exhibits genuine quantum chaos with a positive quantum Lyapunov exponent. However, the system as a whole cannot be called quantum chaotic since the system was analysed in a mixed classical-quantum description. In our opinion only fully quantized systems that show chaos and exponential sensitivity deserve to be called quantum chaotic. It is often asserted that there is a problem with defining quantum chaos. This is not so. The following statement is a perfectly legitimate and natural definition of quantum chaos: A quantum system is chaotic if it shows exponential instabilities and a positive Lyapunov exponent in analogy to the quantum subsystem of a type II quantum chaotic system. But the problem is not in *defining* what quantum chaos is, but, once a definition has been adopted, to show that the definition is not empty. Therefore, a most important question is whether quantum chaotic systems that comply with the above definition exist at all. Although candidate systems are discussed in the literature (Chirikov (1991), Weigert (1993)), the question of the existence of truly quantum chaotic systems is still open for discussion.

We contribute to the discussion with a simple but illustrative example (Blümel (1994a)). This system, too, is still in the discussion stage (Blümel (1995a), Schack (1995)), but it may nevertheless indicate what a truly quantum chaotic type III system looks like. The system consists of a spin-1/2 particle that is passed through a chain of two different magnets A and B as shown in Fig. 4.10. Since the magnets in general induce spin precession, the set-up shown in Fig. 4.10 is also called a *spin-precession*

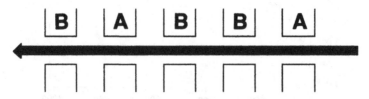

Fig. 4.10. A beam of spin-1/2 particles (arrow) passing through the spin-precession apparatus M_4 consisting of five magnets arranged according to a Fibonacci recurrence.

apparatus. It reminds us strongly of a Stern-Gerlach apparatus (Feynman *et al.* (1965)), but works with a homogeneous magnetic field since the aim is to influence the spin state of the particles, and not to deflect them. The magnets are arranged according to the recursive rule

$$M_{n+1} = M_n \circ M_{n-1}, \quad M_0 = A, \quad M_1 = B. \tag{4.3.1}$$

The symbol $\circ$ in (4.3.1) denotes concatenation, i.e. the operation of chaining the magnets into a string of magnets, and the symbol M_n denotes the string of magnets obtained after n steps of application of the rule (4.3.1). In order to illustrate the rule (4.3.1), we construct the first few chains explicitly. The chain M_2 consists of only two magnets according to $M_2 = M_1 \circ M_0 = BA$. The chain M_3 consists of three magnets arranged according to $M_3 = M_2 \circ M_1 = BAB$. The chains M_4 and M_5 consist of five and eight magnets, respectively. They are given explicitly by $M_4 = M_3 \circ M_2 = BABBA$ and $M_5 = M_4 \circ M_3 = BABBABAB$. It is important to note that the law (4.3.1) is completely deterministic and that the successive stages of M, i.e. M_2, M_3, ..., can in principle be built in the lab.

The propagator U of a spin-1/2 particle, propagating from right to left through M, satisfies the recurrence relation

$$\hat{U}_{n+1} = \hat{U}_n \hat{U}_{n-1}. \tag{4.3.2}$$

The recursion (4.3.2) can be started as soon as the propagators U_A and U_B for passage through the magnetic field sections A and B, respectively, are known. Choosing the z axis as the quantization axis, the propagator U_n at the nth step of (4.3.2) can be parametrized by (Messiah (1979))

$$\hat{U}_n = e^{-i\alpha_n \sigma_z} e^{-i\beta_n \sigma_y} e^{-i\gamma_n \sigma_z}, \tag{4.3.3}$$

where $\sigma_{x,y,z}$ are the Pauli matrices. The general case with α, β and $\gamma \neq 0$ is difficult to treat. We confine ourselves here to a discussion of the trivial case where the magnetic fields of all the magnets are aligned along the y direction. In this case the propagator $\hat{U}_n$ is represented by the matrix

$$\hat{U}_n = \begin{pmatrix} \cos(\beta_n) & -\sin(\beta_n) \\ \sin(\beta_n) & \cos(\beta_n) \end{pmatrix}. \tag{4.3.4}$$

The matrix U_n represents a simple rotation with rotation angle β_n. Using (4.3.4) in (4.3.2), we obtain a recurrence relation for the rotation angles β,

$$\beta_{n+1} = \beta_n + \beta_{n-1} \quad (\text{mod } 2\pi). \tag{4.3.5}$$

For $\beta_0 = 1$ and $\beta_1 = 1$ the recursion (4.3.5) produces the sequence of Fibonacci numbers (Schroeder (1991)) 1,1,2,3,5,8,... (mod 2π). Introducing the reduced angles $b_n = \beta_n/2\pi$, the following recurrence relation is obtained:

$$b_{n+1} = b_n + b_{n-1} \qquad (\text{mod } 1). \qquad (4.3.6)$$

This recurrence can be written as a mapping

$$Q: \quad \vec{w}_{n+1} = \begin{pmatrix} 1 & 1 \\ 1 & 0 \end{pmatrix} \vec{w}_n \qquad (\text{mod } 1) \qquad (4.3.7)$$

with $\vec{w}_n = (b_n, b_{n-1})$. The mapping Q is very similar to the cat map C defined by

$$C: \quad \vec{w}_{n+1} = \begin{pmatrix} 1 & 1 \\ 1 & 2 \end{pmatrix} \vec{w}_n \qquad (\text{mod } 1). \qquad (4.3.8)$$

The mapping C is called "cat map" because of an illustration in the book by Arnol'd and Avez (1968). Together with the baker's map discussed in Section 2.2, the cat map is a favourite textbook example for discussing the stretching and folding mechanisms that lead to chaos (see, e.g., Ott (1993)). Computing the eigenvalues and eigenvectors of C it is seen that any subset S of the unit square is stretched by C in the direction $\vec{v}_1^{(C)} = (1, \bar{g})$ and compressed in the direction $\vec{v}_2^{(C)} = (1, -g)$, where $\bar{g} = (1 + \sqrt{5})/2$ is the inverse of the golden mean $g = (\sqrt{5} - 1)/2$ defined in (2.1.3). A point P of the unit square which would be mapped out of the unit square by the stretching action of C is "folded back" into the unit square by the action of the modulo operation in (4.3.8), which is an integral part of the definition of C. The stretching factor of C is the larger one of its two eigenvalues and is given by $e_1^{(C)} = 1 + \bar{g}$. The compression factor is $e_2^{(C)} = 2 - \bar{g}$. The product of e_1 and e_2 is 1, which means that C is area preserving. Since $e_1^{(C)} > 1$, the mapping C is exponentially sensitive. As discussed in Section 2.2 in connection with the baker's map, the Lyapunov exponent σ is defined as the logarithm of the stretching factor. For the cat map C we have $\sigma_{cat} = \ln(e_1^{(C)}) > 0$. This implies that C is chaotic.

The mapping Q shares many of the properties of the cat map. In particular, Q possesses a stretching direction $\vec{v}_1^{(Q)} = (1, g)$ with a corresponding eigenvalue $e_1^{(Q)} = \bar{g} > 1$ such that its Lyapunov exponent $\sigma_Q > 0$. Therefore, the mapping Q exhibits exponential sensitivity and chaos in complete analogy to the cat map C. As a consequence, the quantum dynamics of spin-1/2 particles in the magnetic chain M is truly chaotic. Moreover, since Q is chaotic, the sequence of rotation angles $\beta_n \pmod{2\pi}$ is chaotic. Also, if the spin-1/2 particles are prepared in a pure spin state polarized in the $+z$ direction, the occupation probability in the $|+z\rangle$ state after section number n is given by $\cos^2(\beta_n)$. Therefore, since β_n is a chaotic sequence, the population in the $|+z\rangle$ state is chaotic too. Therefore, measuring the occupation probability in the $| +z \rangle$ state provides an experimental test for the occurrence of quantum chaos in this system.

Several comments are in order now.

(1) For given n the apparatus shown in Fig. 4.10 is not chaotic. This is a simple consequence of the unitarity of quantum mechanics. An initial state $|\psi_0\rangle$ is mapped into $|\psi_0'\rangle$ under the action of $\hat{U}_n$. An initial state $|\psi_1\rangle$ close to $|\psi_0\rangle$, i.e. $\langle\psi_0|\psi_1\rangle \approx 1$, will stay close to $|\psi_0\rangle$, i.e. $\langle\psi_0'|\psi_1'\rangle = \langle\psi_0|\hat{U}^\dagger\hat{U}|\psi_1\rangle = \langle\psi_1|\psi_0\rangle \approx 1$. Therefore, there is no exponential instability for fixed n. Chaos and instabilities occur as a function of the discrete time n.

(2) A practical question concerns the physical size (length) of the apparatus required to see the quantum chaos effect discussed above. It is well known that the Fibonacci numbers form an exponentially diverging sequence. Since (4.3.1) is closely related to the recurrence relation for Fibonacci numbers, the number of magnets grows exponentially from stage to stage in the construction of the magnetic chain. On the basis of this observation, the action of the magnetic spin-precession apparatus is equivalent with the free motion of a particle on a ring whose position is checked at the end of exponentially growing time intervals (Schack (1995)). The ring analogue is as chaotic as the shift map discussed in Section 2.2. It provides an alternative way for understanding the origin of chaos in the spin-precession apparatus.

(3) Because of the exponential proliferation of magnets, the physical flight time of the particles through the apparatus grows exponentially for a linear growth of n. Spacing the magnets exponentially close together in n, the natural flight time grows only linearly and the spin-precession apparatus is also chaotic with respect to the natural flight time of the particles.

(4) Transforming to the rest-frame of the moving beam-particles the quantum mechanics of a spin-1/2 particle traversing the spin-precession apparatus is equivalent to the quantum mechanics of a stationary spin-1/2 particle perturbed by a sequence of external field pulses. Such a system was investigated by Luck *et al.* (1988). It was found that the quantum dynamics of a spin-1/2 particle perturbed by a sequence of pulses arranged according to a Fibonacci sequence is indeed very complicated.

Although the investigation of the above example is far from complete (only the "trivial" case with aligned magnetic fields has so far been investigated), it is a candidate for a fully quantized type III quantum chaotic system.

With this section we finish our general survey of chaos in classical and quantum mechanics and turn to a discussion of specific examples of the manifestations of chaos in atomic physics.

5

The kicked rotor: paradigm of chaos

In this and the following chapters we will encounter various time dependent and time independent atomic physics systems whose classical counterparts are chaotic. All the systems discussed in the remaining chapters of this book are examples of type I systems, i.e. examples for quantized chaos. This is natural since quantized chaos is by far the most important type of quantum chaos relevant in atomic and molecular physics. In the category "time dependent systems", we discuss the rotational dynamics of diatomic molecules (Section 5.4), the microwave excitation of surface state electrons (Chapter 6), and hydrogen Rydberg atoms in strong microwave fields (Chapters 7 and 8). All these systems are driven by externally applied microwave fields. For strong fields none of these three systems can be understood on the basis of quantum perturbation theory, as the involved multi-photon orders are very high. Processes of multi-photon orders 100 to 300, typically, have to be considered. It is important to realize that in this day and age, with powerful super-computers at hand, there is no problem in implementing a perturbation expansion of such high orders. But the emphasis is on *understanding* the processes involved. Although multi-photon perturbation theory provides valuable insight into the physics of low order multi-photon processes important in the case of weak applied fields (an example is discussed in Section 6.3), not much insight is gained from a perturbation expansion that has to be carried along to the 100th order and beyond in order to converge. But, as we will see below, the behaviour of many atomic and molecular physics systems, involving strong fields and high multi-photon orders, is understood quite naturally in the language of nonlinear dynamics and chaos. Moreover, we will discover and predict a wealth of new physics, including the occurrence of new photo-electric peaks (Casati *et al.* (1986)) that can be understood on the basis of classical nonlinear resonances (Chapters 6 and 7).

Time dependent systems offer the easiest access to chaos and quantum chaology in atomic physics. The reason is that there is a class of time dependent perturbations of very short temporal duration, "kicks", for which the dynamics of these systems can be formulated with the help of discrete mappings. This is important since, as we saw in Chapter 3, discrete mappings illustrate the essentials of chaos without too much formal ballast. We illustrate the essence of kicked systems with the help of a simple atomic physics system, the *kicked molecule*. The kicked molecule set-up is shown in Fig. 5.1. A polar diatomic molecule (we choose CsI as an example) is located between the two plates of a capacitor. The capacitor is connected to a pulse generator that periodically charges and discharges the two electrodes of the capacitor. This process creates a time varying spatially homogeneous electric field such that the molecule experiences a train of electric field pulses that couple to its dipole moment. It is important to notice that this situation is simple and completely deterministic. We assume that there is no noise present. Therefore, given the initial state of the molecule, there is no problem in solving Schrödinger's equation to obtain the wave function of the molecule for all times. A straightforward solution of the Schrödinger equation is certainly a valid approach to the solution of this molecular excitation problem, and its usefulness should not be underestimated. We make extensive use of direct numerical methods throughout this book. We interpret them as "numerical experiments" (Percival (1973)) to provide us with the bare facts (the "experimental data") of the behaviour of the system. This "brute-force" technique, however, although capturing the essential physical features of the system, does not explain them. Physical insight is obtained only with the help of a detailed analysis based on simplified models.

The plan of Chapter 5 is the following. In order to get a feeling for the dynamics of the kicked molecule, we approximate it by a one-dimensional schematic model by restricting its motion to rotation in the (x, z) plane and ignoring motion of the centre of mass. In this approximation the kicked molecule becomes the *kicked rotor*, probably the most widely studied model in quantum chaology. This model was introduced by Casati *et al.* in 1979. The classical mechanics of the kicked rotor is discussed in Section 5.1. Section 5.2 presents Chirikov's overlap criterion, which can be applied generally to estimate analytically the critical control parameter necessary for the onset of chaos. We use it here to estimate the onset of chaos in the kicked rotor model. The quantum mechanics of the kicked rotor is discussed in Section 5.3. In Section 5.4 we show that the results obtained for the quantum kicked rotor model are of immediate relevance for the kicked CsI molecule. At the time of writing the CsI system is still under proposal. To our knowledge it has never been investigated experimentally in the chaotic strong-field regime. Therefore, some questions pertaining to experimental feasibility are also discussed. An interesting recent development is the influence of symmetries on the quantum behaviour of classically chaotic systems. This topic is discussed in Section 5.5. Based on the results obtained in Sections 5.4 and 5.5, we propose to detect the quantum signatures of chaos and their modifications by symmetries in an actual molecular physics experiment with CsI molecules (see also Blümel *et al.* (1986)).

5.1 The classical kicked rotor

At first glance, the kicked molecule experiment sketched in Fig. 5.1 does not appear to be a system worthy of much attention. The set-up is simple, there are no complicated boundary conditions, and noise effects are neglected. But its simplicity notwithstanding, it turns out that the classical as well as the quantum dynamics of the system sketched in Fig. 5.1 are very complicated, and cannot in either case be solved analytically in the presence of a strong driving field.

In order to extract the essence of the dynamics of the kicked molecule, we consider a simple model constructed by replacing the molecule with a two-dimensional dipole (see Fig. 5.2). Moreover, we replace the sequence of finite-width pulses provided by the pulse generator (see Fig. 5.1) by a train of zero-width δ-function kicks. Thus, the dipole is perturbed by a

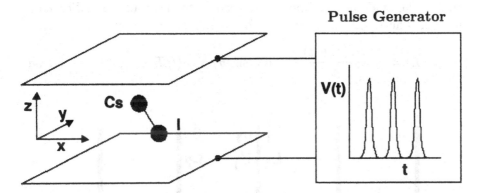

Fig. 5.1. The kicked CsI molecule: schematic sketch of a proposed experimental set-up. The electric field driving the molecule is generated by a voltage $V(t)$ provided by a pulse generator.

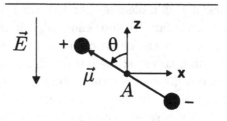

Fig. 5.2.   The one-dimensional kicked rotor.

sequence of electric field bursts according to

$$E(t) = \mathcal{E}_0 \sum_{n=-\infty}^{\infty} \delta(t/T - n), \qquad (5.1.1)$$

where $\mathcal{E}_0$ is proportional to the strength of the field and T is the time interval between kicks.

A time diagram of the electric field bursts (5.1.1) is shown in Fig. 5.3. A planar rotor perturbed by a sequence of delta kicks according to (5.1.1) is known as the *kicked rotor*. In order to familiarize ourselves with the physics of kicked systems, we focus in this section on the classical mechanics of the kicked rotor. If we denote by L the angular momentum of the rotor, the Hamiltonian function of the kicked rotor is given by

$$H(L, \theta; t) = \frac{L^2}{2I} + \mu\mathcal{E}_0 \cos(\theta) \sum_n \delta(t/T - n), \qquad (5.1.2)$$

where I is the moment of inertia of the rotor, μ is its dipole moment and θ is the angle of the dipole with respect to the z axis as indicated in Fig. 5.2. The equations of motion for (5.1.2) are given by:

$$\dot{L} = -\frac{\partial H(L, \theta; t)}{\partial \theta} = \mu\mathcal{E}_0 \sin(\theta) \sum_n \delta(t/T - n), \qquad (5.1.3a)$$

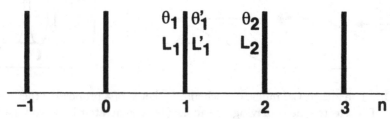

Fig. 5.3.   Time diagram of the electric field bursts of the kicked rotor model.

$$\dot{\theta} = \frac{\partial H(L, \theta; t)}{\partial L} = \frac{L}{I}. \tag{5.1.3b}$$

The solution of (5.1.3) is straightforward. According to Fig. 5.3 there are periods of free motion, during which the external field is switched off, followed by a "kick" at which the angular momentum of the rotor changes abruptly by an amount that depends on the value of the rotation angle θ at the moment at which the kick occurs. If we denote the values of L and θ immediately before kick number n by L_n and θ_n, the solution of (5.1.3) can be stated in the form of a mapping that relates L_n and θ_n immediately before kick number n to L_{n+1} and θ_{n+1} immediately before kick number $n + 1$. If this mapping is known, then, given the initial conditions L_0 and θ_0 at kick number 0, the values of L and θ at kick number n can be computed simply by iterating the mapping. In order to derive the explicit form of the mapping, we start with L_j and θ_j immediately before kick number j and compute, as a first step, the values L'_j and θ'_j immediately after kick number j. Since the rotation angle θ is continuous at the kick, we have immediately: $\theta'_j = \theta_j$. The angular momentum L'_j immediately after kick number j can be obtained by integrating (5.1.3a) in the infinitesimally short time interval $[jT - \epsilon, jT + \epsilon]$, $\epsilon << T$, across kick number j. We obtain

$$L'_j - L_j = \int_{jT-\epsilon}^{jT+\epsilon} \mu \mathcal{E}_0 \sin(\theta) \sum_n \delta(t/T - n)\, dt = \mu \mathcal{E}_0 T \sin(\theta_j). \tag{5.1.4}$$

In the follwing period of free motion the angular momentum is conserved, i.e. $L_{j+1} = L'_j$. The rotation angle increases according to (5.1.3b) to $\theta_{j+1} = \theta_j + TL'_j/I = \theta_j + TL_{j+1}/I$. Therefore we obtain the following mapping as a solution of the dynamics defined by (5.1.3):

$$L_{n+1} = L_n + \mu \mathcal{E}_0 T \sin(\theta_n),$$

$$\theta_{n+1} = \theta_n + TL_{n+1}/I. \tag{5.1.5}$$

Measuring the angular momentum in units of I/T, such that $L_n = Il_n/T$, and introducing the control parameter $K = \mu \mathcal{E}_0 T^2/I$, we obtain the following set of dimensionless mapping equations:

$$l_{n+1} = l_n + K \sin(\theta_n), \tag{5.1.6a}$$

$$\theta_{n+1} = \theta_n + l_{n+1}. \tag{5.1.6b}$$

The mapping (5.1.6) is known as the *standard mapping* (Chirikov (1979), Percival (1987) and references therein). It is one of the paradigms of nonlinear dynamics and chaos theory. One distinguishing mark of the mapping (5.1.6) is that it cannot in general be solved explicitly. This means that in general it is impossible to derive an analytical expression that would predict l_n and θ_n for all n and all initial conditions l_0 and θ_0. In general, i.e. for initial conditions that are not too special, the shortest way of predicting l_n and θ_n is by iterating the mapping (5.1.6).

The analysis so far shows that in the classical approximation the dynamics of the kicked rotor depends only on the single control parameter K. The question is whether a transition to chaos can be observed as a function of K. The easiest way of examining the dynamics of the rotor for the existence of chaos is by means of Poincaré sections introduced in Section 3.2. The phase space of the kicked rotor is the two-dimensional (θ, l) space. Topologically this phase space is a cylinder, since θ is a 2π periodic angle variable and l is unbounded. A portion of the cylindrical rotor phase space ($-2\pi \leq l \leq 4\pi$) is shown in each of the four panels of Fig. 5.4. An initial condition θ_0, l_0 is represented by a point (θ_0, l_0) in the two-dimensional phase space. A phase-space portrait emerges if together with (θ_0, l_0) its iterates (θ_j, l_j), $j = 1, 2, ...$, are plotted. The sequence of points (θ_j, l_j), $j = 0, 1, 2, ...$, the orbit of the starting point (θ_0, l_0), reveals whether the motion is simple or complex. The type of motion depends on the precise values of the initial condition. Therefore, an overview of the motion in the phase space, a complete phase-space portrait, is generated by plotting the first N iterates of a whole family of initial conditions. We choose to plot the $N = 500$ iterates of a two-parameter family of initial

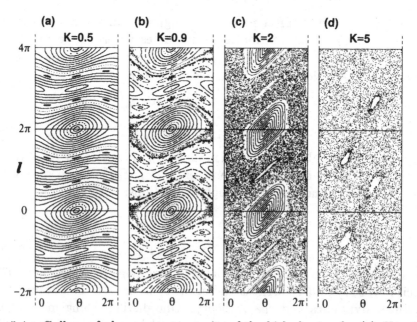

Fig. 5.4.   Gallery of phase-space portraits of the kicked rotor for (a) $K = 0.5$, (b) $K = 0.9$, (c) $K = 2$ and (d) $K = 5$.

conditions $(\theta_0^{(mq)}, l_0^{(mq)})$, $m = 1, 2, ..., 39$, $q = -1, 0, 1$, with $\theta_0^{(mq)} = \pi$, $l_0^{(mq)} = 2m\pi/40 + 2q\pi$, for four different values of the control parameter, $K = 0.5$, $K = 0.9$, $K = 2$ and $K = 5$. The result is shown in Fig. 5.4. The point patterns in Fig. 5.4 indicate that a transition to chaos indeed takes place as a function of increasing control parameter K. We will now discuss the four phase-space portraits shown in Fig. 5.4 one by one in detail.

Our first impression of Fig. 5.4(a) is one of regularity and periodicity. The phase-space structures seem to be periodic in l with a period of 2π. This is indeed so and is easily proved by inspection of the mapping (5.1.6). Since (5.1.6*b*) refers to an angle variable, it is to be understood as an equation modulo 2π. Changing l by multiples of 2π clearly leaves (5.1.6*b*) invariant. The same transformation also leaves (5.1.6*a*) invariant, since the 2π increments in l cancel on both sides. Thus, the rotor phase space is 2π periodic in l. The periodicity in l is often used in the literature to reduce the study of the kicked rotor to the "unit cell" (θ, l) $\in [0, 2\pi] \times [0, 2\pi]$. We do not use this technique here since the phase space appropriate for our particular atomic physics problem, the kicked CsI molecule, is the phase-space cylinder with unrestricted values of l, and not the restricted phase space consisting only of a single unit cell.

The regularity in Fig. 5.4(a) manifests itself in the fact that the iterates of the 3×39 different initial conditions either lie on closed quasi-elliptical curves, or they lie on lines winding around the phase-space cylinder. Both phase-space features are called *invariant curves* because all iterates of an initial condition chosen from such a curve will be mapped onto the curve. It is possible to show mathematically that continuous phase-space features, such as invariant lines, are not allowed to cross. This is a reflection of the idea that because of classical determinism (see Section 1.3) distinct classical trajectories cannot intersect in phase space, since otherwise, two *distinct* "futures" would have the *same* past (the point of intersection). This has an immediate and important physical consequence: for $K = 0.5$ the rotational energy of the kicked rotor cannot grow indefinitely. The reason is the following. Measuring energy in units of I/T^2, the kinetic energy of the kicked rotor is given by $E_{kin} = l^2/2$. Therefore, the rotor energy can increase only if the magnitude of the angular momentum increases. But, as Fig. 5.4(a) shows, a monotonic increase of l is impossible without crossing invariant curves. The angular momentum, and therefore the kinetic energy of the rotor, are bounded. Therefore, the invariant curves stretching across the phase space from $\theta = 0$ to $\theta = 2\pi$ in Fig. 5.4(a) are also called "sealing" curves, because each of them divides the phase space into two parts that are dynamically disconnected. Provided our polar diatomic molecule is well modelled by the kicked rotor, the boundedness of the kicked rotor energy translates directly into an inhibition of molecular excitation for physical parameters that correspond to $K = 0.5$. Therefore, the observation of limited energy growth in the kicked rotor has important consequences for the excitation of molecular rotation in the case of actual diatomic molecules. This subject will be discussed further in Section 5.4 where we study the dynamics of CsI molecules exposed to strong electric field pulses.

The action of the sealing invariant curves can be demonstrated directly with the help of a Monte-Carlo calculation. Consider an ensemble of M rotors with initial conditions $(\theta_0^{(m)}, l_0^{(m)})$, $l_0^{(m)} = 0$ and $\theta_0^{(m)} = 2(m - 1/2)\pi/M$, $m = 1, 2, ..., M$. We choose $M = 10\,000$ for the computations reported below. This ensemble of rotors can also be interpreted as a classical approximation to the quantum mechanical ground state of the kicked rotor. This is so since the angular momentum of the rotor is specified exactly. According Heisenberg's uncertainty principle, the angle variable is then completely undetermined. This can be represented by an ensemble of classical rotors equi-distributed in the angle variable. The ensemble average of the kinetic energy immediately before kick number n is defined as

$$\bar{E}_n = \frac{1}{2M} \sum_{m=1}^{M} \left[l_n^{(m)} \right]^2. \tag{5.1.7}$$

Fig. 5.5(a) shows the mean energy as defined in (5.1.7) for $K = 0.5$. The most important feature of Fig. 5.5(a) is that, as predicted, the energy does not grow at all. It stays bounded for all n.

What happens at larger values of K? Fig. 5.4(b) gives the answer for $K = 0.9$. Compared with Fig. 5.4(a) many of the sealing invariant lines are destroyed. Chaotic regions have appeared. Extrapolating this trend, we expect that at some critical value of K, called K_c, the last sealing invariant curve is broken, and the angular momentum is no longer confined. Detailed calculations reveal that this critical value is $K_c \approx 1$ (for more details see Lichtenberg and Lieberman (1983)).

Fig. 5.4(c) shows the phase-space portrait for $K = 2 > K_c$. Indeed, as expected, all the sealing lines are now broken and the angular momentum is free to roam in the l direction. In this case we expect that the molecule is able to absorb energy out of the external field. This is verified by Fig. 5.5(b), which shows $\bar{E}_n$ for $K = 2$. In this case the energy grows linearly in time and without limit.

Assuming that all the regular structures in the phase space are destroyed, the linear growth of the energy of the kicked rotor can be understood analytically by calculating the mapping of the ensemble averaged

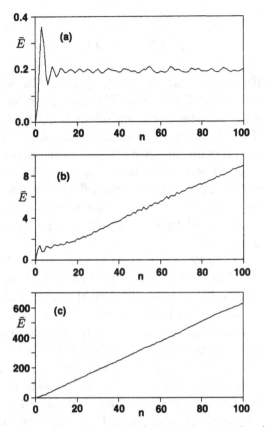

Fig. 5.5. Mean energy of an ensemble of classical rotors for (a) $K = 0.5$, (b) $K = 2$ and (c) $K = 5$.

energy from kick to kick:

$$\bar{E}_{n+1} = \frac{1}{2M} \sum_{m=1}^{M} l_{n+1}^{(m)\,2} = \frac{1}{2M} \sum_{m=1}^{M} [l_n^{(m)} + K \sin(\theta_n^{(m)})]^2$$

$$= \frac{1}{2M} \sum_{m=1}^{M} [l_n^{(m)2} + 2K l_n^{(m)} \sin(\theta_n^{(m)}) + K^2 \sin^2(\theta_n^{(m)})]. \qquad (5.1.8)$$

Assuming that $l_n^{(m)}$ and $\theta_n^{(m)}$ are uncorrelated with zero average, (5.1.8) yields approximately:

$$\bar{E}_{n+1} = \bar{E}_n + \frac{1}{4}K^2. \qquad (5.1.9)$$

The recursion (5.1.9) can be solved explicitly and yields

$$\bar{E}_n = D\,n, \qquad (5.1.10)$$

with

$$D(K) \;=\; K^2/4. \tag{5.1.11}$$

The result (5.1.10) means that the energy grows diffusively. Comparing this result with the result displayed in Fig. 5.5(b), we see that the linear growth of the energy is indeed well reproduced, but that the magnitude of D is far off. This is not hard to explain. According to Fig. 5.4(c) a large number of initial conditions are trapped in the regular structures at $\theta \approx \pi$, $l \approx 0 \bmod 2\pi$. Therefore, a large fraction of initial conditions do not participate in the angular momentum diffusion process, which results in the observed reduction of the magnitude of the diffusion constant D. This problem can be cured by starting trajectories only in the chaotic parts of phase space.

Fig. 5.4(d) shows that at $K = 5$, apart from small regular islands, most of the kicked rotor phase space is chaotic. In this case we expect the energy diffusion constant D to be very close to the analytical result (5.1.11). Fig. 5.5(c) shows that this is indeed the case. The ensemble average of the energy grows nearly linearly in time. The diffusion constant is given by $D \approx 6$ and can be compared with the prediction $D = 6.25$ according to (5.1.11). But the near agreement between the "experimental" and the predicted diffusion constant for $K = 5$ is to a certain extent fortuitous. Going beyond the simple analytical analysis presented above, Rechester and White (1980) predicted that (5.1.11) is only the lowest order approximation to the exact diffusion constant $D(K)$ which oscillates in K around its mean value $\bar{D} = K^2/4$ with an amplitude that can be up to 1.5 times the expected mean value $\bar{D}$. Also, "accelerator modes" (Chirikov (1979)) influence the exact value of the diffusion constant. According to Rechester and White, the amplitude of the oscillations die with $K \to \infty$. Therefore, we expect that (5.1.11) is a good approximation for $K \gg 1$.

Summarizing, we predict on the basis of the one-dimensional classical kicked rotor that, for $K < K_c \approx 1$, the molecule in Fig. 5.1 cannot absorb any substantial quantities of energy from the driving field, no matter how long the molecule is exposed to the field. Strong absorption of energy with a linear growth as a function of exposure time is expected to occur for $K > K_c$.

The magnitude of the critical control parameter K_c can be estimated analytically with the help of Chirikov's criterion. This criterion is naturally derived in the context of the kicked rotor, and is introduced in the following section. The Chirikov criterion is also the basis for estimating the onset of chaos in many other chaotic atomic physics systems. Examples are presented in Chapters 6 and 7.

5.2 Chirikov's overlap criterion

The mapping equations (5.1.6) can be derived from the Hamiltonian

$$H = \frac{1}{2}l^2 + K\cos(\theta)\sum_n \delta(t-n).$$ (5.2.1)

The drive term in (5.2.1) is a sum of 1-periodic δ functions whose Fourier transform is given by

$$\sum_n \delta(t-n) = \sum_m \exp(im\omega t),$$ (5.2.2)

where $\omega = 2\pi$. Whenever the driving frequency ω and the natural frequency $\dot{\theta}$ of the rotor are rationally related ($\dot{\theta} = p\omega/q, p, q \in N$ relatively prime) a *resonance* occurs. The most important resonances are the *primary resonances* which occur for $q = 1$. In this case a harmonic of (5.2.2) equals the angular frequency $\dot{\theta}$ of the free rotation of the rotor. Therefore, the condition for a primary resonance is given by

$$\dot{\theta} = l = m\omega = 2m\pi.$$ (5.2.3)

Thus, the resonance angular momenta are given by $l_m = 2m\pi$. The resonance angular momenta correspond exactly to the large elliptical structures seen in the phase-space plots shown in Fig. 5.4. These phase-space structures, by inference called resonances too, are of prime importance for Chirikov's criterion.

In order to estimate the onset of global chaoticity in the kicked rotor, Chirikov (1979) made the following observation. According to Fig. 5.4(b), chaos is already well developed, but it is prohibited from spreading to other regions of phase space by the existence of the sealing invariant lines that exist between the resonances in Fig. 5.4(b). In order to eliminate the sealing lines, Chirikov argued that by increasing the control parameter K the widths of the resonances in Figs. 5.4(a,b) can be increased to the point where the resonances "touch" each other, effectively squeezing the invariant lines out of phase space. The touching point defines the critical control parameter K_c. For $K > K_c$ the resonances overlap, resulting in global chaos and energy diffusion. Hence the name of the criterion: Chirikov's *overlap criterion*. This criterion is one of the most useful analytical tools in nonlinear dynamics. Although only accurate to within a factor of 2, the criterion gives valuable "ball park" estimates for the transition to global stochasticity. The application of Chirikov's criterion is not restricted to the kicked rotor. In the context of atomic and molecular physics it has been applied in estimating critical fields for driven surface state electrons (Chapter 6), Rydberg atoms (Chapter 7) and diatomic molelcules (Section 5.4).

In order to develop the criterion more quantitatively, consider the sequence of phase-space portraits shown in Figs. 5.4(a) – (d). This sequence suggests that, as the control parameter K increases, the diameter of the resonance islands at $l = 0$ mod 2π grows in action. In order to predict the touching point of the resonances, we need the widths of the resonances as a function of K. The width of the resonances is derived on the basis of the Hamiltonian (5.2.1). Since the dynamics induced by H is equivalent to the chaotic mapping (5.1.6), the Hamiltonian H itself cannot be treated analytically and has to be simplified. One way is to consider only the average effect of the periodic δ kicks in (5.2.1). The average perturbation strength is given by $\int_{-1/2}^{1/2} \delta(t)\, dt = 1$. Therefore, we define the average Hamiltonian

$$\bar{H} = \frac{1}{2} l^2 + K \cos(\theta). \tag{5.2.4}$$

This is the Hamiltonian of a pendulum shown in Fig. 5.6.

The pendulum phase space is two-dimensional. A qualitative sketch is shown in Fig. 5.7. There is a fixed point at $l = 0$, $\theta = \pi$, which corresponds to the lowest pendulum energy $E = -K$. For slightly higher energy the pendulum librates (oscillates). The corresponding phase-space trajectory is a closed curve around the fixed point, as shown in Fig. 5.7. Closed orbits occur for energies up to $E = K$. At energies $E > K$ the pendulum "turns over" and libration turns into rotation. The two qualitatively different types of motion are separated by a one-dimensional curve at energy $E = K$, the *separatrix*.

The pendulum phase space is not periodic in l as was the case for the standard mapping (see Fig. 5.4). But in the vicinity of $l = 0$ it qualitatively reproduces the phase-space features of the kicked rotor very

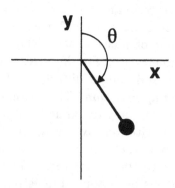

Fig. 5.6.   Pendulum resulting from a time average of the kicked rotor Hamiltonian.

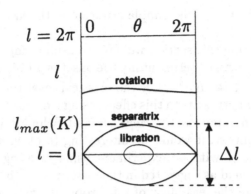

Fig. 5.7. Qualitative sketch of the pendulum phase space indicating libration and rotation regions. The separatrix, in this case a one-dimensional curve, separates the two regions.

well. The resonances in the kicked rotor located in action at multiples of 2π are well modelled by the single pendulum resonance shown in Fig. 5.7.

Estimating the width of the pendulum resonance is straightforward. According to Fig. 5.7 the pendulum resonance has its largest width at the top of the separatrix at $\theta = \pi$. Inserting $E = K$ and $\theta = \pi$ into the averaged Hamiltonian function (5.2.4), we obtain

$$E = K = \frac{1}{2} l_{max}^2 - K. \tag{5.2.5}$$

This relation yields

$$l_{max}(K) = 2\sqrt{K} \tag{5.2.6}$$

for the maximal excursion in angular momentum. The width Δl of the pendulum resonance is twice the maximal excursion (see Fig. 5.7). We obtain

$$\Delta l = 4\sqrt{K}. \tag{5.2.7}$$

Therefore, on the basis of the averaged Hamiltonian (5.2.4), a resonance of the standard mapping located at $l = 2m\pi$ overlaps with its next replica located at $l = 2(m+1)\pi$ if $\Delta l = 2\pi$. This yields

$$K_c^{(a)} = \frac{\pi^2}{4} \tag{5.2.8}$$

as the analytical estimate for the critical control parameter K_c according to the Chirikov overlap criterion. The numerical value of $K_c^{(a)}$ is $K_c^{(a)} \approx 2.47$. This result is about a factor 2 too large compared with the numerical results obtained above. Various elaborations on the Chirikov criterion are reported in the literature (see, e.g., Chirikov (1979), Lichtenberg and Lieberman (1983)). For order-of-magnitude estimates for the onset of global diffusion, however, the simple criterion outlined above is entirely adequate.

Before we conclude this section we have to ask an important question: Is it possible to observe experimentally the onset of global chaos predicted to occur at $K = K_c \approx K_c^{(a)}$? On the classical level the answer to this question is yes. Experiments to this effect, using a driven magnetic dipole, have already been reported by Meissner and Schmidt (1986). In the atomic physics context, it would be interesting to observe the predicted onset of global chaos with an experiment using rotating molecules in a pulsed microwave field as suggested in the introduction to this chapter. In order to investigate the feasibility of this proposal, we have to investigate the quantum kicked rotor model.

5.3 The quantum kicked rotor

The quantum Hamiltonian of the classical kicked rotor, defined by the classical Hamiltonian function (5.1.2), is easily obtained by canonical quantization. On replacing the classical angular momentum L by the quantum angular momentum operator $\hat{L}$ according to

$$L \to \hat{L} = -i\hbar \partial/\partial\theta, \tag{5.3.1}$$

we obtain

$$\hat{H} = \frac{\hat{L}^2}{2I} + \mu\mathcal{E}_0 \cos(\theta) \sum_j \delta(t/T - j). \tag{5.3.2}$$

In the context of the classical kicked rotor, discussed in the previous section, it was the impulsive nature of the driving force that allowed us to solve the classical equations of motion by means of a simple discrete mapping. The same technique works in the case of the quantum kicked rotor. In complete analogy to the classical case, the quantum dynamics of (5.3.2) can be solved by means of a quantum mapping (Berry *et al.* (1979)). The mapping operator is the time evolution operator of the rotor over one period of the driving force. It is constructed in the following way. We start with the time dependent Schrödinger equation for the time dependent rotor states $| \psi(t) \rangle$ given by

$$i\hbar \frac{\partial}{\partial t} | \psi(t) \rangle = \hat{H} | \psi(t) \rangle. \tag{5.3.3}$$

In the vicinity of a kick, say, kick number n, the kinetic energy operator $\hat{L}^2/(2I)$ can be neglected. In this case the Schrödinger equation reads

$$i\hbar \frac{\partial}{\partial t} | \psi(t) \rangle = \mu\mathcal{E}_0 \cos(\theta)\delta(t/T - n) | \psi(t) \rangle. \tag{5.3.4}$$

If we denote by $|\psi_n\rangle$ the state of the rotor immediately before kick number n, and by $|\psi'_n\rangle$ the state of the rotor immediately after kick number n, then $|\psi'_n\rangle$ can be calculated by a simple quadrature. It is given by:

$$|\psi'_n\rangle = \exp[-i\mu\mathcal{E}_0 T \cos(\theta)/\hbar] |\psi_n\rangle. \qquad (5.3.5)$$

Between kicks the external force is zero and the state of the rotor evolves freely:

$$|\psi_{n+1}\rangle = \exp[-i\hat{L}^2 T/2I\hbar] |\psi'_n\rangle. \qquad (5.3.6)$$

Defining the complete propagator over one kick as

$$\hat{U} = \exp[-i\hat{L}^2 T/2I\hbar] \exp[-i\mu\mathcal{E}_0 T \cos(\theta)/\hbar], \qquad (5.3.7)$$

the quantum mapping of the rotor wave function takes the form

$$|\psi_{n+1}\rangle = \hat{U} |\psi_n\rangle. \qquad (5.3.8)$$

All the information about the long-time behaviour of the rotor states is contained in the properties of the one-cycle propagator $\hat{U}$. Introducing the dimensionless angular momentum $\hat{l}$ according to

$$\hat{L} = \hbar\hat{l} \qquad (5.3.9)$$

and the dimensionless control parameters

$$\tau = \hbar T/I, \qquad k = \mu\mathcal{E}_0 T/\hbar, \qquad (5.3.10)$$

the mapping operator $\hat{U}$ can be written as

$$\hat{U} = \exp\left[-i\frac{\tau}{2}\hat{l}^2\right] \exp[-ik\cos(\theta)]. \qquad (5.3.11)$$

The classical analogue of the quantum mapping (5.3.11) is obtained from (5.1.5) by using the definition (5.3.9) and the dimensionless control parameters (5.3.10). The result is

$$l_{n+1} = l_n + k\sin(\theta_n)$$

$$\theta_{n+1} = \theta_n + \tau l_{n+1}, \qquad (5.3.12)$$

where now, i.e. in the classical context, l is a c-number, not an operator. Rescaling the angular momentum according to

$$\tilde{l}_n = \tau l_n, \qquad (5.3.13)$$

the mapping (5.3.12) can be written in the form (5.1.6) according to

$$\tilde{l}_{n+1} = \tilde{l}_n + K\sin(\theta_n)$$

$$\theta_{n+1} = \theta_n + \tilde{l}_{n+1}, \qquad (5.3.14)$$

where

$$K = k\tau. \tag{5.3.15}$$

The relation (5.3.15) allows us to associate a classical kicked rotor characterized by the control parameter K with a quantum kicked rotor characterized by the two quantum control parameters τ and k. It is not possible to reduce the two quantum control parameters to a single control parameter, for instance by means of some appropriate scaling transformation. Thus, contrary to the classical mapping equations (5.3.14) that depend only on a single control parameter K, the quantum mapping operator (5.3.11) depends on *two* essential control parameters τ and k. We will show below that for given k the qualitative features of the quantum rotor dynamics depend decisively on the algebraic properties of the control parameter τ, thus establishing τ as an essential *independent* control parameter.

Because of the simple analytical structure of the quantum mapping operator (5.3.11), numerical computations are very easy to perform. A variety of numerical schemes are known that can be used to propagate the rotor wave function according to (5.3.8). The simplest one is to expand the rotor wave function $| \psi_0 \rangle$ at time $t = 0$ into the complete set of angular momentum eigenstates according to

$$| \psi_0 \rangle = \sum_l A_l^{(0)} \langle l \, | \, \psi_0 \rangle \, | l \rangle, \tag{5.3.16}$$

where the angular momentum eigenstates $| l \rangle$ are given in θ representation according to

$$\langle \theta \, | \, l \rangle = \frac{1}{\sqrt{2\pi}} \exp(il\theta). \tag{5.3.17}$$

The matrix elements of $\hat{U}$ in the representation (5.3.17) can be calculated analytically. They are given by

$$U_{lm} \equiv \langle l \, | \, \hat{U} \, | \, m \rangle = i^{l-m} \exp\left[-i\frac{\tau}{2}l^2\right] J_{|l-m|}(k), \tag{5.3.18}$$

where the functions $J_{|l-m|}(k)$ in (5.3.18) are Bessel functions of the first kind (Abramowitz and Stegun (1964)). The iterates $| \psi_n \rangle$ of a rotor wave function $| \psi_0 \rangle$ can now be obtained via their expansion amplitudes by simple matrix multiplication according to

$$A_l^{(n+1)} = \sum_m U_{lm} A_m^{(n)}. \tag{5.3.19}$$

The matrix multiplication scheme is sufficient for some basic numerical experiments with the quantum kicked rotor. A more efficient scheme based on the fast Fourier transform is discussed, e.g., by Grempel *et al.* (1984).

In analogy to the classical kicked rotor we define the mean kinetic energy of the rotor given by

$$E_n^{(qm)} = \langle \psi_n \mid \frac{1}{2}\hat{l}^2 \mid \psi_n \rangle. \qquad (5.3.20)$$

Using (5.3.17), (5.3.18) and (5.3.19), we are now ready for a first numerical experiment with the quantum kicked rotor. With an eye on the results for the classical kicked rotor, we choose the control parameters to be $\tau = 1$ and $k = 5$. This ensures that the corresponding classical control parameter $K = k\tau = 5$ is well in the chaotic regime. The choice of control parameters also allows a direct comparison with the results for the energy growth of an ensemble of classical kicked rotors displayed in Fig. 5.5(c). For the reader's convenience this result is copied into Fig. 5.8 and labelled "cl, $K = 5$". The classical ensemble of rotors is started at time $t = 0$ with $l = 0$, and is equi-distributed in the rotor angle. This classical situation is best modelled by starting the quantum rotor in the initial state $\mid \psi_0 \rangle = \mid l = 0 \rangle$, since this state has a sharp angular momentum ($l = 0$) and a completely flat distribution in rotation angles, in complete analogy with the classical case. Using (5.3.18) in an angular momentum basis of 401 states ($-200 \le l \le 200$), we propagated the initial rotor function $\mid \psi_0 \rangle = \mid l = 0 \rangle$ over 100 kicks. The resulting $E_n^{(qm)}$ is shown in Fig. 5.8 as the full line marked "qm, $\tau = 1$". The result is completely unexpected. Instead of showing energy diffusion characterized by a persistent linear growth of the energy, the quantum kicked rotor mimics the classical en-

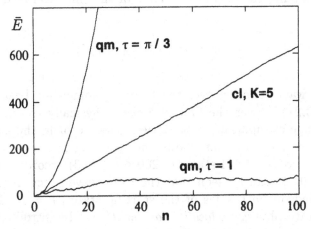

Fig. 5.8.   Comparison between classical and quantum energy growth in the classically chaotic regime.

ergy growth only for a few kicks. Following that, it breaks away abruptly from the classical prediction and stays bounded, well below the classical result, over the whole range of 100 kicks. Thus, the quantum dynamics is completely different from what one would naively expect on the basis of the classical results. The net result can be characterized as a quantum suppression of classical diffusion (see, e.g., Blümel and Smilansky (1990a)).

The quantum suppression of classical diffusion was first noted by Casati and collaborators in their seminal work on the quantum kicked rotor (Casati et al. (1979)). Subsequent work by Fishman et al. (1982) established a connection between the suppression of energy diffusion in the quantum kicked rotor and the phenomenon of Anderson localization (P. W. Anderson (1958)), an important effect that limits the conductance in disordered solids. Following the work of Fishman et al. (1982), quantum localization of chaos was also discovered in the dynamics of hydrogen Rydberg atoms (Casati et al. (1987)). The existence of a quantum suppression effect was experimentally confirmed by studying ionization thresholds in hydrogen Rydberg atoms (Moorman and Koch (1992)). Dynamical localization was also studied theoretically as well as experimentally in the case of rubidium Rydberg atoms (Blümel et al. (1991)).

In the Anderson picture the suppression of classical chaotic diffusion is understood as a destructive phase interference phenomenon that limits the spread of the rotor wave function over the available angular momentum space. The localization effect has no classical analogue. It is purely quantum mechanical in origin. The localization of the quantum rotor wave function in the angular momentum space can be demonstrated readily by plotting the absolute squares of the time averaged expansion amplitudes

$$P_l = \frac{1}{N} \sum_{n=1}^{N} | A_l^{(n)} |^2 \qquad (5.3.21)$$

of the rotor wave function over the angular momentum basis. The average in (5.3.21) is over the first N kicks. Physically the quantities P_l are the average occupation probabilities of the rotor in the angular momentum states $| l \rangle$. Fig. 5.9 shows the resulting P_l for $\tau = 1$, $k = 5$ computed in a basis of 401 states ($-200 \leq l \leq 200$) according to (5.3.21) for $N = 2000$. It can be seen that the rotor wave function is exponentially localized on both sides of the starting state $| l = 0 \rangle$. In analogy with solid state physics, a *localization length* can be introduced. In the case of the quantum kicked rotor it measures the exponential fall-off of the wave function in the angular momentum basis. The time averaged state shown in Fig. 5.9 has a shape that is well approximated by

$$P_l \sim \exp[-2 \, | \, l \, | \, /\lambda], \qquad (5.3.22)$$

where λ is the localization length. For the localized state shown in Fig. 5.9, $\lambda \approx 17$. The factor 2 in (5.3.22) appears because, according to the usual convention, the localization length refers to the *amplitudes* of a quantum state, and not to the occupation probabilities.

Another unexpected feature of the quantum kicked rotor is the existence of resonances that lead to quadratic energy gain. Quantum resonances in the kicked rotor were first analysed by Izrailev and Shepelyanskii (1979). They showed that whenever the quantum control parameter τ is a rational multiple of π, the quantum energy $E_n^{(qm)}$ grows quadratically in the kick number. This phenomenon is illustrated in Fig. 5.8 for $\tau = \pi/3$. The parabolic energy growth is clearly visible. It is surprising that the small change from $\tau = 1$ to $\tau = \pi/3$ results in such a different dynamical behaviour of the quantum kicked rotor.

In the concluding remark of Section 5.2 we asked the question whether the transition from confined chaos to global chaos at $K = K_c$ can be seen in an experiment with diatomic molecules. The technical feasibility of such an experiment is discussed in Section 5.4. Here we ask the more modest question whether, and if so, how, the transition to global chaos manifests itself within the framework of the quantum kicked rotor. Since the transition to global chaos is primarily a classical phenomenon, we expect that we have the best chance of seeing any manifestation of this transition in the quantum kicked rotor the more "classical" we prepare its initial state and control parameters. Thus, we choose a small value

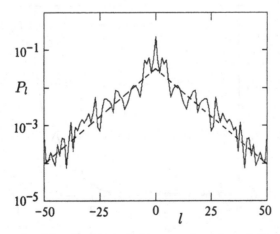

Fig. 5.9.   Exponential localization of the wave function of the quantum kicked rotor in the classically chaotic regime.

of the effective $\hbar$, $\tau = 1/10$, and start the quantum rotor with $\mid l_0\rangle = \mid 10\rangle$. Classically, we saw that for $K < K_c$ the energy of the kicked rotor stays bounded, whereas it diffuses for $K > K_c$. We expect that the average energy of the quantum kicked rotor shows a similar behaviour as a function of the quantum control parameter k as the product $K = k\tau$ is swept by the critical control parameter $K_c \approx 1$. In order to check our expectations, we computed the mean energy $\bar{E}(K)$ of the quantum kicked rotor averaged over 500 kicks for 15 values of the quantum control parameter $k_j = K_j/\tau$, $K_j = 0.2 * j$, $j = 1, 2, ..., 15$. We used the matrix propagation approach with a basis of 1201 rotor states ($-600 \leq l \leq 600$). The result is shown in Fig. 5.10. We see a clear threshold behaviour of the energy at $K \approx 2$, not far from the critical value $K_c \approx 1$. We attribute this threshold to the onset of global chaos in the underlying classical kicked rotor. The shift of the quantum threshold towards higher values of the control parameter K is understandable. Since $\hbar$ is finite, it is not easy for the quantum flux to "squeeze" through the narrow phase-space gaps that just open up at $K = K_c$. Only for K values larger than K_c will the gaps be large enough for the quantum flux to pass through. Therefore, we expect that for $\hbar \to 0$ ($\tau \to 0$) the quantum threshold shown in Fig. 5.10 will move closer to the critical value K_c. Thus, we interpret the threshold in Fig. 5.10 to be a manifestation of classical chaos in the quantum kicked rotor.

The relatively small value of $l_0 = 10$ was chosen with an eye on the experimental verification of the quantum threshold. It is not expected that any diatomic molecule is a good rotor for angular momenta exceeding $100\hbar$. Therefore, l_0 has to be chosen large enough to enter the "semi-

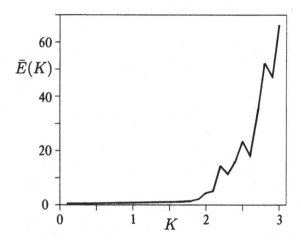

Fig. 5.10.   Average energy of the quantum kicked rotor as a function of $K = k\tau$ for $\tau = 1/10$. The rotor was prepared at $t = 0$ in the rotational state $\mid l_0 = 10\rangle$.

classical" regime, but small enough to leave some room for the controlled excitation of rotational states with $l > l_0$ that are still well below the critical angular momentum for which the molecule under investigation ceases to be a good rotor. It was suggested above to obtain better approximations to K_c for decreasing values of τ. But again, there are experimental limits. A decreasing τ means the coupling of many rotational states (by effectively increasing k for fixed K) that may not be experimentally accessible (nor even existent) for a given diatomic molecule. On the other hand the excitation of high-lying rotational states may lead to the disruption of the molecule, an effect which might be used to advantage. Instead of measuring the excitation probabilities of high-lying rotational states (which may be a formidable job to accomplish), the disruption of the molecule may be used as an experimental signature for the onset of chaos. This is in analogy to similar ideas that have been advanced for surface state electrons and hydrogen Rydberg atoms in strong radiation fields (see Chapters 6 – 8). For these systems the onset of ionization is the tell-tale signal for the transition to chaos.

Although interesting in their own right, the quantum kicked rotor results would only be of academic interest if they did not relate to a real physical situation. Therefore, we now have to address the question of whether an analogue of the quantum kicked rotor can be implemented in the laboratory, and whether this implementation allows for the experimental confirmation of the central features of the quantum kicked rotor, such as (i) the existence of a chaos threshold, (ii) suppression of chaotic diffusion, (iii) exponential localization, and (iv) resonant energy growth. As suggested in the introduction to this chapter, a diatomic molecule in a pulsed electric field (see Fig. 5.1) may be a candidate for a direct implementation of the quantum kicked rotor. The purpose of the following section is to discuss this system quantitatively in the light of the quantum kicked rotor results and to motivate our choice of the CsI molecule as the most promising candidate for a laboratory implementation.

5.4 The kicked molecule

In this section we investigate the physics of impulsively perturbed diatomic molecules. These provide a real physical system whose characteristics are expected to be close to the ones established for the kicked rotor. We establish that the dynamical effects exhibited by the quantum kicked rotor can be demonstrated experimentally with diatomic molecules within the possibilities of present day technology.

Following Townes and Schawlow (1975), we characterize the orientation of a diatomic molecule by two angles, θ and φ. Then, the stationary

Schrödinger equation for the unperturbed molecule reads

$$-\frac{\hbar^2}{2I}\Delta_\Omega\,\Psi(\theta,\varphi)\;=\;E\,\Psi(\theta,\varphi),\tag{5.4.1}$$

where

$$\Delta_\Omega\;=\;\frac{1}{\sin(\theta)}\frac{\partial}{\partial\theta}\left[\sin(\theta)\frac{\partial}{\partial\theta}\right]\;+\;\frac{1}{\sin^2(\theta)}\frac{\partial^2}{\partial\varphi^2}.\tag{5.4.2}$$

On separating the wave function into

$$\Psi(\theta,\varphi)\;=\;\psi(\theta)\,\exp(iM\varphi),\quad M=0,1,2,...,\tag{5.4.3}$$

we obtain

$$-\frac{\hbar^2}{2I}\left\{\frac{1}{\sin(\theta)}\frac{d}{d\theta}\left[\sin(\theta)\frac{d\psi(\theta)}{d\theta}\right]\;-\;\frac{M^2\psi(\theta)}{\sin^2(\theta)}\right\}\;=\;E\,\psi(\theta).\tag{5.4.4}$$

The conditions of normalizability and single-valuedness require that the spectrum of (5.4.4) is discrete. It is given explicitly by

$$E_J\;=\;\frac{\hbar^2}{2I}J(J+1).\tag{5.4.5}$$

The corresponding wave functions are

$$\Psi_{JM}(\theta,\varphi)\;=\;\langle\theta,\varphi|JM\rangle\;=$$

$$\frac{1}{\sqrt{2\pi}}\left\{\frac{(2J+1)\,(J-M)!}{2\,(J+M)!}\right\}^{1/2}P_J^{(M)}\left[\cos(\theta)\right]\exp(iM\varphi),$$

$$J=0,1,2,...,\quad |M|\leq J,\tag{5.4.6}$$

where the functions $P_J^{(|M|)}(x)$ are the associated Legendre polynomials (Abramowitz and Stegun (1964)). The transition frequencies are given by

$$\nu_J\;=\;\frac{1}{h}\left[E_{J+1}-E_J\right]\;=\;2B\,(J+1),\quad J=0,1,2,...,\tag{5.4.7}$$

where the rotational constant

$$B\;=\;\frac{\hbar}{4\pi I}\tag{5.4.8}$$

was introduced. For three different diatomic molecules the rotational constants B as well as the electric dipole moments μ are listed in Table 5.1. The table was compiled from Townes and Schawlow (1975). The dipole moments are quoted in units of debye. One debye equals $3.335\,64\times10^{-30}\,\mathrm{C\,m}$ (H. L. Anderson (1989)). With the help of B and μ, the dimensionless control parameters τ and k defined in (5.3.10) can now be translated into physical estimates for the frequency ν and amplitude $\mathcal{E}_0$ of the driving field. We obtain:

$$\nu = \frac{1}{T} = \frac{4\pi B}{\tau} \qquad (5.4.9)$$

and

$$\mathcal{E}_0 = \frac{4\pi B \hbar}{\mu}\left(\frac{k}{\tau}\right). \qquad (5.4.10)$$

For $\tau = 1$ we can calculate the resulting driving frequencies for the three molecules listed in Table 5.1. We obtain the following numerical values: $\nu_{\text{HCl}} \approx 5000\,\text{GHz}$, $\nu_{\text{KCl}} \approx 50\,\text{GHz}$ and $\nu_{\text{CsI}} \approx 9\,\text{GHz}$. Only the CsI molecule falls into a convenient frequency range that is already widely used by many experimenters in atomic quantum chaos research (see, e.g., van Leeuwen et al. (1985), Blümel et al. (1991), Moorman and Koch (1992)). The expression (5.4.10) shows that, for keeping the electric field strength to manageable levels, the CsI molecule is again favoured by its small rotational constant and large dipole moment. Therefore, the CsI molecule is a promising candidate for experiments with dynamically perturbed diatomic molecules.

A further argument in favour of using CsI molecules is provided by the following consideration. In order to demonstrate convincingly the dynamical effects of the quantum kicked rotor, in particular the dynamical localization effect, we need as many rotational states as possible. A diatomic molecule ceases to be a good rotor if the rotational excitation energy equals the excitation energy of the first vibrational level. In order to estimate the critical angular momentum $J^{(cr)}$ where this condition occurs, we need the excitation energy of the first vibrational state. The vibrational spectrum of a diatomic molecule is approximately given by (H. L. Anderson (1989))

$$\nu_m = c\omega_e(m + 1/2), \quad m = 0, 1, 2, ..., \qquad (5.4.11)$$

Table 5.1. Rotational constant B, vibrational constant ω_e and dipole moment μ for three diatomic molecules.

Molecule	B(MHz)	ω_e(1/cm)	μ(D)
HCl	317 510	2990	1.2
KCl	3856	300	10.5
CsI	708	120	12.1

where c is the velocity of light and ω_e is the vibrational constant, traditionally stated in units of $1/\text{cm}$. It is listed in Table 5.1. With (5.4.11) we obtain for the critical angular momentum

$$J^{(cr)} = \left[\frac{c\omega_e}{B}\right]^{1/2}. \tag{5.4.12}$$

For the three molecules listed in Table 5.1 we obtain $J^{(cr)}_{\text{HCl}} = 17$, $J^{(cr)}_{\text{KCl}} = 48$ and $J^{(cr)}_{\text{CsI}} = 71$. Clearly, CsI has the largest available range of rotor states.

In a realistic experiment we cannot work with zero-width "kicks". Any realistic electric field pulse has a finite width. Therefore, we replace the periodic δ function drive in (5.3.2) by an array of finite-width pulses according to

$$\hat{H} = \frac{\hat{L}^2}{2I} + \mu\mathcal{E}_0\cos(\theta)\sum_n \Delta\left(\frac{t}{T} - n\right), \tag{5.4.13}$$

where $\Delta(\xi)$ is the form factor of the pulses. It is dimensionless and normalized to 1 according to

$$\int_{-\infty}^{\infty} \Delta(\xi)\,d\xi = 1. \tag{5.4.14}$$

We will now integrate the time dependent Schrödinger equation for the wave function $\Psi(t)$ of a CsI molecule. We expand the wave function according to

$$|\Psi(t)\rangle = \sum_{JM} A_J^{(M)}(t)\,|JM\rangle. \tag{5.4.15}$$

It is seen immediately from the structure of the wave functions $\langle\theta,\varphi|JM\rangle$ in (5.4.6) that states with different M do not couple since the perturbation term in (5.4.13) is independent of φ. Therefore, we can classify the wave functions $|\Psi(t)\rangle$ according to the M quantum number in the following way:

$$|\Psi^{(M)}(t)\rangle = \sum_J A_J^{(M)}(t)\,|JM\rangle. \tag{5.4.16}$$

Inserting this expansion into the time dependent Schrödinger equation with the Hamiltonian (5.4.13) yields the following set of coupled equations for the expansion amplitudes $A_J^{(M)}(t)$

$$i\hbar\,\dot{A}_J^{(M)} = E_J A_J^{(M)} + \mu\mathcal{E}_0\sum_{nJ'}\Delta\left(\frac{t}{T} - n\right)C_{JJ'}^{(M)}A_{J'}^{(M)}, \tag{5.4.17}$$

where the coupling matrix elements $C_{JJ'}^{(M)}$ are given by

$$C_{JJ'}^{(M)} = \langle JM|\cos(\theta)|J'M\rangle = C_J^{(M)}\delta_{J,J-1} + C_{J+1}^{(M)}\delta_{J,J+1}. \tag{5.4.18}$$

with

$$C_J^{(M)} = \left[\frac{(J-M)(J+M)}{(2J-1)(2J+1)} \right]^{1/2}. \tag{5.4.19}$$

Since the rotational energies E_J grow quadratically in J, it is advantageous to define the amplitudes

$$B_J^{(M)}(t) = A_J^{(M)}(t) \exp(it E_J / \hbar). \tag{5.4.20}$$

With this definition the system of coupled equations for the expansion amplitudes reads:

$$i\hbar \dot{B}_J^{(M)}(t) =$$

$$\mu \mathcal{E}_0 \sum_{nJ'} \Delta\left(\frac{t}{T} - n\right) \exp[it(E_J - E_{J'})/\hbar] C_{JJ'}^{(M)} B_{J'}^{(M)}(t). \tag{5.4.21}$$

Introducing the dimensionless time

$$x = t/T, \tag{5.4.22}$$

the set of equations (5.4.21) can be expressed as

$$i\frac{dB_J^{(M)}(x)}{dx} = k\,\Delta(x) \sum_{J'} \exp(i\omega_{JJ'}x)\, C_{JJ'}^{(M)}\, B_{J'}^{(M)}(x). \tag{5.4.23}$$

We used the dimensionless kicked rotor control parameter k defined in (5.3.10) and introduced the frequencies $\omega_{JJ'}$ according to

$$\omega_{JJ'} = \frac{1}{\hbar}(E_J - E_{J'})\,T. \tag{5.4.24}$$

Since the coupling matrix elements $C_{JJ'}^{(M)}$ are nonzero only for $J' = J \pm 1$, we need only

$$\omega_{J,J-1} = \tau J, \qquad \omega_{J,J+1} = -\tau(J+1), \tag{5.4.25}$$

where τ is the dimensionless kicked rotor control parameter defined in (5.3.10). The final form of the coupled amplitude equations can now be stated as follows:

$$i\frac{dB_J^{(M)}(x)}{dx} = k\,\Delta(x)\Big\{ \exp[i\tau Jx]\, C_J^{(M)}\, B_{J-1}^{(M)}(x) +$$

$$\exp[-i\tau(J+1)x]\, C_{J+1}^{(M)}\, B_{J+1}^{(M)}(x) \Big\}. \tag{5.4.26}$$

The molecule amplitude equations (5.4.26) together with (5.4.25) show that the analogy between the schematic kicked rotor model and a real diatomic molecule is nearly complete. The essential control parameters in both models, τ and k, are the same. The only difference is the finite width of the perturbation form factor Δ and the numerical values of the coupling matrix elements $C_{JJ'}^{(M)}$. The width of the form factor does not pose a limitation in principle. If the width is made smaller than the inverse exposure time of the molecules, there is no difference between kicks and finite-width pulses. Nothing can be done about the coupling matrix elements. They are given by the geometry of the problem and cannot be changed. The matrix elements (5.4.18) can be compared with the corresponding planar kicked rotor matrix elements

$$\Gamma_{mm'} = \langle m| \cos(\theta)|m'\rangle = \begin{cases} 1/2 & \text{for } |m - m'| = 1 \\ 0 & \text{otherwise,} \end{cases} \tag{5.4.27}$$

where we used the planar rotor states (5.3.17). Especially for $M = 0$ the difference between (5.4.18) and (5.4.27) is very small. The difference vanishes for large J. The difference in the coupling matrix elements can be interpreted as a small renormalization of the coupling constant k. Since the sensitivity in the dynamics of the quantum kicked rotor occurs in τ, and not in k, a small change of k is not expected to do any harm. In fact, as far as the experimental distinction between localized dynamics and resonant dynamics is concerned, we expect that a small renormalization of k is of no consequence.

In the following numerical experiment we show that it is possible to demonstrate the difference between localized and resonant rotor dynamics with CsI molecules. At time $t = 0$ the molecules are prepared in their rotational ground state ($|\Psi(t = 0)\rangle = |J = 0, M = 0\rangle$). For $t > 0$ they are exposed to a string of microwave pulses. The control parameters τ and k determine the repetition frequency and the strength of the pulses. In order to be able to compare with the results for the planar kicked rotor discussed in Section 5.3, we choose $k = 5$ and $\tau = 1$ (for the nonresonant case) and $\tau = \pi/3$ (for the resonant case). This choice of control parameters translates into a driving frequency of $\nu = 1/T \approx$ 9 GHz and a field strength of $\mathcal{E}_0 \approx 1\,\text{kV/cm}$. For the pulse shape we choose

$$\Delta(\xi) = 1 + 2 \sum_{m=1}^{m=7} \cos\left[2m\pi(\xi - 1/2)\right]. \tag{5.4.28}$$

The shape of the pulse (5.4.28) is shown in Fig. 5.11. The shape (5.4.28) was chosen because it may be possible to actually synthesize pulses resembling (5.4.28) with an array of phase-locked microwave generators (see discussion below). "Kicking" the CsI molecules with the pulses (5.4.28) results in the excitation of rotational states.

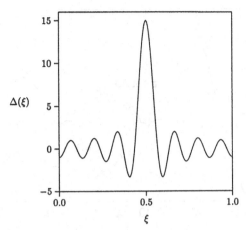

Fig. 5.11. Shape of a microwave pulse consisting of seven harmonics of the fundamental driving frequency.

The occupation probabilities of the rotational states can be computed as a function of the exposure time (characterized by the number of pulses n) by numerically integrating the system (5.4.26). We choose $M = 0$ and the initial condition $B_J^{(M=0)}(x = 0) = \delta_{J0}$. We work in a basis of 70 angular momentum states. In analogy to (5.3.20) we define the dimensionless rotational energy by

$$E_n^{(M)} = \frac{1}{2} \sum_J J(J+1) P_J^{(M)}(n), \qquad (5.4.29)$$

where

$$P_J^{(M)}(n) = |\, B_J^{(M)}(x = n)\,|^2 . \qquad (5.4.30)$$

For the above choices of field parameters and the pulse shape (5.4.28) the resulting excitation energy $E_n^{(M=0)}$ is shown in Fig. 5.12 in the resonant ($\tau = \pi/3$) and nonresonant ($\tau = 1$) cases. It can be seen that the behaviour of the excitation energies is very similar to the analogous cases studied in connection with the planar kicked rotor. The result shown in Fig. 5.12 proves that the difference between a control parameter τ that is rationally related to π and a control parameter that is irrational with respect to π can be demonstrated experimentally with a CsI molecule. The energy in the rational (resonant) case is distinctly higher than the energy in the nonresonant case.

The next question is whether dynamical localization can be demonstrated with the help of CsI molecules. This is indeed the case. In

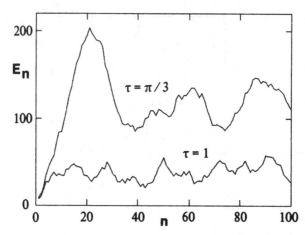

Fig. 5.12. Excitation energy of a CsI molecule exposed to n microwave field pulses in the resonant ($\tau = \pi/3$) and nonresonant ($\tau = 1$) cases.

Fig. 5.13 we plot the time averaged occupation probabilities

$$\bar{P}_J^{(M)} = \frac{1}{100} \sum_{n=1}^{100} P_J^{(M)}(n) \qquad (5.4.31)$$

of the $M = 0$ rotor states as a function of the angular momentum J for $\tau = 1$ and $k = 5$. The occupation probabilities decay exponentially with J, proving exponential localization of the rotational states. It should be emphasized here that the experimental demonstration of exponential (Anderson) localization is extremely difficult to perform in solid state physics. Therefore, a demonstration with CsI molecules would be a promising alternative.

Although only a technical problem, the generation of microwave pulses is one of the major obstacles to performing the CsI experiments. Contrary to continuous wave (cw) microwave sources in the 1 – 20 GHz regime, microwave pulses with a duration much shorter than their repetition frequency cannot be obtained "off the shelf". Therefore we discuss three possible methods for the generation of microwave pulses. The first method is the most straightforward one. It consists of generating the microwave pulses as a superposition of the microwave output of several (in the above example seven) cw microwave generators. The main problem here is to phase-lock the generators. This does not seem to be a major problem. Phase-locking of two microwave sources was already successfully demonstrated in a recent microwave experiment on bichromatically excited hydrogen Rydberg atoms (Haffmans et al. (1994)). The only drawback of this method is the required microwave power. Since a wave guide is used and field strengths in the kV/cm regime are required, the required microwave power has to be on the order of several kW.

The large power requirement can be avoided when working with a microwave cavity. Several frequencies have to be excited simultaneously in the cavity in order to synthesize a pulse resembling (5.4.28). The main problem here is not the simultaneous excitation of many frequencies in a cavity. In fact two-frequency excitation, e.g., has already been successfully demonstrated (see, e.g., Moorman and Koch (1992)). The main problem is to shape the cavity in such a way that the cavity excitation frequencies are harmonically related to the basic mode in the cavity. Only in this case can one synthesize a periodic pulse. Otherwise, a quasi-periodic signal results. It is an interesting mathematical problem to find the shape of a "harmonic cavity" that possesses m consecutive harmonically related eigenfrequencies. For large m the cavity may have a "nonintegrable" shape, and the problem may be solved by using the expertise accumulated in quantum chaos research with nonintegrable stadium shaped enclosures (see, e.g., Heller and Tomsovic (1993)).

A third possibility of obtaining narrow microwave pulses is to use a fast beam of CsI molecules that is passed between a periodic array of sharp edges, as shown in Fig. 5.14. If the edges are spaced at a distance d and the molecules fly with a velocity v, then the molecule experiences microwave pulses of frequency $\nu = v/d$ in its rest-frame. If 10 GHz pulses are desired, and the edges are spaced a distance $d = 1\,\mathrm{mm}$ apart, then $v = 10^7\,\mathrm{m/s}$ is required. This is on the order of 10% of the speed of light, but should be technically feasible.

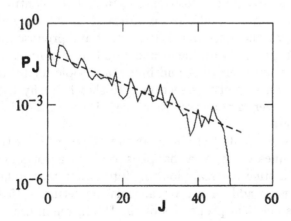

Fig. 5.13. Dynamical localization of the time averaged excitation probability of a CsI molecule in a pulsed microwave field.

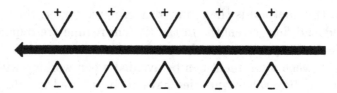

Fig. 5.14. A beam of CsI molecules passed between an array of sharp charged edges: a method for generating electric field pulses in the molecule's rest-frame.

Close to relativistic velocities can be avoided by reducing d. This can be done by using the new technology of laser focussed deposition of atoms. This nano technology was pioneered by Timp et al. (1992) and was developed by McClelland et al. (1993). The latter authors demonstrated fabrication of a chromium grid consisting of parallel lines spaced $d \approx$ 200 nm apart. Using this grid, a velocity of only $v \approx$ 2000 m/s is needed to produce 10 GHz pulses. This velocity is on the order of thermal velocities of atoms at room temperature.

In summary, CsI molecules appear to be the most promising candidates for an experimental verification of the dynamical predictions of the quantum kicked rotor. The most significant result would be a demonstration of Anderson localization with the help of diatomic molecules. Our results indicate that this demonstration is within technical reach.

5.5 Symmetry and localization

Recently, interesting symmetry effects have been discovered in connection with the quantum kicked rotor (see, e.g., Dittrich and Smilansky (1991a,b), Blümel and Smilansky (1992), Thaha et al. (1993), Thaha and Blümel (1994)). The issue concerns the influence of symmetries and their destruction on the localization length of the quantum kicked rotor. Can these symmetry effects be observed in atomic and molecular physics? We think that this issue is important and propose the search for the influence of symmetry on the localization length of dynamically localizing systems as an interesting and important topic for future research. Therefore, the purpose of this section is to sketch briefly the essence of these recent discoveries as an incentive for their application in atomic and molecular physics.

The Hamiltonian (5.3.2) of the quantum kicked rotor is time reversal invariant. The question is, what happens to the localization length of the kicked rotor if a time reversal violating interaction is switched on? The difficulty here is to add time reversal violating terms to (5.3.2) in such a way that the classical properties of the Hamiltonian are not affected. This is important, since otherwise a change in the localization length is not surprising since it can always, at least partially, be "blamed" on the

changed dynamical properties of the system, whereas our aim is to prove
that changes in the localization length are due to the change in symmetry
alone. A suitable Hamiltonian is

$$\hat{H} = \frac{1}{2}\tau\hat{l}^2 + k\Big[\cos(\pi p/2)\cos(\pi q/2)\cos(\theta + \pi r/2)$$

$$+\frac{1}{2}\cos(\pi p/2)\sin(\pi q/2)\sin(2\theta)\sigma_x + \sin(\pi p/2)\sin(\theta)\sigma_z\Big]\,\delta_p(t), \quad (5.5.1)$$

where σ_x, σ_y and σ_z are the Pauli matrices. The parameters τ and k
are the control parameters (5.3.10) of the quantum kicked rotor, and
$0 \le p, q, r \le 1$ are symmetry breaking parameters. The Hamiltonian
(5.5.1) includes the case of spin-1/2 particles. It is usually referred to as
the Hamiltonian of the symplectic kicked rotor. A Hamiltonian similar
to (5.5.1) was first studied by Scharf (1989). The Hamiltonian (5.5.1)
contains the standard kicked rotor as a special case. It is obtained for
$p = q = r = 0$. In analogy to (5.3.11) a one-step propagator $\hat{U}$ exists
for (5.5.1) whose symmetries can be switched between the three different
universality classes (see Section 4.1.1), orthogonal, unitary and symplectic, by changing the control parameters p, q and r. Since $\hat{U}$ is a unitary
operator, its eigenvalues lie on the unit circle. Therefore, its symmetry
classes are denoted by circular orthogonal (COE), circular unitary (CUE)
and circular symplectic (CSE), respectively. The circular ensembles were
introduced by Dyson (1962a). Defining the scaled parameters

$$x = \ln\left[\frac{p}{1-p}\right], \quad y = \ln\left[\frac{q}{1-q}\right], \quad (5.5.2)$$

which were introduced for graphical purposes, Fig. 5.15 shows that even
for $r = 0$ the Hamiltonian (5.5.1) is capable of exhibiting all three symmetry classes in the (x, y) plane of symmetry breaking parameters. Fig. 5.15
is a chart in the x, y plane assigning the three different symmetry classes
to regions in x, y space. The labelled plateaus in Fig. 5.15 correspond to
well developed symmetry regions. The region labelled "COE + COE" in
Fig. 5.15 corresponds to parameter combinations where (5.5.1) exhibits
an additional unitary symmetry causing its spectrum to decompose into
two distinct classes, each of which is COE. The ridges in Fig. 5.15 are the
transition regions between well developed symmetry regions. Fig. 5.15
shows that the straight line $y = 1$ in the x, y plane defines a path which
intersects with all three symmetry regions in the following order: CUE →
CSE → COE. Thaha and Blümel (1994) computed the localization length
of the quantum kicked rotor on this straight line path as a function of x.
With the help of the golden mean $g = (\sqrt{5} - 1)/2$ defined in (2.1.3) the

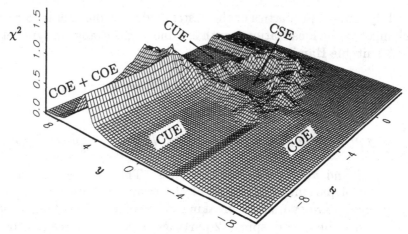

Fig. 5.15. Chart of symmetry regimes in the x, y plane of the symplectic kicked rotor for $r = 0$. Well developed symmetry regions (plateaus) are separated by transition regions (ridges). (From Thaha and Blümel (1994).)

control parameters were chosen as follows: $K = 7$, $\tau = 0.05 \times 2\pi/g$ and $k = K/\tau$.

Normalizing the localization length λ to $\lambda \approx 2$ in the CUE region, Fig. 5.16 shows the resulting scaled localization length λ as a function of x. The localization length is (on average) approximately constant in each of the symmetry regions and changes abruptly when crossing the

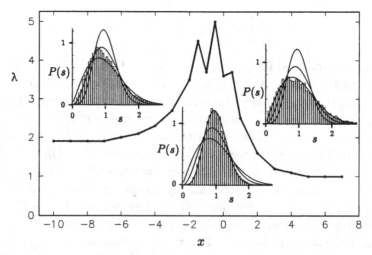

Fig. 5.16. Scaled localization length of the symplectic kicked rotor in the three different symmetry regimes COE, CUE and CSE. The three symmetry regimes are visited as a function of the control parameter x with $y = 1$ and $r = 0$ fixed. (From Thaha and Blümel (1994).)

transition regions. The insets in Fig. 5.16 show typical histograms of the quasi-energy statistics of $\hat{U}$ obtained in the three symmetry regions, respectively. The histograms prove that the path $y = 1$ indeed sweeps through all three symmetry regions. Fig. 5.16 shows that the localization length of the quantum kicked rotor changes roughly by factors 2 and 4 when crossing different symmetry regions. These factors are universal. They also occur in solid state physics were they were predicted by Pichard *et al.* (1990) for the change of the localization length of disordered solid state samples. In generating Fig. 5.16, great care was taken to refer the localization lengths in the different symmetry regimes to approximately the same dynamical situation. Therefore, the change in localization length can be attributed to the change in symmetry alone. This is an astonishing effect which may also occur in dynamically localizing atomic and molecular physics systems.

6

Microwave-driven
surface state electrons

What happens if one sprinkles electrons onto the surface of liquid helium? Surprisingly the electrons are not absorbed into the bulk of the fluid, but form a quasi-two-dimensional sheet of electrons concentrated at some distance above the helium surface. In general, electrons hovering above the surface of a dielectric are called *surface state electrons*. An excellent review of surface state electrons is that by Cole (1974).

Surface state electrons are especially interesting in the context of chaos and quantum chaos. This is so because driven by strong microwave fields, their classical dynamics shows a transition to chaos. The investigation of microwave-driven surface state electrons as a testing ground for quantum chaos was first proposed by Jensen in 1982. So far, and to the best of our knowledge, a successful surface-state-electron (SSE) microwave ionization experiment was never carried out in the chaotic regime. This is mainly due to the formidable experimental difficulties in controlling the fragile SSE system. Electric stray fields, residual helium vapour pressure and interactions with the quantized surface modes ("ripplons") of the liquid helium substrate make it very difficult to reach the high quantum numbers necessary for a quantum chaos experiment. It was, however, realized early on (see, e.g., Shepelyansky (1985)) that the dynamics of low angular momentum hydrogen Rydberg atoms is very similar to the dynamics of surface state electrons. Therefore, building on the knowledge accumulated in the field of surface state electrons, the focus of research shifted to the investigation of microwave-driven hydrogen and alkali Rydberg atoms. This does not mean, however, that the SSE system is now obsolete. The SSE system is essentially one-dimensional and therefore much more easily analysed theoretically than the atomic physics analogues. Therefore, we are convinced that research on surface state electrons will see a renaissance once the experimental difficulties are overcome.

There are three main reasons for presenting the SSE system in this book. (i) Due to its low dimensionality it is a fundamental system. It is described with the help of analytical tools that are widely used in the analysis of Rydberg atoms in strong radiation fields. (ii) The classical version of the SSE system is chaotic. (iii) Surface state electrons possess an ionization continuum, an important feature in all of atomic physics. Since the SSE system is both simple and physical, it is a natural candidate for research in classically chaotic driven systems that allow ionization to take place.

This chapter presents a discussion of some of the central ideas concerning the physics of driven surface state electrons with special emphasis on their chaotic dynamics. In Section 6.1 we discuss the physics of surface state electrons and introduce a one-dimensional model which reproduces some of the most important features of the SSE system. Surface state electrons by themselves are not chaotic. At least not in the low density regime where the mutual influence of the electrons can be neglected. In this low density limit a single-particle description is justified which leads to simple integrable dynamics both on the classical and the quantum level. Chaotic dynamics arises if the electrons are exposed to a strong microwave field. In this situation the quantum dynamics of even a single isolated surface state electron is complicated. Many quantum states are strongly coupled and participate in the dynamics. Simple perturbation theory is no longer sufficient for describing the excitation and ionization dynamics of strongly perturbed surface state electrons, and numerical schemes are necessary to predict their behaviour. Three numerical schemes for solving the time dependent Schrödinger equation are suggested in Section 6.2. A particularly simple scheme is based on analytical one-photon decay rates. It is derived in Section 6.2.3. The weak-field regime is investigated in Section 6.3. In this regime the classical dynamics is not chaotic and a simple multi-photon picture is sufficient to understand the ionization dynamics. For strong fields this is no longer the case, and large fractions of the SSE phase space are chaotic. In Section 6.4 the strong-field ionization process is analysed in a classical picture on the basis of Chirikov's resonance overlap criterion derived in Section 5.2. Critical ionization fields are predicted and linked with the onset of chaos. The predicted values of the critical fields may be tested in future experiments with microwave-driven surface state electrons.

6.1 The SSE model

Consider the situation shown in Fig. 6.1(a). It shows an electron (charge $-e$) located at position $(x, y = 0, z = 0)$ above the planar surface S of a dielectric. The electron induces a surface charge density σ in S. The surface charge density can be calculated explicitly by elementary means, for instance with the tools provided in Section 4.4 of Jackson's *Classical Electrodynamics* (1975). Expressed in SI units, the surface charge density is given by

$$\sigma(\vec{r}) \;=\; \frac{e}{8\pi^2\epsilon_0}\left(\frac{\epsilon-1}{\epsilon+1}\right)\frac{x}{(x^2+r^2)^{3/2}}, \tag{6.1.1}$$

where ϵ_0 is the electric permittivity of the vacuum, ϵ is the dielectric constant of the medium and $r = |\vec{r}| = \sqrt{x^2+y^2}$.

The induced surface charge attracts the electron toward S with a force $\vec{F}(x) = F(x)\hat{x}$, where $\hat{x}$ is a unit vector in the x direction and $F(x)$ is given by

$$F(x) \;=\; \int_S \frac{(-e)\,(\sigma dS)}{x^2+r^2}\frac{x}{(x^2+r^2)^{1/2}} \;=\; -\frac{1}{16\pi\epsilon_0}\left(\frac{\epsilon-1}{\epsilon+1}\right)\frac{e^2}{x^2}. \tag{6.1.2}$$

The force $F(x)$ can be derived from the potential

$$v(x) \;=\; -\frac{Ze^2}{4\pi\epsilon_0 x} \tag{6.1.3}$$

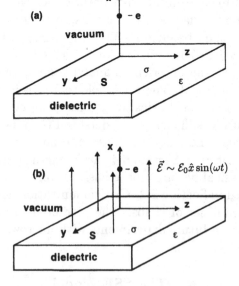

Fig. 6.1.   A surface state electron hovering at $(x, y = 0, z = 0)$ above the surface S of a dielectric with dielectric constant ϵ. The electron induces a surface charge density σ in S. (a) No external fields present. (b) A homogeneous microwave field parallel to the x axis is switched on.

via $F(x) = -\partial v(x)/\partial x$. The effective charge Z in (6.1.3) is given by

$$Z = \frac{1}{4}\left(\frac{\epsilon - 1}{\epsilon + 1}\right). \qquad (6.1.4)$$

For $\epsilon > 1$ the force (6.1.2) is attractive and accelerates the electron toward S. But the electron cannot attach to S, nor can it intrude into the bulk of the dielectric, since there are no vacant states. This is a direct consequence of Pauli's exclusion principle. Therefore S acts like a barrier between the inside and the outside of the dielectric. This situation is effectively modelled by the potential

$$V(x) = \begin{cases} \infty, & \text{for } x \leq 0 \\ -\dfrac{Ze^2}{4\pi\epsilon_0 x}, & \text{for } x > 0. \end{cases} \qquad (6.1.5)$$

The potential (7.1.5) is sketched in Fig. 6.2. The potential $V(x)$ forms a one-dimensional pocket. The shape of this pocket is that of a one-dimensional Coulomb problem in the half-space, which establishes its connection with atomic physics. Depending on the electron energy E, two types of motion are possible in $V(x)$. For $E > 0$ the electron escapes from S. For $E < 0$ the electron is trapped. Classically, it executes an oscillatory bouncing motion between the surface and the turning point $x_t = Ze^2/4\pi\epsilon_0 \mid E \mid$, if we assume perfectly elastic reflection from the surface S at $x = 0$. Quantum mechanically the potential $V(x)$ supports an infinite sequence of bound states, the "surface states". This is due to the long range nature of $V(x)$ which is of the same form as the familiar Coulomb potential of atomic physics, but with a rather smaller coupling constant. In the following we analyse both the classical and the quantum mechanics of surface state electrons.

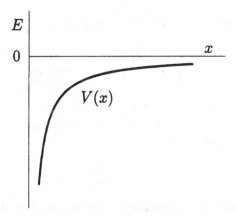

Fig. 6.2. Effective potential for surface state electrons.

With the help of a typical velocity

$$v_0 = \frac{Ze^2}{4\pi\epsilon_0\hbar} \tag{6.1.6}$$

and m denoting the electron mass, we define the units of momentum, energy, time and length according to

$$p_0 = mv_0, \quad E_0 = \frac{p_0^2}{m}, \quad t_0 = \frac{\hbar}{E_0}, \quad l_0 = \frac{\hbar}{p_0}, \tag{6.1.7}$$

respectively. In these units the classical equations of motion in the potential $V(x)$ are given by

$$\dot{x} = p, \quad \dot{p} = -1/x^2, \quad x > 0. \tag{6.1.8}$$

The equations of motion (6.1.8) supplemented with a perfectly elastic reflection condition at $x = 0$ can be derived as the canonical equations of motion from the Hamiltonian

$$H_0 = \begin{cases} \infty, & \text{for } x \leq 0 \\ \dfrac{p^2}{2} - \dfrac{1}{x}, & \text{for } x > 0. \end{cases} \tag{6.1.9}$$

The equations (6.1.8) are solved by a canonical transformation to action and angle variables in the following way. Since (6.1.9) is autonomous, the energy E is conserved. For bounded motion E is negative. Therefore,

$$H_0 = \frac{p^2}{2} - \frac{1}{x} = E = -|E|. \tag{6.1.10}$$

For oscillatory motion the action variable is defined by

$$I = \frac{1}{2\pi} \oint p\, dx, \tag{6.1.11}$$

where the integral is extended over one period of the motion. In our case we have

$$I = \frac{1}{\pi} \int_0^{1/|E|} p\, dx = \frac{1}{\sqrt{-2E}}. \tag{6.1.12}$$

For bounded motion E ranges from $-\infty$ to 0 (see Fig. 6.2). Therefore, I ranges from 0 to ∞. The relation (6.1.12) provides a connection between the action and the energy:

$$E = -\frac{1}{2I^2}. \tag{6.1.13}$$

The variable canonically conjugate to the action I is the angle variable θ. According to (3.1.36) it can be obtained from the generating function

$F_2(I, x)$ via $\theta = \partial F_2/\partial I$. The relation $p = \partial F_2/\partial x$ (see (3.1.36)) suggests construction of F_2 according to

$$F_2(I, x) = \int_0^x p\, dx. \tag{6.1.14}$$

The integral (6.1.14) can be evaluated analytically. We obtain:

$$F_2(I, x) = I\, \theta(E, x), \tag{6.1.15}$$

where

$$\theta(E, x) = 2\eta(E, x) - \sin\left[2\eta(E, x)\right] \tag{6.1.16}$$

and

$$\eta(E, x) = \begin{cases} \arcsin\sqrt{x|E|}, & \text{for } p > 0 \\ \pi - \arcsin\sqrt{x|E|}, & \text{for } p < 0. \end{cases} \tag{6.1.17}$$

The canonical transformation to action-angle variables can now be stated explicitly. With $p = \partial F_2/\partial x$ and $\theta = \partial F_2/\partial I$ we obtain

$$p = \frac{1}{I}\cot(\eta)\ ;\quad x = 2I^2\sin^2(\eta). \tag{6.1.18}$$

Since the canonical transformation is time independent, the new Hamiltonian equals the old Hamiltonian expressed in the new variables. We obtain

$$H_0 = -\frac{1}{2I^2}. \tag{6.1.19}$$

In the new variables the canonical equations of motion are

$$\dot{I} = -\frac{\partial H_0}{\partial \theta} = 0,\quad \dot{\theta} = \frac{\partial H_0}{\partial I} = 1/I^3. \tag{6.1.20}$$

From (6.1.20) we obtain immediately

$$I(t) = const.,\quad \theta(t) = \frac{t}{I^3} + \theta_0. \tag{6.1.21}$$

Thus, in (I, θ) phase space, the motion governed by H_0 is on straight lines, as shown in Fig. 6.3.

We turn now to a discussion of the quantum mechanics of surface state electrons. Written in the units (6.1.7) the quantum Hamiltonian is given by

$$\hat{H}_0(x) = \begin{cases} \infty, & \text{for } x \le 0 \\ -\dfrac{1}{2}\dfrac{d^2}{dx^2} - \dfrac{1}{x}, & \text{for } x > 0. \end{cases} \tag{6.1.22}$$

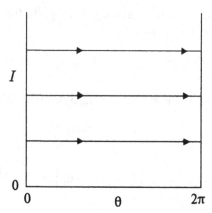

Fig. 6.3.   Phase-space portrait of unperturbed surface state electrons.

The spectrum of $\hat{H}_0$ is obtained from the stationary Schrödinger equation

$$\hat{H}_0 \mid \psi\rangle = E \mid \psi\rangle. \qquad (6.1.23)$$

It divides into two categories. For $E < 0$ we obtain an infinite discrete sequence of bounded states; for $E > 0$ we obtain a continuous spectrum. Due to the similarity of (6.1.22) to the $l = 0$ radial Hamiltonian of the hydrogen atom, the wave functions and energies can be computed analytically. For the bound state wave functions we obtain

$$\varphi_n(x) = \frac{1}{n^{3/2}} \left(\frac{2x}{n}\right) L_{n-1}^{(1)}\left(\frac{2x}{n}\right) \exp\left(-\frac{x}{n}\right), \quad n = 1, 2, ..., \qquad (6.1.24)$$

where $L_n^{(1)}(x)$ are associated Laguerre polynomials as defined, e.g., in the *Handbook of Mathematical Functions* (Abramowitz and Stegun (1964)). The quantized energies corresponding to the wave functions $\varphi_n(x)$ are

$$E_n = -\frac{1}{2n^2}. \qquad (6.1.25)$$

The squares of the first three wave functions φ_1^2, φ_2^2 and φ_3^2, are shown in Fig. 6.4.

The continuum wave functions are given by

$$\varphi_k(x) = \frac{2xk^{1/2}}{\sqrt{1 - \exp(-2\pi/k)}} e^{-ikx} {}_1F_1\left(1 + \frac{i}{k}, 2; 2ikx\right), \qquad (6.1.26)$$

where ${}_1F_1$ is the confluent hypergeometric function (see, e.g., Abramowitz and Stegun (1964)). The corresponding continuum energies are given by

$$E_k = \frac{1}{2} k^2. \qquad (6.1.27)$$

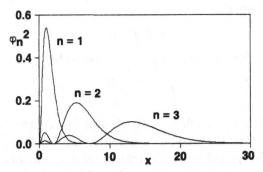

Fig. 6.4. Squares of the ground state wave function ($n = 1$) and the first two excited states ($n = 2, 3$) of the SSE system.

The bound state wave functions are normalized according to

$$\langle n \mid m \rangle = \int_0^\infty \varphi_n(x)\, \varphi_m(x)\, dx = \delta_{nm}; \qquad (6.1.28)$$

the continuum wave functions are normalized according to

$$\langle k \mid k' \rangle = \int_0^\infty \varphi_k(x)\, \varphi_{k'}(x)\, dx = \delta(k - k'). \qquad (6.1.29)$$

Bound and continuum states are orthogonal, i.e.

$$\langle n \mid k \rangle = \int_0^\infty \varphi_n(x)\, \varphi_k(x)\, dx = 0. \qquad (6.1.30)$$

It is interesting to note that straightforward Bohr-Sommerfeld quantization of the action (6.1.11) yields the exact result (6.1.25) for the bound state energies. In our units the Bohr-Sommerfeld condition results in $I = n$, $n = 1, 2, \ldots$. Inserting this result into (6.1.13) indeed reproduces (6.1.25) exactly. This is the same happy accident which allowed Bohr (1913) to obtain the Balmer formula from a simple solar system model of a one-electron atom.

We are now in a position to discuss the physical properties of surface state electrons. We choose liquid helium for the dielectric. According to H. L. Anderson (1989) the dielectric constant for liquid helium is given by

$$\epsilon_{He} = 1.057. \qquad (6.1.31)$$

According to (6.1.4) this results in the effective charge

$$Z_{He} = 6.93 \times 10^{-3}. \qquad (6.1.32)$$

With this value of the effective charge, the unit of energy becomes

$$E_0 = 2.1 \times 10^{-22}\,\mathrm{J} = 1.3 \times 10^{-3}\,\mathrm{eV} \sim 320\,\mathrm{GHz}, \qquad (6.1.33)$$

and the unit of length is

$$l_0 = 76 \, \text{Å}. \tag{6.1.34}$$

With (6.1.33) the SSE spectrum is predicted to be

$$E_n = -\frac{R}{n^2}, \tag{6.1.35}$$

where $R = 160 \, \text{GHz}$ is the effective Rydberg constant for the SSE system.

The spectrum (6.1.35) strongly resembles the bound spectrum of hydrogen. But while typical low-lying transition frequencies in hydrogen are in the optical or ultra-violet regime, the SSE transition frequencies are conveniently located in the microwave regime. Moreover, since the wave functions (6.1.24) do not have a good parity, as this is a half-space, rather than a radial Coulomb problem, there are no dipole selection rules and all transitions $n \leftrightarrow m$, $n, m = 1, 2, ..., n \neq m$ are allowed. According to the Rydberg-Ritz combination principle (French and Taylor (1978)), the following frequencies are the theoretical prediction for the SSE transition frequencies that should be observable in an SSE microwave absorption experiment:

$$\nu_{nm}^{(th)} = R \left(\frac{1}{n^2} - \frac{1}{m^2} \right), \quad n < m. \tag{6.1.36}$$

For the $1 \to 2$ and the $1 \to 3$ transitions, e.g., we obtain $\nu_{12}^{(th)} = 120 \, \text{GHz}$ and $\nu_{13}^{(th)} = 142 \, \text{GHz}$, respectively. In 1974 Grimes and Brown performed a microwave absorption experiment for surface state electrons on liquid helium. They obtained $\nu_{12}^{(exp)} = 126 \, \text{GHz}$ and $\nu_{13}^{(exp)} = 147 \, \text{GHz}$. The experimental transition frequencies agree well with the theoretical predictions.

Because of the small transition frequencies, the SSE system is very delicate. Great care has to be taken to provide a cold environment. Since the Boltzmann constant, expressed in frequency, is given by $k_B = 21 \, \text{GHz/K}$, the energy gap to the continuum is bridged easily at a temperature on the order of $10 \, \text{K}$. This means that a quasi-two-dimensional electron gas localized in $\varphi_1(x)$ can exist only at liquid helium temperatures. This is one of the reasons why liquid helium is a preferred dielectric for SSE experiments.

The requirement of a low temperature becomes even more stringent if meaningful quantum chaos experiments are to be conducted with surface state electrons. In this case, as will become apparent later, many SSE levels have to be excited simultaneously and coherently. Suppose we aim at an experiment involving only the coherent excitation of a modest five levels. This experiment can be conducted only if we can exclude thermally induced transitions between the fifth and the sixth levels. Thus, this experiment has to be conducted at a temperature $T \ll \nu_{5\to6}/k_B \approx 0.1\,\text{K}$. Temperatures in this regime are within the reach of modern experiments, but are still very hard to obtain. This illustrates the level of the technical difficulties involved in conducting controlled SSE experiments.

The condition on the temperature may be relaxed if the SSE system is irradiated with microwave pulses of short duration only. In this case we may work at considerably higher temperatures, arguing that, owing to the short interaction times, the thermal radiation does not have enough time to destroy quantum coherence. This argument is used and found to be valid in atomic fast beam experiments (see, e.g., Moorman and Koch (1992)) that are conducted at room temperature.

Surface state electrons can be interpreted as model hydrogen atoms scaled down to the microwave regime. The resemblance of the two systems can be exploited for a qualitative investigation of the physics of the hydrogen atom in strong microwave fields. It is experimentally very difficult to ionize the ground state of a hydrogen atom with electromagnetic radiation. Excessively strong laser fields not easily available with present day technology are required. In the microwave regime, however, there is no shortage of power, and even the ground state of the SSE system can be ionized in a controlled way. The SSE system, therefore, has recently acquired the status of an experimentally accessible model system for the investigation of atoms in super-strong radiation fields.

The spatial extent of SSE wave functions is remarkable. From (6.1.24) we obtain the SSE ground state wave function

$$\varphi_1(x) = 2x \exp(-x). \tag{6.1.37}$$

Its maximum occurs at $x_{max} = 1$, which, according to (6.1.34), corresponds to $l_0 x_{max} \approx 76\,\text{Å}$. This means that the maximum occupation probability of the electrons is not directly at the surface but some $76\,\text{Å}$ removed from the surface.

The expectation value of the position is given by

$$\bar{x} = \langle 1 \mid x \mid 1 \rangle = 3/2, \tag{6.1.38}$$

which corresponds to $\approx 114\,\text{Å}$. Therefore, even in its ground state the SSE system is much larger than the size of a typical atom. Quantum systems which maintain their phase coherence over a spatial extent approaching macroscopic sizes ($\sim 1\,\mu\text{m}$) are called mesoscopic systems because they are intermediate between the microscopic quantum world and the macroscopic "classical" world. Mesoscopic systems are currently at the centre of intensive research (see, e.g., Altshuler *et al.* (1991)), especially in connection with quantum chaos issues (see, e.g., Baranger *et al.* (1993a,b), Marcus *et al.* (1993)).

Since the SSE system by itself is essentially one-dimensional and autonomous, there is no chaos in this system. Even exposing the surface state electrons to weak microwave radiation does not change the situation. As a matter of fact, exposure to weak microwave fields was discussed above in connection with the spectroscopy experiments by Grimes and Brown (1974). In these experiments the microwave fields were deliberately chosen to be weak in order not to disturb the SSE system too much. Weak-field irradiation does not produce chaos, but results in regular absorption lines.

The situation changes drastically, however, if we irradiate the surface state electrons with a sufficiently strong microwave field. Strong fields change the physics of the SSE system profoundly, eventually driving it into chaos. That chaos can indeed occur in the SSE system was first demonstrated by Jensen in 1982. The system he proposed is shown in Fig. 6.1(b). It is an extension of the system shown in Fig. 6.1(a). A microwave field is applied perpendicular to the helium surface S such that the field direction is parallel to the x direction. The resulting classical Hamiltonian of the combined SSE plus microwave field can be written as

$$H(x,t) = H_0 - \mathcal{E}x\sin(\omega t), \qquad (6.1.39)$$

where H_0 is defined in (6.1.9), $\mathcal{E}$ is the dimensionless microwave field strength in units of

$$\mathcal{E}_0 = \frac{m^2(Ze^2)^3}{e\hbar^4(4\pi\epsilon_0)^3} = \frac{m^2}{e\hbar}v_0^3 \qquad (6.1.40)$$

and ω is the dimensionless microwave frequency in units of

$$\Omega_0 = \frac{1}{t_0} = \frac{m(Ze^2)^2}{\hbar^3(4\pi\epsilon_0)^2} = \frac{m}{\hbar}v_0^2. \qquad (6.1.41)$$

For liquid helium, $\mathcal{E}_0$ and Ω_0 are given by

$$\mathcal{E}_0 = 1.7 \times 10^3 \,\text{V/cm} \qquad (6.1.42)$$

and

$$\Omega_0 = 2.0 \times 10^{12} \,/\text{s}. \qquad (6.1.43)$$

The equations of motion for the Hamiltonian H are given by

$$\dot{x} = p, \quad \dot{p} = -\frac{1}{x^2} + \mathcal{E}\sin(\omega t). \qquad (6.1.44)$$

Written in action-angle variables, the Hamiltonian H reads

$$H = -\frac{1}{2I^2} - 2\mathcal{E}I^2 \sin^2(\eta)\sin(\omega t). \qquad (6.1.45)$$

From (6.1.16) we obtain

$$\frac{d\eta}{d\theta} = \frac{1}{4\sin^2(\eta)}. \tag{6.1.46}$$

Together with $\partial\eta/\partial I = 0$, the result (6.1.46) can be used to obtain

$$\dot\theta = \frac{\partial H}{\partial I} = \frac{1}{I^3} - 4\mathcal{E}I\sin^2(\eta)\sin(\omega t),$$

$$\dot I = -\frac{\partial H}{\partial\theta} = \mathcal{E}I^2\cot(\eta)\sin(\omega t). \tag{6.1.47}$$

Unlike the equations of motion (6.1.8), the equations of motion (6.1.47) cannot be solved by a simple globally integrating canonical transformation and we have to resort to numerical methods.

The $\dot I$ equation in (6.1.47) is singular and has to be regularized. This can be done by introducing a fictitious time τ which results in the following set of coupled equations:

$$\frac{dt}{d\tau} = \sin^2(\eta),$$

$$\frac{dI}{d\tau} = -\frac{1}{2}\mathcal{E}I^2\sin(2\eta)\sin(\omega t), \qquad \frac{d\eta}{d\tau} = \frac{1}{4I^3} + \mathcal{E}I\sin^2(\eta)\sin(\omega t). \tag{6.1.48}$$

The price for regularization is one additional differential equation. An approximate way of regularization which results in only two equations of motion is obtained in the following way. Since according to (6.1.18) x is a periodic function of η, which itself is a periodic function of θ, the variable x can be expanded in a Fourier series. Following Landau and Lifschitz (1977), we expand the position variable $x(I,\theta)$ in (6.1.18) into a Fourier-cosine series:

$$x(I,\theta) = \sum_{m=0}^{\infty} x_m(I)\cos(m\theta). \tag{6.1.49}$$

The expansion coefficient x_0 is given by

$$x_0 = \frac{1}{2\pi}\int_0^{2\pi} x(I,\theta)\,d\theta = \frac{I^2}{\pi}\int_0^\pi \sin^2(\eta)\frac{d\theta}{d\eta}\,d\eta. \tag{6.1.50}$$

With (6.1.46) we have

$$x_0 = \frac{4I^2}{\pi}\int_0^\pi \sin^4(\eta)\,d\eta = \frac{3I^2}{2}. \tag{6.1.51}$$

For the $m \neq 0$ coefficients we obtain

$$x_m = \frac{1}{\pi}\int_0^{2\pi} x(I,\theta)\cos(m\theta)\,d\theta = \frac{8I^2}{\pi}\int_0^\pi \sin^4(\eta)\cos(m\theta)\,d\eta. \tag{6.1.52}$$

The integral in (6.1.52) can be expressed with the help of the derivative of the Bessel functions, and we obtain

$$x_m = -\frac{2I^2}{m} J'_m(m).$$

(6.1.53)

The final result is

$$x(I,\theta) = \frac{3}{2}I^2\left\{1 - \frac{4}{3}Q(\theta)\right\},$$

(6.1.54)

where

$$Q(\theta) = \lim_{N\to\infty} Q_N(\theta), \quad Q_N(\theta) = \sum_{k=1}^{N} \frac{J'_k(k)}{k}\cos(k\theta).$$

(6.1.55)

For $k = 1, 2, 3$ we have $J'_k(k) = 0.325,\ 0.224,\ 0.177$, respectively. For $k > 3$ we can use

$$J'_k(k) = 0.411/k^{2/3}$$

(6.1.56)

as a good approximation (Abramowitz and Stegun (1964)). The Hamiltonian (6.1.45) becomes

$$H = -\frac{1}{2I^2} + \frac{3}{2}I^2\mathcal{E}\left\{1 - \frac{4}{3}Q(\theta)\right\}\sin(\omega t)$$

(6.1.57)

and the equations of motion are

$$\dot{I} = 2\mathcal{E}I^2 Q'(\theta)\sin(\omega t), \quad \dot{\theta} = \frac{1}{I^3} + 3I\mathcal{E}\left\{1 - \frac{4}{3}Q(\theta)\right\}\sin(\omega t), \quad (6.1.58)$$

where Q' denotes the derivative of Q with respect to θ. Replacing Q in (6.1.58) by Q_N, N finite, results in a set of two regularized equations. From the numerical point of view, the sums Q_N and Q'_N can be computed very efficiently using the algorithm of Cooley and Tukey (1965), which is the basis of the fast Fourier transform. From the analytical point of view, the expansion (6.1.55) is also the basis for the analytical calculation of resonance widths, since, close to resonances, the truncation of the sum Q after a finite number of terms is an excellent approximation. The analytical investigation of resonance widths is discussed in Section 6.4. It is also the starting point for the analytical calculation of ionization thresholds for microwave-driven hydrogen atoms presented in Section 7.3. Here we use numerical solutions of the regularized equations (6.1.48) to demonstrate the occurrence of chaos in the SSE system.

Numerical solutions of (6.1.48) show that already for moderate microwave field strengths the smooth straight line phase-space orbits shown in Fig. 6.3 start to deform. At higher field strengths they break up entirely. This scenario strongly resembles the analogous scenario already encountered in Chapter 5 in connection with the kicked rotor. Once the last invariant curve is broken, the iterates of phase-space points may be transported chaotically toward the ionization threshold at $E = 0$. We illustrate this scenario with the help of a phase-space portrait which shows the location of phase-space trajectories as points at times $t_j = 2j\pi/\omega$, $j = 1, 2, \dots$. We choose $\mathcal{E} = 0.03$, $\omega = 1$ and integrate the set of equations (6.1.48) numerically for 62 different initial conditions $\theta_0^{(ml)} = l\pi/4$, $I_0^{(ml)} = 1 + 0.02 \times m$, $m = 0, \dots, 32$, $l = 1, 2$ with the help of a fourth order Runge-Kutta scheme (see, e.g., Milne (1970)). Every single one of the 62 resulting trajectories was followed over 200 cycles of the microwave field, and the values of I and θ after every completion of a full cycle of the microwave field were plotted as dots in a (θ, I) phase-space diagram. The result is shown in Fig. 6.5. Regular and chaotic regions are clearly visible.

Because of the apparent chaos in Fig. 6.5, simple analytical solutions of the driven SSE system probably do not exist, neither for the classical nor for the quantum mechanical problem. Therefore, if we want to investigate the quantum dynamics of the SSE system, powerful numerical schemes have to be devised to solve the time dependent Schrödinger equation of the microwave-driven SSE system. While the integration of classical trajectories is nearly trivial (a simple fourth order Runge-Kutta scheme, e.g., is sufficient), the quantum mechanical treatment of microwave-driven surface state electrons is far from trivial. In the chaotic regime many SSE bound states are strongly coupled, and the existence of the continuum and associated ionization channels poses additional problems. Numerical and approximate analytical solutions of the quantum SSE problem are proposed in the following section.

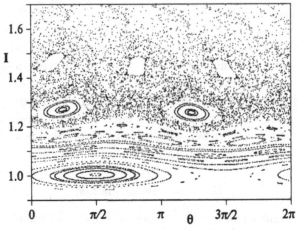

Fig. 6.5. Poincaré section of a microwave-driven surface state electron in the chaotic regime.

6.2 Quantum dynamics: technical tools

The quantum mechanical description of perturbed surface state electrons
starts with the time dependent Schrödinger equation

$$i\frac{\partial}{\partial t}\mid \Psi(t)\rangle \; = \; \hat{H}\mid \Psi(t)\rangle, \qquad (6.2.1)$$

where

$$\hat{H} \; = \; \hat{H}_0 - x \cdot f(t). \qquad (6.2.2)$$

Here $\hat{H}_0$ is the unperturbed SSE Hamiltonian defined in (6.1.22) and $f(t)$
is an arbitrary force function. Since no analytical solutions of (6.2.1) are
known in the classically chaotic regime, we have to resort to numerical
methods. It is desirable that the numerical methods to be devised are
applicable for any shape of the force function $f(t)$. In this case it is possible to apply the numerical methods to a type of experiment in which
the microwave field is switched on and off during a relatively short period
of time (Bayfield and Sokol (1988), Blümel *et al.* (1991), Moorman and
Koch (1992)). It was shown by Breuer *et al.* (1988) and Breuer and
Holthaus (1989) that switch-on and switch-off processes are very important for the SSE dynamics. Two numerical methods, powerful enough
to handle any time dependent form factor $f(t)$, are discussed in the following two subsections. Section 6.2.1 presents a method based on square
normalizable basis functions, the *Sturmian method*. The method is equivalent to discretizing the continuum. In Section 6.2.2 we present a method
that does not discretize the continuum. It leads to a system of integro-
differential equations of the Fredholm type. Therefore, this approach is
called the *Fredholm approach*. In the weak-field one-photon regime the
integro-differential equations can be solved approximately with the help
of analytical methods. This approach is presented in Section 6.2.3. Apart
from computing one-photon rates, the analytical approach can be used
to check the numerical methods. Moreover, in the weak-field regime the
analytical one-photon rates derived in Section 6.2.3 can be used as approximations for the decay widths of unperturbed SSE states. Introducing complex energies computed on the basis of the analytical one-photon
decay rates defines a "pocket version" of the Fredholm approach. This
method is also discussed in Section 6.2.3.

6.2.1 Sturmian method

The most straightforward way to solve the time dependent Schrödinger
equation (6.2.1) with the Hamiltonian (6.2.2) is to expand the wave function $\mid \Psi(t)\rangle$ in a complete set of square normalizable states. This method

is called the Sturmian approach, and the square normalizable wave functions are called Sturmian functions. The advantage of this method is that it is technically simple. The disadvantage is that square normalizable functions do not approximate continuum states very well, and many basis functions have to be taken into account in order to represent properly bound and continuum states. In order to reduce the number of Sturmian basis states required for a converged calculation, it is advantageous to choose a set of states that is adjusted to the physical size and the structure of the problem at hand. In the SSE case, e.g., we may construct a set of Sturmian states from the unperturbed SSE states (6.1.24) in the following way:

$$\langle x|n; \alpha \rangle \;=\; \varphi_n^{(\alpha)}(x) \;=$$

$$\left[\frac{2}{\alpha n(n+1)}\right]^{1/2} \left(\frac{2x}{\alpha}\right) L_{n-1}^{(2)}\!\left(\frac{2x}{\alpha}\right) \exp(-x/\alpha), \qquad n = 1, 2, 3, ..., \quad (6.2.3)$$

where α is a fixed real number (the "Sturmian label"), and the functions $L_n^{(2)}(x)$ are the associated Laguerre polynomials. The set (6.2.3) is orthonormal according to

$$\int_0^\infty dx \; \varphi_n^{(\alpha)}(x) \; \varphi_m^{(\alpha)}(x) \;=\; \delta_{nm}. \qquad (6.2.4)$$

Defining the time dependent expansion amplitudes

$$a_n^{(\alpha)}(t) \;=\; \langle n, \alpha | \Psi(t) \rangle, \qquad (6.2.5)$$

a set of coupled first order differential equations can be derived for the expansion amplitudes by inserting the expansion

$$|\Psi(t)\rangle \;=\; \sum_{n=1}^{\infty} a_n^{(\alpha)}(t) \, |n, \alpha\rangle \qquad (6.2.6)$$

into the time dependent Schrödinger equation (6.2.1). The following set of coupled equations results:

$$i \dot{a}_n^{(\alpha)}(t) \;=\; \sum_{m=1}^{\infty} \langle n, \alpha | \hat{H}_0 \,-\, f(t)x | m, \alpha \rangle \, a_m^{(\alpha)}(t). \qquad (6.2.7)$$

Because of the simple analytical structure of the Sturmian states (6.2.3) the matrix elements of $\hat{H}_0$ and x can be computed analytically. For the dipole matrix elements we obtain

$$
\langle n, \alpha | x | m, \alpha \rangle = \frac{\alpha}{2} \times \begin{cases} -\sqrt{n^2 - 1} & \text{for } m = n - 1 \\ 2n + 1 & \text{for } m = n \\ -\sqrt{n(n+2)} & \text{for } m = n + 1 \\ 0 & \text{for } |m - n| > 1. \end{cases} \tag{6.2.8}
$$

The $\hat{H}_0$ matrix elements are composed of the matrix elements of the potential energy $1/x$ and the kinetic energy $-\frac{1}{2}d^2/dx^2$. For the potential matrix elements we obtain

$$
\langle n, \alpha | \frac{1}{x} | m, \alpha \rangle = \frac{1}{\alpha} \times \begin{cases} \left[\dfrac{m(m+1)}{n(n+1)} \right]^{1/2} & \text{for } m < n \\ 1 & \text{for } m = n \\ \left[\dfrac{n(n+1)}{m(m+1)} \right]^{1/2} & \text{for } m > n \end{cases} \tag{6.2.9}
$$

and the matrix elements of the kinetic energy operator are given by

$$
\langle n, \alpha | - \frac{1}{2} \frac{d^2}{dx^2} | m, \alpha \rangle =
$$

$$
\frac{1}{\alpha^2} \times \begin{cases} \left[\dfrac{m(m+1)}{n(n+1)} \right]^{1/2} \left\{ -\dfrac{1}{2} \delta_{nm} + \dfrac{2m+1}{3} \right\} & \text{for } m \le n \\ \left[\dfrac{n(n+1)}{m(m+1)} \right]^{1/2} \left\{ -\dfrac{1}{2} \delta_{nm} + \dfrac{2n+1}{3} \right\} & \text{for } m > n. \end{cases} \tag{6.2.10}
$$

The matrix elements known, the set (6.2.7) can now be integrated with any of the standard methods for the numerical solution of coupled differential equations (see, e.g., Milne (1970)). In order to be able to interpret the results obtained by the Sturmian method in the physical space of SSE states, we need the overlaps between the Sturmian states (6.2.3) and the SSE states (6.1.24). The result is

$$
T_{nm}^{(\alpha)} = \int_0^\infty \varphi_n^{(\alpha)}(x)\, \varphi_m(x)\, dx =
$$

$$
\frac{4\sqrt{2}}{\gamma_{\alpha m}^3 [\alpha^3 n(n+1) m^5]^{1/2}} \int_0^\infty y^2\, e^{-y}\, L_{n-1}^{(2)}\left(\frac{2y}{\alpha \gamma_{\alpha m}} \right) L_{m-1}^{(1)}\left(\frac{2y}{m \gamma_{\alpha m}} \right) dy, \tag{6.2.11}
$$

where

$$
\gamma_{\alpha m} = \frac{1}{\alpha} + \frac{1}{m}. \tag{6.2.12}
$$

Although it is, in principle, possible to evaluate (6.2.11) analytically, it is numerically more advantageous to evaluate the integral in (6.2.11) with the help of a suitable Gauss-Laguerre integration formula (see, e.g., Stroud and Secrest (1966)).

The Sturmian method as defined here suffers from serious convergence problems. For any finite number of basis states the Sturmian method corresponds to a discretization of the continuum. This means that for a fixed and finite basis size, ionization probabilities computed with the Sturmian method will show spurious resonance structures as a function of the microwave frequency. This phenomenon is demonstrated and discussed in Section 6.3. The problem with the spurious resonances can be mended in two ways: (i) by taking a very large number of Sturmian states into account or (ii) by using the method of complex rotation (see, e.g., Chu and Reinhardt (1977), Reinhardt (1982)). This method was used successfully by Buchleitner (1993) and Buchleitner and Delande (1993) for the computation of ionization rates of the microwave-driven hydrogen atom. Over the past 20 years the method of complex rotation has also been successfully applied in the context of many other atomic and molecular physics problems, especially for the computation of ionization probabilities, resonance lifetimes and dissociation cross-sections. The method of complex rotation is discussed in Section 10.4.1 and is applied to the computation of the locations and widths of resonances in the one-dimensional helium atom.

6.2.2 Fredholm approach

We now turn to a completely different method for solving the time dependent Schrödinger equation (6.2.1). The central idea of the method is to neglect transitions between states that correspond to continuum states of the unperturbed SSE Hamiltonian, but to take transitions from the bound space into the continuum (and back) accurately into account. We start with the unperturbed Hamiltonian $\hat{H}_0$ defined in (6.1.22). Its bound spectrum is given by

$$\hat{H}_0 \mid n\rangle = E_n \mid n\rangle, \quad E_n = -\frac{1}{2n^2}. \tag{6.2.13}$$

The spectrum of the continuum is

$$\hat{H}_0 \mid k\rangle = E_k \mid k\rangle, \quad E_k = \frac{1}{2}k^2. \tag{6.2.14}$$

The bound and continuum wave functions $\langle x|n\rangle = \varphi_n(x)$ and $\langle x|k\rangle = \varphi_k(x)$ are defined in (6.1.24) and (6.1.26), respectively. The time dependent wave function in (6.2.1) is expanded in the set of unperturbed states

(6.2.13) and (6.2.14) according to

$$| \Psi(t) \rangle = \sum_n a_n(t) | n \rangle + \int_0^\infty dk \, a_k(t) | k \rangle. \tag{6.2.15}$$

Inserting this result into the time dependent Schrödinger equation (6.2.1) we obtain a set of coupled first order differential equations for the amplitudes $a_n(t)$ and $a_k(t)$ in the form

$$i\dot{a}_n(t) = E_n a_n(t) - f(t) \left\{ \sum_{n'} x_{nn'} a_{n'}(t) + \int_0^\infty dk' \, x_{nk'} a_{k'}(t) \right\} \tag{6.2.16}$$

and

$$i\dot{a}_k(t) = E_k a_k(t) - f(t) \left\{ \sum_{n'} x_{kn'} a_{n'}(t) + \int_0^\infty dk' \, x_{kk'} a_{k'}(t) \right\} \tag{6.2.17}$$

with the matrix elements

$$x_{nn'} = \langle n|x|n' \rangle, \quad x_{nk} = \langle n|x|k \rangle, \quad x_{kk'} = \langle k|x|k' \rangle. \tag{6.2.18}$$

Explicit expressions for the dipole matrix elements (6.2.18) were computed, e.g., by Blümel and Smilansky (1987) and Susskind and Jensen (1988). At this point we make the crucial approximation. We neglect all coupling between the continuum states by dropping the integral over k' in (6.2.17). By substituting

$$a_k = e^{-iE_k t} b_k \tag{6.2.19}$$

we get from (6.2.17)

$$b_k(t) = i \int_0^t dt' \, e^{iE_k t'} f(t') \sum_{n'} x_{kn'} a_{n'}(t'). \tag{6.2.20}$$

Inserting this result into (6.2.16) we get

$$i\dot{a}_n(t) = E_n a_n(t) -$$

$$f(t) \sum_{n'} \left\{ x_{nn'} a_{n'}(t) + i \int_0^t K_{nn'}(t - t') f(t') a_{n'}(t') dt' \right\}, \tag{6.2.21}$$

where we introduced the memory kernels

$$K_{nm}(\tau) = \int_0^\infty dk \, x_{nk} e^{-iE_k \tau} x_{km}. \tag{6.2.22}$$

We solve the integro-differential equation (6.2.21) by fitting the memory kernels to a function consisting of a sum of decaying exponents

$$K_{nm}(\tau) = \sum_{\lambda=1}^\Lambda \left\{ A_{nm}^{(\lambda)} e^{-B_{nm}^{(\lambda)} \tau} - i\tau \tilde{A}_{nm}^{(\lambda)} e^{-\tilde{B}_{nm}^{(\lambda)} \tau} \right\}. \tag{6.2.23}$$

Decaying exponents were chosen because in this representation the memory kernels (6.2.22) separate and the integro-differential equation (6.2.21) can be reduced to a set of ordinary coupled first order differential equations. This is achieved by introducing the functions:

$$y_{nm}^{(\lambda)}(t) = A_{nm}^{(\lambda)} e^{-B_{nm}^{(\lambda)} t} \int_0^t a_m(t') f(t') e^{B_{nm}^{(\lambda)} t'} \, dt', \qquad (6.2.24)$$

$$\tilde{y}_{nm}^{(\lambda)}(t) = \tilde{A}_{nm}^{(\lambda)} e^{-\tilde{B}_{nm}^{(\lambda)} t} \int_0^t a_m(t') f(t') e^{\tilde{B}_{nm}^{(\lambda)} t'} \, dt', \qquad (6.2.25)$$

$$\tilde{z}_{nm}^{(\lambda)}(t) = \tilde{A}_{nm}^{(\lambda)} e^{-\tilde{B}_{nm}^{(\lambda)} t} \int_0^t a_m(t') t' f(t') e^{\tilde{B}_{nm}^{(\lambda)} t'} \, dt' \qquad (6.2.26)$$

and

$$\tilde{w}_{nm}^{(\lambda)}(t) = t\tilde{y}_{nm}^{(\lambda)}(t) - \tilde{z}_{nm}^{(\lambda)}(t). \qquad (6.2.27)$$

With the help of (6.2.24) – (6.2.27) we arrive at the following set of first order coupled differential equations:

$$i\dot{a}_n(t) = E_n a_n(t) - f(t) \sum_{n'} \left\{ x_{nn'} a_{n'}(t) + i \sum_{\lambda=1}^{\Lambda} [y_{nn'}^{(\lambda)}(t) - i\tilde{w}_{nn'}^{(\lambda)}(t)] \right\}, \qquad (6.2.28)$$

$$\dot{y}_{nm}^{(\lambda)}(t) = -B_{nm}^{(\lambda)} y_{nm}^{(\lambda)}(t) + A_{nm}^{(\lambda)} f(t) a_m(t), \qquad (6.2.29)$$

$$\dot{\tilde{y}}_{nm}^{(\lambda)}(t) = -\tilde{B}_{nm}^{(\lambda)} \tilde{y}_{nm}^{(\lambda)}(t) + \tilde{A}_{nm}^{(\lambda)} f(t) a_m(t) \qquad (6.2.30)$$

and

$$\dot{\tilde{w}}_{nm}^{(\lambda)}(t) = -\tilde{B}_{nm}^{(\lambda)} \tilde{w}_{nm}^{(\lambda)}(t) + \tilde{y}_{nm}^{(\lambda)}(t). \qquad (6.2.31)$$

In the case where $f(t)$ is a periodic function, (6.2.28) – (6.2.31) is a system of first order equations with periodic coefficients. In this case Floquet's theorem (Ince (1956)) can be applied and a one-cycle propagator exists for the system (6.2.28) – (6.2.31). This propagator can be constructed immediately by solving the system (6.2.28) – (6.2.31) directly with the help of a standard numerical integration method for ordinary differential equations. However, in the case of a strong perturbation, many bound states $|n\rangle$ have to be taken into account, making this method rather inefficient. The reason is that the higher states $|n\rangle$ in the basis introduce fast oscillations in the system (6.2.28) – (6.2.31) forcing the solver of differential equations to spend most of the integration time following fast oscillations of small amplitude. The following "algebraic" method eliminates this problem. This method was inspired by Gordon's algorithm (1969), which has already been applied successfully in scattering theory.

The method consists in approximating the force function $f(t)$ by a sequence of step-functions as shown in Fig. 6.6. The width of the steps $\Delta t_j = t_j - t_{j-1}$ is not necessarily constant and the height of the steps is $f_j = f((t_j + t_{j-1})/2)$. According to the approximation the perturbation in the time interval $I_j = [t_{j-1}, t_j]$ is now constant, and the Hamiltonian is given by

$$\hat{H} = \hat{H}_0 - f_j x. \tag{6.2.32}$$

The propagation of the full wave function $| \Psi(t) \rangle$ over the time interval I_j consists of two steps. In the first step we calculate an approximation for the bound state amplitudes $a_n(t)$ in I_j by neglecting all coupling to the continuum. On the basis of this approximation we evaluate the functions $y^{(\lambda)}(t)$, $\tilde{y}^{(\lambda)}(t)$ and $\tilde{z}^{(\lambda)}(t)$ in I_j. In the spirit of a corrector method (Stoer and Bulirsch (1978)) the bound state amplitudes are then re-evaluated interpreting the functions $y^{(\lambda)}(t)$, $\tilde{y}^{(\lambda)}(t)$ and $\tilde{z}^{(\lambda)}(t)$ as known "source functions". More about the technical details of this method can be found in Blümel and Smilansky (1990b). We emphasize that this method, although "perturbative" over a given interval I_j, produces an exact solution of (6.2.28) – (6.2.31) in the limit $I_j \to 0$.

6.2.3 Analytical decay rates

In order to check our numerical calculations we need analytical results. Let us assume that the surface state electrons are exposed to a weak homogeneous cw microwave field characterized by $f(t) = -\epsilon \sin(\omega t)$, where ϵ is the strength of the microwave field. In this case it is possible to calculate one-photon ionization rates analytically. We start with the amplitude equations (6.2.16) and (6.2.17). Neglecting the continuum-

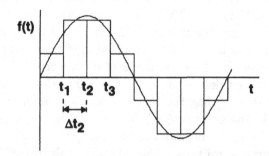

Fig. 6.6. Approximation of the force function $f(t)$ by a sequence of step-functions.

continuum coupling term in (6.2.17) we obtain

$$i\dot{a}_n = -\frac{1}{2n^2}a_n + \epsilon\sin(\omega t)\sum_{m=1}^{\infty} x_{nm}a_m + \epsilon\sin(\omega t)\int_0^{\infty} dk\, x_{nk}\, a_k$$

$$(6.2.33)$$

and

$$i\dot{a}_k = \frac{1}{2}k^2 a_k + \epsilon\sin(\omega t)\sum_{m=1}^{\infty} x_{km}\, a_m. \qquad (6.2.34)$$

For weak fields ($\epsilon \ll 1$) and a microwave frequency tuned slightly above the one-photon threshold of a state $|\,n\rangle$, the states $|\,m\rangle \neq |\,n\rangle$ do not play a significant role. Neglecting all states $|\,m\rangle \neq |\,n\rangle$ in (6.2.33) and (6.2.34) we obtain

$$i\dot{a}_n = -\frac{1}{2n^2}a_n + \epsilon\sin(\omega t)\,x_{nn}a_n + \epsilon\sin(\omega t)\int_0^{\infty} dk\, x_{nk}\, a_k \qquad (6.2.35)$$

and

$$i\dot{a}_k = \frac{1}{2}k^2 a_k + \epsilon\sin(\omega t)x_{kn}\, a_n. \qquad (6.2.36)$$

Defining

$$a_k(t) = \exp\left(-\frac{1}{2}ik^2 t\right) b_k(t), \qquad (6.2.37)$$

the equation (6.2.36) for the continuum amplitudes is transformed into

$$i\dot{b}_k = \epsilon\sin(\omega t)\exp\left(\frac{1}{2}ik^2 t\right) x_{kn}\, a_n. \qquad (6.2.38)$$

This equation can be solved immediately and results in

$$b_k(t) = -i\epsilon\int_0^t dt'\,\sin(\omega t')\exp\left(\frac{1}{2}ik^2 t'\right) x_{kn}\, a_n(t'). \qquad (6.2.39)$$

For weak fields the self-coupling of a_n via the permanent dipole matrix element x_{nn} in (6.2.35) can be neglected. Defining

$$a_n(t) = \exp\left(\frac{it}{2n^2}\right) b_n(t), \qquad (6.2.40)$$

the amplitude equation (6.2.35) is solved by quadratures according to

$$i b_n(t) = -i\epsilon^2\int_0^t dt'\,\sin(\omega t)\sin(\omega t')\exp\left[-\frac{i}{2n^2}(t-t')\right] K_{nn}(t-t')\, b_n(t'),$$

$$(6.2.41)$$

where we used the memory kernels K_{nn} defined in (6.2.22). We assume now that $K_{nn}(t-t')$ is peaked at $t \approx t'$. As a consequence, we can take b_n out of the integral in (6.2.41) since we can assume that as a result of the substitution (6.2.40) $b_n(t)$ is a slowly varying function of t. In

the spirit of a rotating wave approximation we replace $\sin(\omega t)\sin(\omega t')$ by
the term $\exp[i\omega(t - t')]/4$, since this term is approximately resonant with
$\exp[-i(t - t')/2n^2]$. All together we obtain

$$\frac{\dot{b}_n}{b_n} = -\frac{\epsilon^2}{4}\int_0^\infty dk \mid x_{nk}\mid^2 \int_0^t d\sigma \, \exp\left[i\left(\omega - \frac{1}{2n^2} - \frac{1}{2}k^2\right)\sigma\right],$$

(6.2.42)

where we substituted $t - t' = \sigma$. What we really need is the decay of the
staying probability

$$P_n^B(t) = \mid b_n(t)\mid^2,$$

(6.2.43)

i.e. the probability of staying bounded in the state $\mid n\rangle$. We have

$$\dot{P}_n^B = \dot{b}_n b_n^* + b_n \dot{b}_n^* = 2\,\Re(\dot{b}_n b_n^*).$$

(6.2.44)

Therefore,

$$\frac{\dot{P}_n^B}{P_n^B} = \frac{2\,\Re(\dot{b}_n b_n^*)}{P_n^B} = 2\,\Re\left(\frac{\dot{b}_n b_n^*}{b_n b_n^*}\right) = 2\,\Re\left(\frac{\dot{b}_n}{b_n}\right).$$

(6.2.45)

Using (6.2.42) in (6.2.45) together with

$$\Re \int_0^t d\sigma \, \exp(i\nu\sigma) = \pi\,\delta(\nu), \quad \text{for } t \to \infty$$

(6.2.46)

we obtain

$$\frac{\dot{P}_n^B}{P_n^B} = -\lambda_n(\omega),$$

(6.2.47)

where

$$\lambda_n(\omega) = -\frac{\epsilon^2\pi}{2k}\mid x_{nk}\mid^2, \quad k = \sqrt{2\omega - 1/n^2}.$$

(6.2.48)

For a given microwave frequency ω, the right hand side of (6.2.47) is con-
stant. Thus, (6.2.47) can be solved immediately with the result $P_n^B(t) = \exp(-\lambda_n t)$. Therefore, λ_n has the physical meaning of a one-photon decay rate to the continuum. The expression (6.2.48) is a form of Fermi's
golden rule.

It was shown by Blümel and Smilansky (1987) that the bound-conti-
nuum coupling matrix element x_{nk} can be written as

$$x_{nk} = \sqrt{k}\, g_n(k),$$

(6.2.49)

where the functions $g_n(k)$ can be calculated analytically. The function
$g_1(k)$, e.g., is given by

$$g_1(k) = \frac{-8}{\sqrt{1 - \exp(-2\pi/k)}} \frac{1}{(1 + k^2)^2}\exp\left[-\frac{2}{k}\arctan(k)\right].$$

(6.2.50)

At the one-photon threshold, i.e. $k \to 0$, we have

$$g_1(k) \approx -8 \exp(-2). \tag{6.2.51}$$

This results in

$$\lambda_1(\omega) \approx 32\pi \exp(-4)\epsilon^2, \quad \text{for } \omega \gtrsim 1/2. \tag{6.2.52}$$

For weak fields and short exposure to microwave radiation the ionization probability

$$P_n^I = 1 - P_n^B \tag{6.2.53}$$

is small. In this case $P_n^I(t)$ can be approximated by $P_n^I(t) \approx \lambda_n t$. Using the estimate (6.2.52) and irradiating the surface state electrons with N cycles of the microwave field, we obtain the one-photon ionization probability $P_1^I(N)$ after N cycles of the field as

$$P_1^I(N) \approx 128\pi^2 N n^2 \exp(-4)\epsilon^2, \quad \text{for } \omega \gtrsim 1/2. \tag{6.2.54}$$

With the help of an asymptotic result derived by Susskind and Jensen (1988) we calculate the decay rates λ_n in (6.2.48) approximately for arbitrary n. We obtain

$$g_n(k \to 0) \approx -\frac{2^{4/3}\, 3^{1/6}}{\pi} \Gamma\left(\frac{2}{3}\right) n^{11/6}. \tag{6.2.55}$$

Using this result in (6.2.48), we obtain

$$\lambda_n = \frac{1}{\pi} 2^{5/3}\, 3^{1/3}\, \Gamma^2\left(\frac{2}{3}\right) \epsilon^2\, n^{11/3} \approx 2.67\, \epsilon^2\, n^{11/3}. \tag{6.2.56}$$

This result can be compared with the one-photon ionization rates derived by Casati *et al.* (1987)

$$\lambda_n = 0.265\, \epsilon^2 / (\omega^{10/3}\, n^3). \tag{6.2.57}$$

Substituting $\omega = 1/2n^2$ in (6.2.57), we reproduce (6.2.56). Compared with (6.2.52) the result (6.2.56) is about 30% off. It is expected that (6.2.56) improves rapidly for larger n.

The one-photon decay rates λ_n can be used to implement a "short cut" for computing numerical ionization rate. We start from (6.2.16), but approximate the coupling to the continuum by the single decay term $\lambda_n/2$ according to

$$i\dot{a}_n(t) = \left[E_n - i\frac{\lambda_n}{2}\right] a_n(t) - f(t) \sum_m x_{nm}\, a_m(t). \tag{6.2.58}$$

The method defined in (6.2.58) turns the real bound state energies E_n into complex resonances, where the width of the resonances, $\lambda_n/2$, is assumed to contain only the one-photon contribution. For weak microwave fields this approximation is acceptable.

So far we have considered the case of weak external fields. In this limit a one-photon description of ionization is applicable and useful. Next we consider the regime of microwave fields of intermediate strength. In this regime one-photon processes are no longer the most efficient pathways to the continuum, and higher order multi-photon, multi-step processes dominate the decay modes to the continuum. We define the "intermediate strength regime" as a regime of microwave field strengths strong enough to render a one-photon description useless, but still weak enough not to lead to global chaos in the associated classical phase space.

6.3 The multi-photon regime

For microwave field stengths in the intermediate regime the quantum dynamics of the SSE system is still simple. It can be described in a multi-photon picture. For fixed microwave field strength we expect that the ionization probability exhibits a pronounced peak or threshold structure with large amounts of ionization occurring whenever the microwave frequency is in resonance with unperturbed SSE levels, or tuned to the ionization threshold. A schematic sketch of the first four SSE levels is shown in Fig. 6.7 together with possible ionization routes to the continuum.

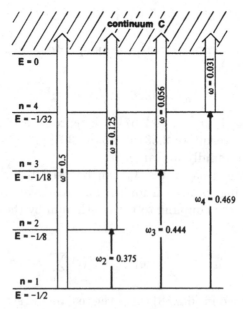

Fig. 6.7.   Sketch of possible ionization routes to the continuum.

For $\omega > 1/2$ direct one-photon ionization from the ground state to the continuum is possible. We denote this process by $1 \to C$, where "C" stands for "continuum". At lower frequencies, an infinity of two-step processes is possible according to $1 \to n \to C$, $n = 2, 3, \dots$. For every n we expect to see an associated peak in the ionization probability. The peaks are expected to occur at

$$\omega_n = \frac{1}{2} - \frac{1}{2n^2}, \quad n = 2, 3, \dots. \tag{6.3.1}$$

How many of these structures can be resolved depends on the microwave power which controls the widths of these transitions. The lowest resonance frequencies are given by $\omega_2 = 0.375$, $\omega_3 = 0.444$ and $\omega_4 = 0.469$. We check this simple picture with the help of numerical solutions of the time dependent Schrödinger equation (6.2.1). We choose a dimensionless microwave field strength of $\epsilon = 0.01$ and compute the ionization probabilities after 100 cycles of the microwave field for 70 microwave frequencies chosen as $\omega_j = 0.2 + (j-1)\Delta\omega$, $j = 1, \dots, 70$, and $\Delta\omega = 0.006$. The initial state is chosen to be the SSE ground state $| n = 1 \rangle$.

First, we try the Sturmian method outlined in Section 6.2.1. The computations are performed in a basis of 50 Sturmian states as defined in (6.2.3) for three different choices of the Sturmian label, $\alpha = 3, 4, 5$. The resulting ionization probabilities after 100 cycles of the field, $P^I(100)$, are shown in Figs. 6.8(a) – (c) for the three different Sturmian labels, respectively. Our first impression is that the results depend strongly on the choice of the Sturmian label. This dependence can in principle be reduced, but only at the cost of increasing the basis size substantially.

A heuristic method, however, yields qualitative results faster. We average $P^I(100)$ over the three different labels and in addition perform a frequency average over three neighbouring ionization probabilities on our frequency mesh. The resulting doubly averaged ionization signal is shown in Fig. 6.9. The ionization threshold at $\omega = 1/2$ is clearly visible together with three peaks at the expected positions ω_2, ω_3 and ω_4. For driving frequencies $\omega > 1/2$, but close to the one-photon threshold, (6.2.54) predicts an ionization probability of $P_1^I(100) = 0.23$. This is in good agreement with the ionization probability close to threshold shown in Fig. 6.9.

We now use the method defined in (6.2.58) to compute microwave ionization probabilities for the same field strength and frequencies as were used above in connection with the Sturmian method. Using the exact decay rates $\lambda_n = \pi\epsilon^2 g_n^2/2$ according to (6.2.48) and (6.2.49), and retaining only the first five SSE states in (6.2.58), we obtain the ionization probabilities shown in Fig. 6.10. They compare favourably with the probabilities

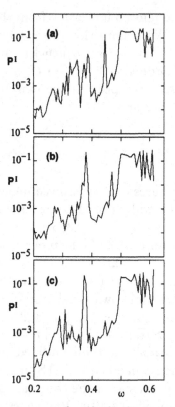

Fig. 6.8.   Ionization rates computed with the Sturmian method for three different Sturmian labels. (a) $\alpha = 3$, (b) $\alpha = 4$, (c) $\alpha = 5$.

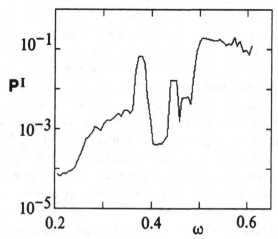

Fig. 6.9.   Ionization rates computed with the Sturmian method, doubly averaged
over frequency and the Sturmian label.

shown in Fig. 6.9, which were obtained from the Sturmian approach retaining 50 coupled Sturmian states. But while the Sturmian method took several hours to execute on a powerful work-station, the data in Fig. 6.10 were generated in a few minutes on a personal computer. The ionization probabilities shown in Figs. 6.9 and 6.10 can be interpreted within the simple $1 \rightarrow n \rightarrow C$ two-step picture illustrated in Fig. 6.7.

The data shown in Fig. 6.9 and Fig. 6.10 confirm our suspicion that for weak microwave fields no chaos mechanisms have to be invoked for an adequate physical understanding of microwave ionization data. The situation, however, is quite different in the case of strong microwave fields. In this case the ionization routes are very complicated, and the multiphoton picture loses its attractiveness. It has to be replaced by a picture based on chaos. Chaos provides a simpler description of the ionization process and consequently a better physical insight. The discussion of the chaotic strong-field regime is the topic of the following section.

6.4 The chaotic regime

With Fig. 6.5 we established that the phase space of the classical version of the SSE system contains chaotic regions. But since the SSE system is a manifestly quantum mechanical system, the central question is whether the classical chaos in the SSE system is at all relevant for the quantum dynamics, and, if yes, what are the signatures?

The SSE phase-space portrait shown in Fig. 6.5 reminds us of the phase-space portraits of the kicked rotor presented in Chapter 5. In Fig. 6.5 we can identify resonances and sealing invariant curves. In Chapter 5 we saw that resonance overlap in the standard mapping defines a sudden "percolation transition" when for $K > K_c$ the sealing invariant

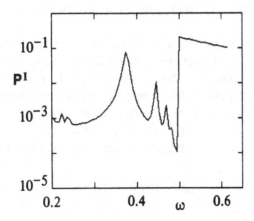

Fig. 6.10. Ionization rates computed by turning the SSE energies into complex resonances with widths corresponding to the one-photon decay rates.

lines are destroyed. In this case a large amount of angular momentum can be excited. Since the phase space of the kicked rotor and the SSE phase space are so similar, we expect that critical fields do exist for the SSE system as well. Furthermore we expect that, in analogy to the kicked rotor, the SSE system is able to absorb large quantities of energy for field strengths exceeding a critical field, while for field strengths below the critical field the SSE energy stays bounded. Because of the existence of an ionization continuum in the SSE system, the diagnostics of the critical field is much simpler than in the rotor case. While the excitation energy of a rotating molecule has to be inferred indirectly by measuring the occupation probabilities of the highly excited rotational states, the onset of global chaos in the SSE system can be measured summarily by measuring ionization probabilities. The rationale is the following. In the chaotic regime a considerable fraction of classical SSE trajectories is free to absorb large amounts of energy. According to (6.1.13) this corresponds to an increase in action, and according to (6.1.18) an increase in action corresponds to an increase in the x expectation value. In other words, the SSE electron will detach from the helium surface, i.e. it ionizes. Classically this ionization channel is open only if the sealing invariant lines are broken. Thus, we answered the question posed at the beginning of this section. The signature of chaos in the quantum SSE system is the occurrence of strong ionization for driving fields with field strengths $\mathcal{E}$ that exceed a critical value $\mathcal{E}_c$. What remains to be done is to predict the critical field $\mathcal{E}_c$.

In the case of the kicked rotor we were able to predict the critical perturbation strength by applying the Chirikov overlap criterion. This criterion does two things for us. First, it provides us with an excellent physical picture which explains qualitatively the mechanism of the chaos transition; secondly, it provides us with an analytical estimate for the critical field.

The central idea of the overlap criterion is to focus on the most important resonances in the phase space and to calculate their widths as a function of the strength of the applied field. In a classical picture the critical field strength is reached when the resonances overlap.

In order to derive a resonance overlap criterion we observe that according to (6.1.39) the potential of the perturbed SSE problem is given by $V(x,t) = -\mathcal{E}x\sin(\omega t)$. Using the expansion (6.1.49) we obtain:

$$V(x,t) = -\frac{1}{2}\mathcal{E}\sum_{m=0}^{\infty} x_m \left\{\sin(m\theta - \omega t) + \sin(m\theta + \omega t)\right\}. \quad (6.4.1)$$

The first term in (6.4.1) is a resonant term; the second term corresponds to fast oscillations. Neglecting the second term, a resonance in the SSE system occurs if the argument of the first term is stationary, i.e. if the frequency of the microwave field equals a multiple of the natural libration frequency of the electron (the frequency of the bouncing motion). Therefore, we get the condition:

$$m\dot{\theta} = \frac{m}{I_m^3} = \omega, \tag{6.4.2}$$

i.e.

$$I_m = \left(\frac{m}{\omega}\right)^{1/3}. \tag{6.4.3}$$

Keeping only the resonance term, the Hamiltonian at the mth resonance can be approximated as

$$H = H_0 - \frac{1}{2}\mathcal{E}\,x_m\,\sin(m\theta - \omega t). \tag{6.4.4}$$

We introduce the width of the mth resonance, $\Delta = I - I_m$, and the resonance angle $\xi = \theta - \omega t/m$. Keeping only terms up to second order in Δ, and dropping a constant term, the canonical transformation

$$F_2(\Delta, \theta; t) = (I_m + \Delta)[\theta - \omega t/m] \tag{6.4.5}$$

transforms the Hamiltonian (6.4.4) into

$$\tilde{H} = -\frac{3\Delta^2}{2I_m^4} - \frac{1}{2}\mathcal{E}x_m\sin(m\xi). \tag{6.4.6}$$

Except for the negative sign in front of the kinetic energy, which can be interpreted as a "negative mass", the Hamiltonian (6.4.6) is the (autonomous) Hamiltonian of a pendulum. The resonance width W_m can immediately be calculated. Using the results derived in Section 5.2 and the approximation (6.1.56) for the derivatives of the Bessel functions we obtain:

$$W_m \approx 1.5\mathcal{E}^{1/2}I_m^3 m^{-5/6}. \tag{6.4.7}$$

The separation between the mth and the $(m+1)$th resonance is given by

$$\delta_m = \frac{I_m}{3m}. \tag{6.4.8}$$

Resonance overlap is achieved for $W_m/\delta_m > 1$, or

$$4.5\,\mathcal{E}^{1/2}\,m^{5/6}\,\omega^{-2/3} > 1. \tag{6.4.9}$$

Therefore the critical field is given by

$$\mathcal{E}_c = 0.05\,\omega^{4/3}\,m^{-5/3}. \tag{6.4.10}$$

For field strengths larger than $\mathcal{E}_c$ we expect efficient ionization to occur. We emphasize that (6.4.10) is valid only for $\omega I_m^3 \gtrsim 1$. Resonance overlap formulae for $\omega I_m^3 < 1$ are much harder to obtain. They are presented in the following chapter.

A clarifying remark is now in order. Since the surface state electrons possess an ionization continuum, ionization *always* occurs as soon as an external microwave field is switched on. In other words, the mere observation of ionization tells us nothing about the chaotic or regular nature of a quantum system. This is because quantum mechanics allows for tunnelling and resonant multi-photon processes to occur, phenomena absent in classical mechanics. The ionization probability in the tunnelling regime, however, is exponentially small, and the probabilities in the multiphoton regimes follow high order powerlaws. The importance of $\mathcal{E}_c$ is that it marks a cross-over from negligible ionization rates below $\mathcal{E}_c$ to very efficient ionization for field strengths above $\mathcal{E}_c$. This definition of $\mathcal{E}_c$ holds for generic driving frequencies. It is clear that, as an exception to this definition of $\mathcal{E}_c$, efficient ionization may occur even below $\mathcal{E}_c$ at special multi-photon/multi-step resonance frequencies. This discussion shows that the concept of a "critical field" in quantum mechanics has to be defined differently, for instance according to the behaviour of the ionization rates. In a quantum treatment of microwave ionization the quantum analogue of $\mathcal{E}_c$ may be defined as the field strength $\mathcal{E}_c^{(qm)}$ at which the microwave ionization *rate* shows a sharp bend in its magnitude corresponding to the cross-over point from small tunnelling/multi-photon ionization to very effective stochastic ionization.

In conclusion we hope that, despite the experimental difficulties discussed in the previous sections, SSE experiments will eventually be performed in the chaotic regime, experiments that put the very existence of critical fields, as well as their numerical values estimated in (6.4.10), to an experimental test.

7

The hydrogen atom in a strong microwave field

The experimental investigation of chaos in atomic physics began with the historic experiment on microwave ionization of Rydberg atoms reported by Bayfield and Koch in 1974. The central result is the existence of an ionization threshold as a function of the microwave *field*. At the time (1974) this result was totally unexpected since ionization thresholds, in analogy to the photo-electric effect, were thought to appear only as a function of the *frequency*. Nowadays, especially in the light of the material presented in Chapters 5 and 6, the existence of an ionization threshold in the microwave *field* is less surprising and may be attributed to the existence of a critical microwave field that marks the onset of global chaos in the classical analogue of the Bayfield-Koch experiments. But at the time the Bayfield-Koch experiment was conducted, a connection with chaos was not suspected. Leopold and Percival (1978, 1979) were the first to investigate the Bayfield-Koch experiments using purely classical mechanics. This is allowed on the basis of the correspondence principle, Leopold and Percival argued, since the quantum numbers involved in the ionization experiments are large. This line of thought turned out to be very fruitful. With the help of Monte Carlo simulations of the time evolution of classical trajectories in phase space, Leopold and Percival were able to reproduce the existence and location of the microwave thresholds established by the Bayfield-Koch experiment. Encouraged by the good performance of purely classical calculations, Meerson *et al.* (1979) went one step further and linked the microwave ionization threshold fields with chaos thresholds. Based on Chirikov's overlap criterion (see Section 5.2), Meerson and collaborators derived an analytical estimate for the critical microwave ionization field strength which agrees well with the experimental results. The pioneering work of the 1970s was continued vigorously throughout the 1980s by many experimental and theoretical groups. As a result of this concentrated effort, the hydrogen atom in a strong microwave field is by now the most thoroughly studied time dependent classically chaotic atomic physics system. A host of experimental data are available which can be compared with theoretical models and predictions. The availability of experimental data puts the chaotic hydrogen atom in a different category than the kicked rotor, or microwave-driven surface state electrons. In the absence of experimental data the kicked rotor and the SSE system currently have the status of schematic models, with corresponding experiments only having been suggested, while in the case of driven hydrogen atoms the theory has to face up to already established experimental results. Given this situation, Chapter 7 is mainly about the comparison between experimental results and their interpretation using concepts of chaos and nonlinear dynamics.

In Section 7.1 we start with a brief survey of experiments and experimental results on microwave-driven hydrogen and alkali Rydberg atoms. In Section 7.2 we interpret the experimental results in the context of chaos. The numerical calculations discussed in Section 7.2 show that, even in the classically chaotic regime, quantum mechanics is the correct theoretical framework for the interpretation of the microwave ionization experiments. This, however, does not mean that we can now dispense with chaos. It serves as an important tool providing valuable physical pictures for the quantum mechanical interpretation of the experimental results. This point of view is illustrated in Section 7.3, where the onset of classical chaos is shown to correlate very accurately with the observed microwave ionization thresholds. This correspondence, however, holds only in the case where the microwave driving frequency ω is small compared with the classical orbiting frequency Ω of the unperturbed Rydberg electron. The orbiting frequency Ω of the unperturbed electron is also called the *Kepler frequency*. For $\omega > \Omega$ the chaos thresholds do not correspond so well with the microwave ionization thresholds. This deviation is due to an important effect, quantum suppression of chaos, already encountered in Chapter 5. The quantum suppression effect, akin to Anderson localization in solids (P. W. Anderson (1958)) is briefly discussed in Section 7.4. Section 7.5 discusses the prediction of a "giant resonance" in the chaotic two-frequency ionization of hydrogen Rydberg atoms. This result is currently only partially verified experimentally. Since extensive reviews on microwave ionization of hydrogen Rydberg atoms are available (see, e.g., Casati *et al.* (1987), Jensen *et al.* (1991), Moorman and Koch (1992)), the focus in this chapter is not on a complete survey of the field. Rather, we will restrict ourselves to a discussion of a few selected topics, with special emphasis on the correspondence between classical and quantum mechanics in the classically chaotic regime.

7.1 Experiment

Fig. 7.1 shows a schematic sketch of a typical microwave ionization experiment. The set-up consists of three parts. In section A hydrogen or alkali Rydberg atoms are prepared in a highly excited state $|n_0\rangle$. They leave section A in the form of a narrow atomic beam to enter section B where they are exposed to electromagnetic radiation. The effect of section B is to induce ionization or transitions to neighbouring bounded states $|n\rangle$. Leaving the interaction region B, the atoms enter section C where their current state is analysed. In an ionization experiment, section C determines the percentage of atoms that lost their electron in the interaction region B. In an excitation experiment section C determines the occupation probabilities p_n of atomic states $|n\rangle$ induced by the electromagnetic fields in section B. Section A may consist of a proton accelerator that generates highly excited hydrogen atoms by means of a charge exchange reaction in a gas cell, or it may consist of an atomic oven that generates a beam of alkali atoms that are excited to a high-lying Rydberg state by means of a carefully tuned laser. The interaction region B typically consists of a wave guide or a cavity with small holes for the atomic beam to pass through, or, for very low frequency irradiation, simply of a set of parallel plates. The most complicated part of the experiment is typically section C where the degree of excitation of the atoms has to be determined. This section may consist of static field ionization regions to determine the occupation probability of a state $|n\rangle$ by selective ionization using an electric field ramp, or, especially for ionization experiments, it may consist of a Faraday cup that collects the total charge arriving at section C in the form of ionized atoms. Actual technical realizations of sections A, B and C have been discussed in detail in the literature. Descriptions of possible experimental set-ups can be found, e.g., in Bayfield and Sokol (1988), Blümel et al. (1991), Moorman and Koch (1992), Koch and van Leeuwen (1995).

Using an apparatus of the type shown in Fig. 7.1, Bayfield and Koch (1974) conducted an ionization experiment with hydrogen Rydberg atoms that were prepared in the band $63 \leq n_0 \leq 69$ and exposed to electromagnetic radiation of three different frequencies $\omega_i = 2\pi f_i$, with $f_1 = 30\,\mathrm{MHz}$,

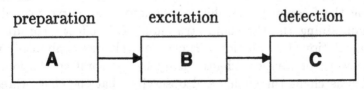

Fig. 7.1.   Schematic sketch of a typical microwave ionization experiment.

$f_2 = 1.5\,\text{GHz}$ and $f_3 = 9.9\,\text{GHz}$. Bayfield and Koch (1974) present their data in the form of an ionization signal versus the *adiabaticity parameter* γ (Keldysh (1965)) defined as

$$\gamma = \omega/n_0\mathcal{E}, \qquad (7.1.1)$$

where ω is the field's frequency, n_0 is the principal quantum number of the starting state and $\mathcal{E}$ is the field strength. Both ω and $\mathcal{E}$ in (7.1.1) are in atomic units. The adiabaticity parameter is chosen because it plays an important role in multi-photon theories of ionization by electromagnetic fields. In the context of chaos it is better to present the ionization data as the measured ionization signal versus the field strength. Moreover, it is advantageous to introduce the scaled frequency

$$\omega_0 = \omega n_0^3 \qquad (7.1.2)$$

and the scaled field strength

$$\mathcal{E}_0 = \mathcal{E} n_0^4. \qquad (7.1.3)$$

Both definitions are natural since ω_0 turns out to be the ratio of the microwave frequency ω and the Kepler frequency Ω of the Rydberg electron, and $\mathcal{E}_0$ is the ratio of the microwave field strength and the field strength experienced by an electron in the n_0th Bohr orbit of the hydrogen atom. Motivated by the above discussion we have redrawn the results obtained by Bayfield and Koch (1974) and present them in Fig. 7.2 as an ionization signal (in arbitrary units) versus the scaled field strength $\mathcal{E}_0$ defined in (7.1.3). For n_0 in (7.1.3) we chose $n_0 = 66$, the centroid of the band of Rydberg states present in the atomic beam.

Fig. 7.2 shows that for all frequencies used in the experiment we obtain a pronounced threshold structure in the scaled field $\mathcal{E}_0$. For $f_1 = 30\,\text{MHz}$ (data points connected by the full line in Fig. 7.2) this result is not surprising since the Kepler frequency of a Rydberg electron in the state $|n_0 = 66\rangle$ is much larger than f_1. Therefore, the situation is quasi-static and we expect a threshold in the ionization data as soon as the peak field strength of the electromagnetic perturbation exceeds the static ionization threshold of hydrogen given by $\mathcal{E}_0 \approx 0.13$. The data displayed in Fig. 7.2 reflect this behaviour. Fig. 7.2 also shows that the effect of increasing the frequency is a shift of the ionization curve toward lower fields maintaining the threshold structure. This effect, reproduced by model calculations (see Section 7.2) can be explained analytically within the framework of classical chaos theory by interpreting the ionization thresholds as chaos thresholds and observing that the chaos thresholds decrease with increasing frequency. This will be shown in Section 7.3.

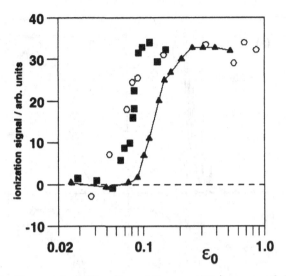

Fig. 7.2. Ionization probability (in arbitrary units) versus field strength for hydrogen Rydberg atoms prepared in states $63 \leq n_0 \leq 69$ and irradiated with three different electromagnetic fields of frequencies 30 MHz (triangles), 1.5 GHz (squares) and 9.9 GHz (circles). The full line connecting the 30 MHz data is drawn to guide the eye. The data for this figure were compiled from Bayfield and Koch (1974).

Since 1974 the experimental groups of Bayfield and Koch have considerably refined and extended the experimental data available on microwave ionization of hydrogen Rydberg atoms. Ionization data are now available covering a large range of principal quantum numbers ranging from $n_0 \approx 30$ to $n_0 \approx 100$. Some sample ionization data at $f = 9.92$ GHz covering the range from $n_0 = 36$ to $n_0 = 40$ are shown as the dashed lines in Fig. 7.3. Since the frequency is the same for each one of the starting states, it is possible to gauge the field axes in Fig. 7.3 conveniently in V/cm. Again, the threshold structure of the data is apparent. The shift of the data toward lower field strengths as a function of increasing n_0 is not surprising, since the atomic states are naturally more fragile, and therefore more susceptible to ionization, the higher the principal quantum number with which they enter the microwave field.

It is demonstrated in the following section that quantum calculations in large basis sets are able to reproduce qualitatively the threshold behaviour of the ionization curves. This demonstrates that there is no "mystery" beyond quantum mechanics hidden in the experimental results. But although numerical quantum calculations reproduce the experimental results adequately, they are limited in that they do not provide us with any insight into the physical mechanisms responsible for the occurrence

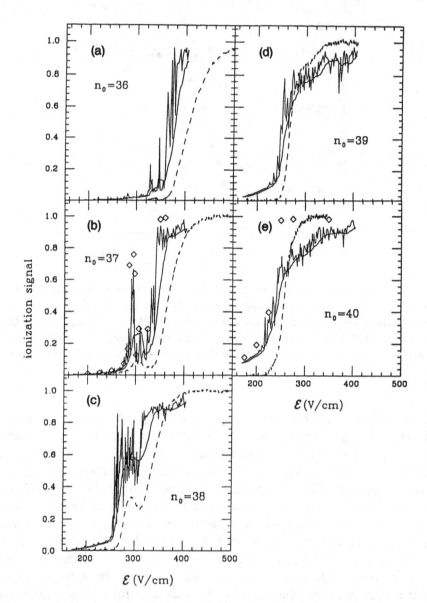

Fig. 7.3. Ionization probability versus field strength for hydrogen Rydberg atoms prepared in the states $36 \leq n_0 \leq 40$ (panels (a) – (e), respectively) and ionized at 9.92 GHz. Dashed lines: experimental data. Full lines: Results of quantum mechanical model calculations discussed in Section 7.2. (From Blümel and Smilansky (1987).)

of the sharp thresholds. This insight is provided by classical trajectory calculations in the spirit of the Monte Carlo calculations of Leopold and Percival mentioned in the previous section. The trajectory calculations show that, in analogy with the results obtained in Chapters 5 and 6, the hydrogen ionization thresholds correspond to chaos thresholds. This means that for given n_0 and frequency ω the motion of the hydrogen Rydberg electron becomes globally chaotic at some critical field $\mathcal{E}_c$ from where on it is free to diffuse toward the ionization continuum. The onset of global chaos, classically sharply defined to occur *exactly* at the critical field $\mathcal{E}_c$ explains the sharpness of the ionization thresholds observed in the microwave ionization experiments. The analytical determination of the critical fields $\mathcal{E}_c$ on the basis of resoance overlap in phase space is the topic of Section 7.3.

7.2 Quantum calculations

The aim of this section is to show that the experimentally obtained ionization thresholds displayed in Fig. 7.3 can be reproduced by numerical quantum calculations conducted within the framework of the SSE model discussed in detail in Chapter 6. Since the SSE model is one-dimensional, it is surprising that it should be possible to use this model with any hope of success for the description of the manifestly "three-dimensional" experiments with real Rydberg atoms in the laboratory. Therefore, the main point here is to motivate and to justify the use of a one-dimensional model for the description of Rydberg atoms in a strong linearly polarized radiation field.

In standard introductory text books, the quantum mechanics of the hydrogen atom is usually discussed in spherical coordinates. In the spherical description, neglecting the electron spin, the hydrogen states are classified with the help of three quantum numbers, the principal quantum number n, the angular quantum number l and the magnetic quantum number m. The hydrogen wave functions are given by

$$\langle \vec{r} | nlm \rangle = R_{nl}(r) Y_{lm}(\theta, \varphi), \qquad (7.2.1)$$

where R_{nl} are the hydrogenic radial functions and Y_{lm} are the spherical harmonics. But the spherical representation is not the only one possible for the hydrogen atom. Especially in a linearly polarized radiation field, such as the microwave fields commonly used in the ionization of Rydberg atoms, the use of parabolic coordinates is more effective. The parabolic coordinates are defined as (Landau and Lifschitz (1971))

$$\xi = r + z; \quad \eta = r - z; \quad \varphi = \arctan(y/x). \qquad (7.2.2)$$

In parabolic coordinates the hydrogen states are known as the Stark states. They are classified with the help of three quantum numbers, the parabolic quantum numbers n_1 and n_2 and the magnetic quantum number m. The Stark states are denoted by

$$|n; n_1, n_2, m\rangle, \qquad (7.2.3)$$

where n in (7.2.3) is the principal quantum number. It is the same as the principal quantum number in the spherical representation (7.2.1). Not all the quantum numbers in (7.2.3) are independent. They are connected by the relation (Landau and Lifschitz (1971))

$$n = n_1 + n_2 + |m| + 1. \qquad (7.2.4)$$

In $\vec{r}$ representation the states (7.2.3) are given by

$$\langle \vec{r} | n; n_1, n_2, m \rangle = \frac{1}{n^2 \sqrt{\pi}} f_{n_1 m}(\xi/n) \, f_{n_2 m}(\eta/n) \, e^{im\varphi}, \qquad (7.2.5)$$

where

$$f_{pm}(x) = \sqrt{\frac{p!}{(p + |m|)!}} \, x^{|m|/2} \, L_p^{|m|}(x) \, e^{-x/2} \qquad (7.2.6)$$

and $L_p^{|m|}(x)$ are the associated Laguerre polynomials. There are two exceptional classes of Stark states, called *extremal* Stark states. They are defined by

$$|S_n^{(\rightarrow)}\rangle = |n; n - 1, 0, 0\rangle \qquad (7.2.7a)$$

and

$$|S_n^{(\leftarrow)}\rangle = |n; 0, n - 1, 0\rangle. \qquad (7.2.7b)$$

From (7.2.6) and the fact that $L_0^0(x) \equiv const.$ it can be seen that $\langle \vec{r} | S_n^{(\rightarrow)} \rangle$ is localized around $r \approx z$, whereas $\langle \vec{r} | S_n^{(\leftarrow)} \rangle$ is localized around $r \approx -z$. Thus, the extremal Stark states are quasi-one-dimensional highly elongated states. This is illustrated in Fig. 7.4, which shows $|\langle \vec{r} | S_{40}^{(\rightarrow)} \rangle|^2$ as a three-dimensional plot in the $x - z$ plane.

In atomic units the hydrogen atom in a linearly polarized microwave field of frequency ω and field strength $\mathcal{E}$ is given by

$$\hat{H} = \hat{H}_0 + \mathcal{E} \, z \, g(t) \, \sin(\omega t), \qquad (7.2.8)$$

where

$$\hat{H}_0 = -\frac{1}{2} \Delta - \frac{1}{r} \qquad (7.2.9)$$

is the unperturbed hydrogen Hamiltonian and $g(t)$ is an envelope function which can be used to describe the switch-on and switch-off stages in a microwave ionization experiment. While $\hat{H}_0$ is diagonal in the Stark

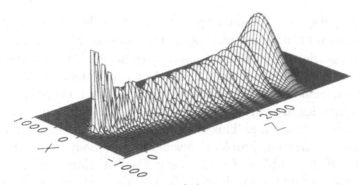

Fig. 7.4. The extremal Stark state $|S_{40}^{(\rightarrow)}\rangle$ as a three-dimensional plot of the spatial probability $|\langle \vec{r}|S_{40}^{(\rightarrow)}\rangle|^2$ in the $x-z$ plane. (From Blümel and Smilansky (1987).)

basis (7.2.3) this is not true for the matrix elements of z. Only m is a good quantum number. Thus (7.2.8) is a time dependent problem in two dimensions. Although such problems can nowadays be solved using powerful computers (see, e.g., Buchleitner (1993), Buchleitner and Delande (1993)) this is not necessary here since at this point we are not interested in high accuracy calculations, but in a demonstration of the basic physical mechanisms underlying the ionization process. A detailed inspection of the magnitudes of the dipole coupling matrix elements $\langle n; n_1, n_2, m|z|n'; n_1', n_2', m\rangle$ shows that the most effective pathway to ionization proceeds via the extremal Stark states $|S_n^{(\leftrightarrow)}\rangle$ (Jaeckel (1988)). Therefore, we restrict the two-dimensional space of Stark states (7.2.3) to the one-dimensional space of extremal Stark states. In order to proceed, we need the dipole coupling matrix elements. The diagonal matrix element is given by

$$\langle S_n^{(\rightarrow)}|z|S_n^{(\rightarrow)}\rangle = \frac{3}{2}n(n-1). \qquad (7.2.10)$$

The dipole matrix element (7.2.10) can be compared with the corresponding dipole matrix element for the SSE model,

$$< n|z|n > = \frac{3}{2}n^2. \qquad (7.2.11)$$

For high quantum numbers n, the matrix element (7.2.11) is very close to the matrix element (7.2.10). This fact is consistent with the elongated nature of the extremal Stark states (see Fig. 7.4), which renders them quasi-one-dimensional as discussed above. This correspondence holds for the off-diagonal matrix elements as well. Therefore, and as far as the dipole matrix elements are concerned, the SSE states are good approximations to the extremal Stark states. Since both the SSE states and the

Stark states diagonalize their respective $\hat{H}_0$ with the result $-1/2n^2$, we can approximate the three-dimensional Hamiltonian $\hat{H}$ defined in (7.2.8) by a one-dimensional SSE Hamiltonian by replacing $\hat{H}_0$ in (7.2.8) with the unperturbed SSE Hamiltonian (6.1.22). As a consequence of this approximation we conclude that for high quantum numbers n the SSE dynamics and the dynamics of hydrogen Rydberg atoms in extremal Stark states are qualitatively the same. This is an important result since it simplifies the hydrogen ionization problem considerably without losing the essential physics of the problem. Consequently the SSE Hamiltonian provides the theoretical basis for much of the analytical and numerical research work done during the 1980s and early 1990s on the microwave ionization problem.

The quality of the SSE approach to the hydrogen ionization problem is demonstrated in Fig. 7.3, where microwave ionization probabilities computed on the basis of the SSE model (full lines) can be compared with the experimental results (dashed lines). The spiky full lines in 7.3 were obtained by solving the time dependent Schrödinger equation of the SSE model with the help of the Fredholm approach discussed in Section 6.2.2 (for more details see Blümel and Smilansky (1987)). Although the numerical results are close to the experimental ionization probabilities, the spikes in the theoretical ionization probabilities are absent in the experimental data, whose behaviour is much smoother. This apparent discrepancy is not yet fully understood, but some of the smoothing observed in the experiments may be due to the higher dimensionality of the experimental situation which makes many more states available for transitions. Part of the smoothing is certainly due to the fact that the electromagnetic fields in the interaction region vary in space, especially if the interaction field is produced by a standing wave in a microwave cavity. Therefore, the finite width of the atomic beam introduces an effective average over the interaction field strength. Taking this effect into account and averaging the spiky results in Fig. 7.3 over the experimental field distribution yields the smooth full lines in Fig. 7.3 as the final theoretical result to be compared with experiment. Fig. 7.3 shows that the smoothed numerical results and the experimental ionization signals are astonishingly close given the simplification introduced by reducing the three-dimensional Hamiltonian (7.2.8) to a one-dimensional problem. Even some prominent structures in the ionization curves, especially for $n_0 = 37$ and $n_0 = 38$, are adequately reproduced. But the most important result is the ability of the one-dimensional SSE model to reproduce the ionization thresholds. Whatever the physical reason for the thresholds, the numerical calculations show that it is contained within the one-dimensional model. In Chapter 6 we saw that the SSE model allows for a transition to global chaos if the applied microwave field exceeds a critical field strength. Since the onset of chaos is the most drastic occurrence in phase space as a function of increased field strength, it is tempting to identify the microwave ionization thresholds with chaos thresholds. It is shown in the following section that this physical picture is consistent with the experimental results.

7.3 Ionization thresholds as chaos thresholds

In this section we show that the ionization thresholds in Fig. 7.3 can be interpreted as chaos thresholds. In order to be specific, we focus on the case $n_0 = 36$ shown in Fig. 7.3(a). For both the experiment and the quantum mechanical computations the ionization signal starts to rise monotonically from about $\mathcal{E} = 350$ V/cm on. There is small (in the quantum mechanical computations) to negligible (in the experiments) ionization at $\mathcal{E} = 330$ V/cm. Strong ionization is observed at $\mathcal{E} = 380$ V/cm. In order to prove that the ionization threshold in Fig. 7.3(a) corresponds to a chaos threshold, we used the classical one-dimensional SSE model and computed Poincaré sections of the classical dynamics of the hydrogen Rydberg electron in the vicinity of the classical action $I \approx 36$ for a driving frequency of 9.92 GHz and (a) $\mathcal{E} = 330$ V/cm, (b) $\mathcal{E} = 350$ V/cm and (c) $\mathcal{E} = 380$ V/cm. The results are shown in Fig. 7.5. Since for the hydrogen atom integer values of I correspond to the principal quantum number n, Fig. 7.5 is of direct relevance for the quantum dynamics. Therefore, in what follows, I and n are used interchangably.

First we analyse Fig. 7.5(a). The phase space is regular for actions in the immediate vicinity of $I = 36$. In particular we notice the existence of invariant lines at actions $I > 36$ that shield a Rydberg electron initially placed in $I = 36$ from ionization. As a result, no classical ionization is possible for $\mathcal{E} = 330$ V/cm. This result is consistent with the behaviour of the ionization curves shown in Fig. 7.3(a).

Next we analyse Fig. 7.5(b). We see that the sealing lines at $I > 36$ are now destroyed and chaotic regions exist at $I = 36$. This means that parts of the phase space of a Rydberg electron prepared in $I = 36$ are now free to diffuse chaotically to the continuum. Thus, $\mathcal{E} = 350$ V/cm marks the onset of ionization in the classical model.

For $\mathcal{E} = 380$ V/cm Fig. 7.5(c) shows that the starting action $I = 36$ is in the middle of a chaotic sea. Practically all the classical phase-space probability started out at $I = 36$ is now free to diffuse to the continuum, resulting in large ionization. This result, again, is consistent with the experimental data.

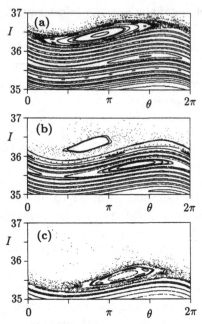

Fig. 7.5. Phase-space portraits of the driven one-dimensional hydrogen atom for $f = 9.92\,\mathrm{GHz}$ and $\mathcal{E} =$ (a) 330 V/cm, (b) 350 V/cm and (c) 380 V/cm.

On the basis of the phase-space portraits presented in Fig. 7.5, we conclude that the ionization thresholds displayed in Fig. 7.3 do indeed correlate with the onset of chaos. The equation "ionization thresholds = chaos thresholds" is therefore justified.

The numerical results give us the confidence to attempt an analytical calculation of critical ionization fields. This can be done with some success by computing the widths of the resonances apparent in Fig. 7.5 and using the widths as input to Chirikov's overlap criterion as discussed in Section 5.2. The analytical method allows us to compute critical ionization fields for many initial conditions n_0 and field parameters $\mathcal{E}$ and ω without the need to inspect a sequence of Poincaré sections in each particular case. Presently, however, the available analytical methods are not very accurate. For rough estimates of classical critical ionization fields, however, the currently available analytical techniques are very useful.

The analytical computation of critical ionization fields starts with an analysis of the widths of resonances in the classical phase space. Resonances occur whenever the ratio of the external driving frequency ω and the unperturbed Kepler frequency Ω of the Rydberg electron is rational, i.e.

$$\frac{\omega}{\Omega} = \omega I_{KM}^3 = \frac{K}{M}, \tag{7.3.1}$$

where K and M are relatively prime integers. The resonance (7.3.1) is called a K/M type resonance. The action I_{KM} in (7.3.1) is the location of the K/M type resonance. Using the methods discussed in Section 5.2 it was shown recently (Blümel (1994b)) that the width w_{KM} of a K/M type resonance is given by

$$w_{KM} = \frac{4 I_{KM}\sqrt{2}}{3}\left\{ x_K J_M(3M\mathcal{E} I_{KM}^4) + \right.$$

$$\left. K^2 \sum_{\gamma(p,q,r,s)} \frac{q s x_p x_r}{[Mp - Kq]^2} J_q\left(\frac{3Mp\mathcal{E} I_{KM}^4}{K}\right) J_s\left(\frac{3Mr\mathcal{E} I_{KM}^4}{K}\right) \right\}^{1/2},$$

(7.3.2)

where

$$x_k = J_k'(k)/k,$$

(7.3.3)

and $J_n(x)$ are Bessel functions of the first kind. The summation in (7.3.2) is over all positive integers p, q, r, s which satisfy the following condition

$$\gamma: \quad \frac{p+r}{q+s} = \omega I_{KM}^3 = \frac{K}{M}, \quad (p,q) \neq (K,M).$$

(7.3.4)

For $\mathcal{E} \to 0$ and fixed K, M (7.3.2) can be expanded to yield the following simpler expression

$$w_{KM} = \frac{4 I_{KM}}{3\sqrt{\pi}}\left\{ x_K \sqrt{\frac{2\pi}{M}} + \right.$$

$$\left. K^2 \left(\frac{M}{K}\right)^M \sum_{\gamma(p,q,r,s)} \frac{x_p x_r \sqrt{qs}}{[pM - qK]^2}\left(\frac{p}{q}\right)^q \left(\frac{r}{s}\right)^s \right\}^{1/2} \left(\frac{3e\mathcal{E} I_{KM}^4}{2}\right)^{M/2}.$$

(7.3.5)

In the adiabatic regime, $\omega n_0^3 < 1$, where detailed experimental data are available, the exponent $M/2$ in (7.3.5) is typically a large number. This feature makes the application of Chirikov's resoance overlap criterion a promising procedure for calculating the classical critical ionization fields . This is so because, due to the high powers of $\mathcal{E}$, the resonances "explode" in size close to the critical field. Therefore, the critical fields in the adiabatic regime appear to be well defined. This feature, again, correlates well with experimental findings. As shown in Fig. 7.3 the ionization thresholds in the adiabatic regime are steep functions of the field strength. In contrast to the behaviour in the adiabatic regime, experiments show that for typical exposure times of a few hundred to a few thousand microwave cycles the ionization curves in the nonadiabatic regime, $\omega n_0^3 > 1$, are very shallow, making it hard to define the "critical field" on the basis of experimental ionization data Even in the adiabatic regime, and for fixed exposure time, the thresholds are expected to become more shallow as the scaled frequency ωn_0^3 increases toward 1. In this case the denominator M of the resonances decreases and the power $M/2$ in the resonance widths (7.3.5) decreases accordingly, with the result that the resoance overlap is "slower" in the field. This feature agrees with numerical calculations and experimental observations.

We discuss now an analytical procedure for the calculation of critical ionization fields that is based on Chirikov's overlap criterion. For a given microwave frequency ω we first compute the locations I_{1M} of all $1/M$ type resonances. We then determine a list of fields, $\{\mathcal{E}_M\}$, defined as the critical fields of overlap between the $1/M$ type resonance and the $1/(M-1)$ type resonance, i.e. $\mathcal{E}_M$ is determined from the condition

$$\frac{1}{2}\left[w_{1M} + w_{1,M-1}\right] = I_{1M} - I_{1,M-1}. \tag{7.3.6}$$

The smallest field $\mathcal{E}_M$ defines the critical field $\mathcal{E}$ for ω. The heavy full line in Fig. 7.6 shows the resulting scaled critical fields $\mathcal{E}n_0^4$ for scaled microwave field strengths ωn_0^3 ranging from 0.1 to 0.5. This result can be compared with 10% ionization fields determined experimentally (full squares) and critical ionization fields determined numerically (bullets) from a detailed analysis of Poincaré sections. There is good qualitative agreement between the analytical model and the numerical and experimental results.

On summarizing the material presented in this section we obtained two major results. (a) With the help of numerical simulations we proved the equivalence of ionization thresholds and chaos thresholds in the regime $\omega n_0^3 < 1$; and (b) we formulated an analytically solvable model which predicts threshold fields in close agreement with experimental results. This model, however, should be considered only as a first step for a more complete understanding of the behaviour of the critical fields. For instance, closer inspection of Fig. 7.6 shows that the critical fields predicted by the model have the wrong slope. In the limit ($\omega \to 0$), e.g., the model predicts a static ionization field of $\mathcal{E}n_0^4 = 2/3e \approx 0.25$. This is about a factor 2 higher than the exact static ionization field $\mathcal{E}_s n_0^4 \approx 0.13$. We are confident that this problem can be solved by a more detailed analysis of phase-space features. Thus, the obvious deficiencies in the present state of the analytical analysis of critical ionization fields should not distract from the central result obtained in this section, namely the relevance of chaos for the physics of hydrogen Rydberg atoms in strong radiation fields.

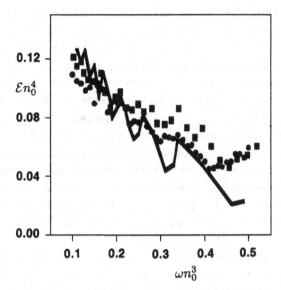

Fig. 7.6. Numerically computed scaled critical fields (bullets) and experimental 10% threshold fields (squares) (from Moorman and Koch (1992)) as a function of scaled frequency. The full line represents critical fields computed on the basis of Chirikov's overlap criterion. (From Blümel (1994b).)

7.4 High frequency localization of chaos

In the previous sections we saw that, at least as far as their ionization behaviour is concerned, hydrogen Rydberg atoms in a strong microwave field behave very similarly to the kicked rotor discussed in Chapter 5. One of the most interesting findings in connection with the kicked rotor is the existence of a localization mechanism that suppresses the chaotic diffusion. This result immediately invites two questions: Does quantum localization exist for microwave-driven Rydberg atoms as well? Can it be measured experimentally? The first question was answered in the affirmative by Casati *et al.* (1987). On the basis of numerical and analytical quantum calculations, Casati and collaborators predicted the existence of localization in the microwave-driven hydrogen atom. In the atomic physics context localization means that the time dependent wave function of a hydrogen Rydberg atom in the diffusive, chaotic regime is not as extended over the principal quantum numbers of the unperturbed atom as would be expected on the basis of the strong chaotic diffusion that occurs in the classical analogue of the atom. In other words, while for a certain microwave field strength the classical atom shows a transition to global chaos and strong diffusion ionization, the wave function of the "quantum atom", due to the inhibiting effect of localization, falls short of reaching the continuum threshold, unable to transport the electron into the continuum. This observation can be used for the prediction of a measurable signature for the presence of localization: compared with the classical predictions for the onset of ionization, the onset of ionization in the presence of localization is *delayed*; i.e. in the presence of localization strong ionization should be observed at considerably higher field strengths than is predicted by classical trajectory calculations. Following the predictions of Casati *et al.* (1987), Galvez *et al.* (1988) and Bayfield *et al.* (1989) were indeed able to measure delayed ionization that is consistent with the presence of localization in microwave-perturbed hydrogen Rydberg atoms. The localization effect is most pronounced in the nonadiabatic regime of hydrogen ionization, i.e. for $\omega > \Omega$, where ω is the microwave frequency and Ω is the natural Kepler frequency of the atom. This answers the second question asked above: yes, the localization effect is indeed measurable experimentally.

Following the discussion in Section 5.5 concerning the influence of symmetry breaking on the localization we can ask yet a third question: Are the effects of symmetry breaking observable in microwave-driven hydrogen Rydberg atoms? At the time of writing neither theoretical nor experimental work exists addressing this question. We feel, however, that this question is important, and recommend it for future theoretical and experimental investigation.

7.5 Bichromatic drive: prediction of a "giant resonance"

We conclude our discussion of the physics of the chaotic microwave-driven hydrogen atom with some recent theoretical and experimental developments in bichromatically driven hydrogen Rydberg atoms. Bichromatic drive is interesting from a theoretical point of view since none of the periodically perturbed quantum systems so far investigated in the literature yield any indication for genuine quantum chaos (see Chapter 4). As a "way out" theorists suggested using two-frequency perturbations of a quantum system in order to destabilize it and drive it into chaos. Two-frequency perturbation of the kicked rotor is suggested and discussed in detail by Chirikov *et al.* (1981). Some important investigations were also reported by Samuelides *et al.* (1986). As the main result, these researchers indeed obtained an enhanced quantum diffusion and a broadening of the frequency response of the kicked rotor.

Two-frequency irradiation of a two-level atom was proposed by Pomeau *et al.* (1986). As compared with one-frequency irradiation, a rapid decay of correlations, indicative of true quantum chaos, was observed. However, for the particular choice of parameters in the paper by Pomeau *et al.* (1986), Badii and Meier (1987) were able to demonstrate that the response is not chaotic, but quasi-periodic, albeit on a very long time scale. Although for the particular parameters chosen by Pomeau *et al.* (1986) a careful analysis of the quasi-periodically driven two-level atom did not reveal a chaotic response, the issue of whether, in general, a quasi-periodic (two-frequency) perturbation of a quantum system may result in a chaotic response is still under investigation.

In order to contribute to this discussion, two-frequency experiments with microwave-driven hydrogen Rydberg atoms were performed by Koch *et al.* (see, e.g., Moorman and Koch (1992)) and were analysed with the help of numerical quantum calculations within the one-dimensional model by Blümel *et al.* (1989b). The focus here was not on investigating the temporal behaviour of the occupation probabilities of high-lying Rydberg states (which may or may not have revealed chaotic features), but to investigate differences in the ionization behaviour of one-frequency and two-frequency perturbations. It was shown theoretically that, over a finite interaction time, a two-frequency perturbation can be approximated effectively by a one-frequency perturbation, since, within the frequency uncertainty resulting from the finite experimental interaction time, any pair of frequencies can be replaced by two frequencies that are rationally related. Therefore, in order to contribute on a fundamental level to the quantum chaos discussion, controlled experiments with much larger exposure times have to be performed in order to reduce the frequency uncertainty and thus render the ratio of the two frequencies more "irrational". It is clear that any experiment is by necessity carried out in a finite time, and the "irrational" ratio remains elusive. Nevertheless, by increasing the exposure time (and at the same time keeping the destructive effects of technical noise under control) the ideal of an irrational ratio can be approached experimentally.

In the following we do not concentrate on the temporal aspect of two-frequency irradiation, but on the prediction of a new observable atomic physics effect that relies essentially on the simultaneous presence of two frequencies and the possibility of chaos occurring in the hydrogen atom's phase space. It is demonstrated that the presence of a second frequency can be used to produce interspersing resonances that "short out" previously nonoverlapping phase-space resonances. Thus, Chirikov's idea of resoance overlap finds here a technical application: resonance engineering. More precisely, this effect works in the following way. The first microwave field sets up two nonoverlapping resonances. Let us call these two nonoverlapping resonances R_1 and R_2. On setting up a third resonance S by the application of a second microwave field with a different frequency, we are able to place S between R_1 and R_2 and thus achieve a global overlap between the three resonances. In this situation the interspersed resonance S acts like an effective bridge for the phase-space flow. The bridge may be turned into a "switch" with the help of which phase-space diffusion, and thus ionization, can be switched on and off. This is done by controlling the location and width of the interspersed resonance S. This is possible since the width of a resonance and its location in action are controlled by the field strength and the frequency, respectively. Because we have two fields at our disposal, we have yet another option: controlling the location of S with respect to R_1 and R_2 with the help of the relative phase between the two fields. This three-parameter control gives a wide variety of possibilities for optimally placing S. The three "degrees of freedom" can be used to define three different ways of achieving switching.

(a) By placing S right between R_1 and R_2, an effective switch can be set up by varying the field strength of the second field. Increasing the field strength connects R_1 and R_2 by "blowing up" the width of S. Decreasing the field again disconnects R_1 and R_2.

(b) Another type of switch may be constructed by choosing the field strength of the second field to be large enough so that the width of S is large enough to connect R_1 and R_2. But because of the "eye-shape" of resonances, by an appropriate choice of the microwave frequency and the relative phase φ, the resonance S may be located between R_1 and R_2 as far as the action is concerned, but shifted in angle to the side of R_1 and R_2 such that no overlap occurs. By varying the relative phase φ between the first and second microwave fields, the resonance S moves horizontally in phase space and may be brought to lie between R_1 and R_2, where it can connect them. Thus, variations in φ can also define a switching action.

(c) A third obvious way to connect R_1 and R_2 is by again preparing a wide resonance that in principle could connect R_1 and R_2. But this time we choose the frequency such that S is located above or below (in action) the pair of resonances R_1 and R_2. By increasing or decreasing the second frequency, S can be brought right in between R_1 and R_2, thus again connecting them. In this way, the frequency, too can be used to define a switch.

The mechanisms (a) and (c) are discussed in more detail below. Theoretical and experimental results are currently available that confirm the existence and validity of the two resonance switching mechanisms. Mechanism (b) has not yet been tested, either theoretically or experimentally. Of course, by simultaneously changing two or all three parameters, any combination of (a) – (c) can be used to define a "generalized" switch. Depending on how fast the field parameters can be changed experimentally, the three fundamental switches (a) – (c) and their generalizations are potentially very fast and may find some interesting technical applications.

The three switches defined above rest on the mechanism of interspersed resonances, which was first suggested theoretically by Howard (1991). In the following we demonstrate the action of such a switch in the case of bichromatically driven hydrogen Rydberg atoms. It results in the prediction of a new kind of ionization peak in the microwave ionization of hydrogen Rydberg atoms. Recently performed experiments indicate that the effect actually exists.

We proceed now to a discussion of this new ionization peak which occurs in the classical and quantum dynamics of bichromatically driven hydrogen Rydberg atoms. As we shall see below, the ionization peak turns out to be rather broad. In analogy to similarly broad structures that can be observed in the photo-absorption of γ rays in heavy nuclei (see, e.g., Ring and Schuck (1980)) we call this structure a "giant resonance". We emphasize that, although the nuclear giant resonances and the giant ionization structure to be discussed here exhibit a striking visual similarity, the underlying physics of both phenomena may be very different.

The prediction of the giant resonance in microwave ionization starts from the one-dimensional Hamiltonian

$$\hat{H}(t) = \hat{H}_0 + xg(t)\Big[\mathcal{E}_1 \sin(\omega_1 t) + \mathcal{E}_2 \sin(\omega_2 t + \varphi)\Big], \qquad (7.5.1)$$

where $\hat{H}_0$ is defined in (6.1.22), φ is the phase difference between the two microwave modes with frequencies ω_1 and ω_2, respectively, and $g(t)$ is the envelope function introduced in (7.2.8). The frequencies $\omega_{1,2}$, as well as the field strengths $\mathcal{E}_{1,2}$ and time t, are in atomic units. For easier comparison with experiment we also introduce the microwave frequencies $\nu_{1,2} = \omega_{1,2}/2\pi$ expressed in GHz and the microwave field strengths $F_{1,2}$ expressed in V/cm that correspond to the fields $\mathcal{E}_{1,2}$ in atomic units.

In order to illustrate the ionization mechanism, we first analyse (7.5.1) within the framework of classical mechanics. We choose actions, frequencies and field strengths in a range that is accessible to existing experiments currently performed at Stony Brook (Moorman and Koch (1992), Koch and van Leeuwen (1995)). Since we are aiming at a comparison of the classical results with quantum computations and experiments, it is advantageous to transform (7.5.1) to action-angle representation, where the action I and the angle variable θ are defined in (6.1.11) and (6.1.16), respectively. The bound-space dynamics of (7.5.1) can be visualized in the (θ, I) phase space by choosing several initial conditions (θ_0^j, I_0^j), $j = 1, 2, ..., N$, and marking them as well as their M iterates after time $t_m = m\Delta t$, $m = 1, 2, ..., M$ by a dot. For $\nu_1 = 25\,\text{GHz}$, $\nu_2 = 35\,\text{GHz}$, $\Delta t = 8.26 \times 10^6$, $N = 27$ and $M = 100$, the result is shown in Fig. 7.7. In this figure we see four resonance islands corresponding to three resonances that are caused by either ω_1 or ω_2 being in resonance with the unperturbed Kepler motion of the Rydberg electron. The resonance condition is $\omega_i I^3 = r_i/s_i$, where r_i and s_i are integers. We see that for $r_2 = s_2 = 1$ the structure at $I_1 \approx 57.3$, as well as the two structures at $I_2 \approx 72.2$ (r_2=2, $s_2 = 1$), are due to the driving frequency ω_2, while the interspersed resonance at $I_I \approx 64.1$ ($r_1 = 1$, $s_1 = 1$) is due to ω_1.

We consider now the case of an arbitrarily chosen frequency ν_1. For ionization to occur it is necessary that the interspersed resoance overlaps with *both* the $I_2 \approx 72.2$ and the $I_1 \approx 57.3$ resonances. The position of the interspersed resonance is given by $I_I = 187.35/\nu_1^{1/3}$. With (7.3.2) we obtain the following two conditions for resoance overlap:

(i) $68.8 \, [g(t)F_1]^{1/2}/\nu_1 + 2.0 \, [g(t)F_2]^{1/2} > I_I - I_1$;

(ii) $68.8 \, [g(t)F_1]^{1/2}/\nu_1 + 2.2 \, [g(t)F_2]^{1/2} > I_2 - I_I$.

Depending on the shape of the envelope function $g(t)$ and the field strengths F_1 and F_2, the conditions (i) and (ii) may, or may not, be simultaneously fulfilled. Ionization is expected to occur only if both (i) and (ii) are fulfilled. The overlap condition depends essentially on ν_1. Thus, as ν_1 is swept from small values to large values, overlap can be achieved, and lost again, giving rise to a broad ionization structure. A first qualitative analysis of this structure has already been achieved on the basis of (i) and (ii). The decay to the continuum is approximated by an exponential decay with decay constants determined from a classical Monte Carlo calculation. The decay is assumed to start as soon as (i) and (ii) are fulfilled. On the basis of this model Haffmans *et al.* (1994) obtained the ionization probabilities as a function of ν_1 shown as the

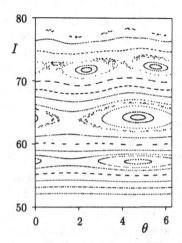

Fig. 7.7. Phase-space resonances for bichromatically driven hydrogen Rydberg atoms within the one-dimensional SSE model. Parameters: $g(t) = 1$, $\varphi = 0$, $\nu_1 = 25 \, \text{GHz}$, $F_1 = 1 \, \text{V/cm}$, $\nu_2 = 35 \, \text{GHz}$, $F_2 = 0.5 \, \text{V/cm}$.

full line in Fig. 7.8. The frequency ν_1 is swept, and ν_2, F_1 and F_2 are fixed at 34.998 GHz, 14.5 V/cm and 8.2 V/cm, respectively. The sharp cut-off of $P_{ion}(\nu_1)$ at $\nu_1 \approx 33$ GHz is due to the fact that according to the model the interspersed resonance no longer overlaps with the resonance at $I_2 \approx 72.2$. The centre of the ionization structure occurs when for the given shape of the envelope function the conditions (i) and (ii) are fulfilled for the longest time. This is the case at $\nu_1 \approx 25$. The left wing of the ionization structure is explained because the overlap time of the resonances decreases toward low ν_1.

Fig. 7.8 also shows the results of a classical calculation and a quantum calculation that both confirm the prediction of the giant resonance based on the simple overlap criterion discussed above. The crosses in Fig. 7.8 are the results of classical Monte Carlo calculations. They were performed by choosing 200 different initial conditions in the classical phase space at $I_0 = 57$. The "ionization" probability in this case was defined as the excitation probability of actions beyond the cut-off action $I_c = 86$. This definition is motivated by experiments that, due to stray fields and the particular experimental procedures, cannot distinguish between excitation above $I_c > 86$ and "true" ionization, i.e. excitation to the field-free hydrogen continuum. The crosses in Fig. 7.8 are close to the full line and thus confirm the model prediction. The open squares are the results of quantum calculations within the one-dimensional SSE model. The computations were performed in the simplest way, i.e. no continuum was

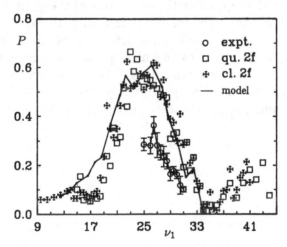

Fig. 7.8. "Ionization" probabilities of bichromatically driven hydrogen Rydberg atoms as a function of ν_1 obtained by various methods. Full line: resoance overlap model; crosses: classical Monte Carlo simulations; open squares: quantum calculations; open circles with error bars: experimental results. (Adapted from Haffmans *et al.* (1994).)

coupled to the SSE bound states and ionization was again defined as excitation above the cut-off action $n_c = I_c = 86$. Fig. 7.8 shows that all three, the quantum calculation as well as the classical Monte Carlo simulation and the simple overlap model, consistently predict the existence of the giant resonance structure.

In order to check the theoretical prediction, an experiment was performed (Haffmans et al. (1994)). A beam of hydrogen atoms with principal quantum number $n_0 = 57$ entered a wave guide in which the atoms interacted with a superposition of two microwave fields with field strengths and frequencies chosen to be equal to the values used in the theoretical analysis. The "ionization" probability, again defined as excitation above the critical action $I_c = 86$ (i.e. above principal quantum number $n_c = 86$, including the field-free continuum) is depicted by the open circles with error bars in Fig. 7.8. The absolute values of the experimental ionization probabilities are below the model predictions. This is expected, since the theoretical analysis was performed within the one-dimensional SSE model, which is known to overestimate the experimental ionization probabilities. This is due to the fact that the SSE model describes the low angular momentum states in the experimental sub-state distribution, and it is those states that are the most easily ionized. But since the experimental initial state also contains higher angular momentum states that are harder to ionize, the SSE model gives the low angular momentum states an undue weight, which results in an overestimation of the ionization probability. Apart from this "renormalization" problem, the experimental results reproduce the right wing of the giant resonance predicted above. Unfortunately, for the time being, the experiments cannot go lower in frequency to explore the left wing of the giant resonance. This, however, is only due to technical problems, and data points for lower frequencies may be forthcoming. We hope that experiments at lower frequencies will confirm qualitatively the shape of the left wing of the resonance, and thus establish the existence of the entire structure.

8

The kicked hydrogen atom

In Chapters 5 – 7 we studied the onset of global chaos and its various manifestations in atomic and molecular systems. It was shown that in the kicked molecule (Section 5.4) the onset of chaos is responsible for population transfer to highly excited rotational states. A similar effect is active in microwave-driven surface state electrons and hydrogen Rydberg atoms where the onset of chaos results in strong ionization. But so far the focus has been on the computation of critical strengths and control parameters, whereas the ionization signal was reduced to play a secondary role as a probe, or an indicator for the onset of chaos. In this chapter we shift the focus to the investigation of the ionization signal itself, especially its time dependence.

The time dependence of weakly ionizing systems that are well described by a multi-photon process of order p has been studied extensively in the literature. In this case the time dependence of the ionization signal does not offer any surprises. We expect exponential decay with a decay rate ρ that is proportional to the pth power of the field intensity $\mathcal{I}$ according to $\rho \sim \mathcal{I}^p$. This prediction of multi-photon theory has been verified in numerous experiments. In fact, experimentalists often use the field dependence of the ionization rates to assign a multi-photon order to an experimentally observed ionization signal. For strong fields, however, especially in situations where the associated classical system is globally chaotic, the ionization signal is no longer expected to be exponential. Many researchers predict an algebraic dependence of the ionization signal as a function of time (see, e.g., Karney (1983), Chirikov and Shepelyansky (1984), Meiss and Ott (1985)). Since the theoretical predictions are based on classical chaos theory, interesting cross-over effects may be observable due to the finite size of Planck's constant.

In this chapter we restrict ourselves to a classical analysis of the ionization process. Within the classical description, we ask the central question:

How does an atom ionize? Obviously, subsets of phase space are leaving for the continuum. While this is undisputed, another, more interesting, question arises: What is the nature of the sets leaving for the continuum, and what is the nature of the sets that stay bounded?

In Section 2.3 we studied the tent map, a schematic model for ionization that was able to produce fractal structures as a result of ionization. An important question is therefore whether the results presented in Section 2.3 are only of academic interest, or whether fractal structures can appear as a result of ionization in physical systems. In order to answer this question we return to the microwave-driven one-dimensional hydrogen atom. As we know from the previous chapter, this model is ionizing and realistic enough to qualitatively reproduce measured ionization data. Therefore this model is expected to be a fair representative for a large class of chaotic ionization processes.

We investigate now whether the microwave ionization process is capable of producing a fractal in the (θ, n) phase space of the one-dimensional model. We choose $\epsilon = 0.04$, $\omega = 1$ and a rectangular region $\mathcal{R}$ in phase space, defined by $\mathcal{R} = [0, 2\pi] \times [0.9, 1.1]$. We cover $\mathcal{R}$ with a regular mesh of 137 500 points defined by $(\theta_j, n_k) = 2\pi(j - 1/2)/550$, $j = 1, 2, ..., 550$, and $n_k = 0.9 + (k - 1)/249$, $k = 1, 2, ..., 250$. In what follows these points will serve as initial conditions for the time evolution of $\mathcal{R}$ under the dynamics induced by the microwave field. We also choose a cut-off action $n_{cut} = 1.2$, and call points with $n > n_{cut}$ "ionized". This kind of ionization can be achieved experimentally by applying a static electric field. We now expose the collection of phase-space points in $\mathcal{R}$ to the microwave field and ask the question: What is the set of points ionized (not ionized) after N cycles of the field? The answer is presented in Fig. 8.1. It consists of three frames, (a), (b) and (c), that correspond to $N = 10$, 50, and 100, respectively. The white regions in Fig. 8.1 show the set of initial conditions that are ionized after N cycles, the black regions correspond to initial conditions not ionized after N cycles. The black phase-space regions shown in Fig. 8.1 appear to be the initial stages of a fractal set. This is verified upon closer inspection. The succession of frames in Fig. 8.1 indicates that the ionization mechanism defines a fractal generating mechanism. For $N \to \infty$ the black regions in Fig. 8.1 are expected to converge to a fractal set representing the points in phase space that never ionize. The set of never-ionizing points was called $\Lambda^{(+)}$ in Section 2.3. While the sets $\Lambda^{(+)}$ discussed in Section 2.3 are relatively simple fractals, Fig. 8.1 shows that $\Lambda^{(+)}$ for the microwave ionization problem is very complicated. It consist of contiguous regions with fractal edges, and of fractal regions of zero measure. Thus, we have the answer

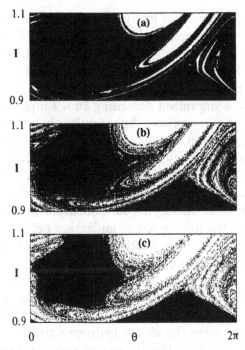

Fig. 8.1. Three successive stages in the generation of a phase-space fractal for the one-dimensional microwave-driven hydrogen atom. The black areas in frames (a)–(c) represent initial conditions in phase space that do not ionize after (a) $N = 10$, (b) $N = 50$ and (c) $N = 100$ microwave cycles.

to our central question. Indeed, physical models exist which show fractal phase-space structures as a result of ionization.

The time dependence of the ionization signal for the microwave-driven hydrogen atom was investigated by Lai *et al.* (1992a,b). Powerlaw decay of the phase-space probability was found to be in agreement with predictions based on model calculations. The reason for the powerlaw decay is that phase-space points sometimes "hide" in the fractal crevices at the edges of stable islands, and are thus "invisible" to the microwave perturbation for some time. A decay slower than exponential results. This mechanism is most effective for long-lived phase-space points. Therefore, the powerlaw behaviour is expected to be most pronounced in the asymptotic region of long exposure times.

Although illustrative, the phase-space structures of the microwave-driven hydrogen atom are very hard to analyse analytically. Therefore, we simplify the problem by replacing the smooth microwave drive by impulsive kicks in analogy to the procedure used to arrive at the kicked rotor model in Chapter 5. The resulting model, investigated in detail in the following sections, is called the kicked hydrogen atom. The model is defined in Section 8.1. It is shown that its dynamics can be reduced to a mapping which is an important simplification that allows various analytical techniques to be applied successfully to the analysis of the model. The decay properties of the kicked hydrogen atom are studied in Section 8.2. Although in certain parameter regimes the phase space does not possess any stable islands and is organized according to a simple fractal, the model ionizes according to a powerlaw. Apparently, the phase-space fractal of the kicked hydrogen atom cannot be a self-similar scaling fractal, since otherwise, as we saw in Section 2.3, exponential decay is expected. It is shown in Section 8.2 that the hydrogen fractal is a nonscaling "exotic" fractal, which explains the powerlaw decay. Thus, the precise nature of the phase-space fractal underlying an ionization process plays a major role in determining the nature of the ionization process.

In Section 8.3 we present an analytical proof for the absence of regular islands for a certain choice of control parameters. The mere possibility of such a proof is astonishing. This is so because many physical models are currently discussed in the literature for which an absence of islands is suspected (see, e.g., Gutzwiller (1990)), but have so far defied rigorous proofs. To our knowledge the kicked hydrogen atom is the only physical model for which a proof for the absence of regularity exists. Finally, we mention that the kicked hydrogen atom is not only of academic interest. It can be realized experimentally according to the same schemes outlined in Section 5.4 in the context of the kicked CsI molecule.

8.1 The model

The Hamiltonian of the kicked one-dimensional hydrogen atom is constructed from the SSE Hamiltonian H_0 defined in (6.1.9) by adding an impulsive drive term according to

$$H(x,t) = H_0(x,p) - \beta x \delta_{2\pi}(\omega t), \qquad (8.1.1)$$

where β is the strength of the drive, ω is the kick frequency and $\delta_{2\pi}$ is the 2π-periodic δ function defined by

$$\delta_{2\pi}(\omega t) = \frac{1}{2\pi} \sum_{m=-\infty}^{\infty} e^{im\omega t} = \frac{1}{2\pi}\left\{1 + 2\sum_{m=1}^{\infty} \cos(m\omega t)\right\}. \qquad (8.1.2)$$

The electron is restricted to move in the half-space $x > 0$. There is a totally reflecting wall at $x = 0$. Since the Hamiltonian (8.1.1) of the kicked hydrogen atom and the Hamiltonian of microwave-driven surface state electrons are so similar, we can use many of the results that were derived in Chapter 6. The most important result is the transformation to action and angle variables I and θ, respectively, defined in (6.1.18). The

Hamiltonian (8.1.1) yields the following equations of motion for I and θ

$$\dot{I} = -\frac{\partial H}{\partial \theta} = \beta\, \delta_{2\pi}(\omega t)\, I^2 \cot(\eta)$$

$$\dot{\theta} = \frac{\partial H}{\partial I} = \frac{1}{I^3} - 4I\beta \sin^2(\eta)\, \delta_{2\pi}(\omega t). \qquad (8.1.3)$$

These equations are invariant under the scale transformation $I \to I/I_0$ provided that frequency, interaction strength and time are scaled according to

$$\omega \to \omega_0 = \omega I_0^3\,; \quad \beta \to \beta_0 = \beta I_0^4\,; \quad t \to t_0 = t/I_0^3. \qquad (8.1.4)$$

Exploiting the invariance of (8.1.3) with respect to the scale transformations (8.1.4), it is advantageous to introduce the scaled action

$$y = \beta I/\omega, \qquad (8.1.5)$$

the control parameter

$$\xi = \beta^3/\omega^4, \qquad (8.1.6)$$

and the scaled time

$$\tau = \xi\omega t. \qquad (8.1.7)$$

In the new variables (8.1.5) – (8.1.7), the equations of motion (8.1.3) take the form

$$\frac{dy}{d\tau} = \frac{1}{\xi} y^2 \cot(\eta)\, \delta_{2\pi}(\tau/\xi)$$

$$\frac{d\theta}{d\tau} = \frac{1}{y^3} - \frac{4y}{\xi} \sin^2(\eta)\, \delta_{2\pi}(\tau/\xi). \qquad (8.1.8)$$

It is helpful to introduce the scaled position and momentum variables X and P, respectively, defined according to

$$X = 2y^2 \sin^2(\eta)\,; \quad P = \cot(\eta)/y. \qquad (8.1.9)$$

In the new variables, and in the bounded space, the Hamiltonian (8.1.1) takes the form

$$h(\theta, y) = -\frac{1}{2y^2} - \frac{2y^2}{\xi} \sin^2(\eta)\, \delta_{2\pi}(\tau/\xi). \qquad (8.1.10)$$

It is worth mentioning that although the original problem (8.1.1) depended on the two field parameters β and ω, the scaled equations of motion (8.1.8) as well as the scaled Hamiltonian (8.1.10) depend only on the single control parameter ξ. This is a major difference compared with the problem of microwave-driven surface state electrons, but is reminiscent of the classical mechanics of the kicked rotor discussed in Chapter

5. For the kicked rotor, too, we were able to scale out one of the control parameters, leaving only the single control parameter K. A further analogy to the kicked rotor is the fact that in the case of the quantized one-dimensional kicked hydrogen atom (see, e.g., Blümel and Smilansky (1984)) a reduction to one control parameter is not possible.

Because of the impulsive nature of the δ-kick drive, the time evolution of a point (θ, y) over one cycle of the external perturbation can be written in the form of an area-preserving mapping

$$T: \quad (\theta, y) \rightarrow (\theta'', y''), \tag{8.1.11}$$

where T can be stated analytically. Since the free motion governed by H_0 is naturally treated in action-angle variables, while the result of a δ-kick is more easily expressed in coordinate-space representation, it is advantageous to present the mapping T in four steps, with two of the steps mediating between the two possible representations via the canonical transformation (8.1.9). The remaining (nontrivial) steps can be derived from the equations of motion (8.1.8), or from the equations of motion for the scaled position and momentum variables (8.1.9) given by

$$\frac{dX}{d\tau} = P, \qquad \frac{dP}{d\tau} = -\frac{1}{X^2} + \delta_{2\pi\xi}(\tau). \tag{8.1.12}$$

Starting with the phase-space point (θ, y), the following are the four steps T_j, $j = 1, ..., 4$ of the mapping $T = T_4 \circ T_3 \circ T_2 \circ T_1$ which transform (θ, y) into the image point (θ'', y'') according to (8.1.11):

(1) *Change of representation (canonical transformation):*

$$T_1: \quad (\theta, y) \rightarrow (X, P). \tag{8.1.13}$$

(2) *Kick:*

$$T_2: \quad X' = X; \quad P' = P + 1. \tag{8.1.14}$$

(3) *Change of representation (inverse canonical transformation):*

$$T_3: \quad (X', P') \rightarrow (\theta', y'). \tag{8.1.15}$$

(4) *Free motion:*

$$T_4: \quad y'' = y'; \quad \theta'' = \theta' + 2\pi\xi/y'^3. \tag{8.1.16}$$

The mapping T is not defined for all points of phase space. Ionization can occur in step (8.1.14) if the energy after the kick (8.1.14) turns out to be positive. In the scaled variables (8.1.9) the energy of the electron is given by

$$E = \frac{1}{2}P^2 - \frac{1}{X}. \tag{8.1.17}$$

With (8.1.17) the energy E' after step (2) of the mapping T is given by

$$E' = E + P + 1/2. \tag{8.1.18}$$

For $E' < 0$ the electron is still bounded and the mapping can proceed with step (3). Otherwise the electron is called "ionized" and the evolution of the trajectory is terminated. For $\xi > 0$ the termination of the trajectory is well justified. In this case an electron, once ionized, stays ionized forever. We encounter here a physical example of the never-come-back property introduced in Section 2.3. The never-come-back property holds only for $\xi > 0$. For $\xi < 0$, the termination of the trajectory of an ionized electron is only an approximation since in this case trajectories once ionized can be trapped back into the bounded space. The possibility of back-trapping complicates the exact analysis of the kicked hydrogen atom enormously. Moreover, the phase space for the case $\xi < 0$ is much more complicated (as a matter of fact "richer") than in the case $\xi > 0$. It is the existence of the never-come-back property together with a simple phase-space structure that makes the "positively kicked" hydrogen atom ($\xi > 0$) amenable to analytical investigation. Therefore, from now on, we focus on the case $\xi > 0$ only. Before proceeding with a more detailed investigation of the mapping T, we prove now the never-come-back property.

Assume that the point (θ, y) is bounded, but is ionized by the next kick. Then, $E = P^2/2 - 1/X < 0$, but $E' = P'^2/2 - 1/X > 0$ with $P' = P + 1$ according to (8.1.14). According to (8.1.18) we have $E' = E + P + 1/2$. Since $E < 0$ and $E' > 0$, $P = E' - E - 1/2 > -1/2 - E > -1/2$. Therefore, $P' = P + 1 > 1/2 > 0$. During the free-motion part following the kick the energy is conserved, but P' will change. But since there are no turning points in the continuum, we have $P > 0$ immediately prior to the next kick. Therefore, according to (8.1.18), the energy after the now following kick will increase at least by $1/2$ beyond E'. The same reasoning applies to all the following kicks. Therefore, following ionization, we have as a lower bound for the energy after N kicks:

$$E_N \geq (N - 1)/2. \tag{8.1.19}$$

Since for a return to the bounded space $E_N < 0$ for some $N > 1$ is necessary, (8.1.19) proves the never-come-back property.

According to Hillermeier et al. (1992) the four steps of the mapping T can be condensed into two steps by writing T as the product of a "kick-mapping" and a "twist-mapping" according to

$$T = T_{twist} \circ T_{kick}. \tag{8.1.20}$$

Explicitly, the kick-mapping is given by

$$
T_{kick} : \quad
\begin{cases}
y' = \dfrac{y}{\sqrt{1 - 2y\cot(\eta) - y^2}} = y\dfrac{\sin(\eta)}{\sin(\eta')} \\[4mm]
\eta' = \begin{cases}
\arcsin\left\{\sin(\eta)[1 - 2y\cot(\eta) - y^2]^{1/2}\right\}, & p' > 0 \\[2mm]
\pi - \arcsin\left\{\sin(\eta)[1 - 2y\cot(\eta) - y^2]^{1/2}\right\}, & p' < 0.
\end{cases}
\end{cases}
$$

$$(8.1.21)$$

The twist-mapping is given by

$$
T_{twist} : \quad
\begin{cases}
y'' = y' \\
\theta'' = \theta' + 2\pi\xi/y'^3.
\end{cases}
\qquad (8.1.22)
$$

There are a variety of reasons for considering the one-dimensional kicked hydrogen atom as a model for studying decay processes. A few reasons are listed here:

(i) The Hamiltonian H_0 is physical: (a) it describes approximately the dynamics of highly excited hydrogen Rydberg atoms, and (b) it is a good description of the physics of surface state electrons discussed in Chapter 6.

(ii) As discussed in Section 5.4, there exists no problem in principle in generating narrow microwave pulses. Therefore, the Hamiltonian (8.1.1) can in principle be implemented in the laboratory.

(iii) As shown above, the classical dynamics induced by the Hamiltonian (8.1.1) is generated by a relatively simple mapping. Thus, in analogy to the kicked rotor, the one-dimensional kicked hydrogen atom can be treated with analytical tools. This is demonstrated in Section 8.3, where the phase-space structure of the kicked hydrogen atom is analysed analytically for the case $\xi > 0$.

(iv) The kicked hydrogen atom is complementary to the kicked rotor in the following sense. The spectrum of the unperturbed rotor is discrete, while the spectrum of the unperturbed one-dimensional hydrogen atom possesses both a discrete and a continuous component. Thus, decay processes can be studied in a natural way.

(v) It is shown in the following section that the one-dimensional kicked hydrogen atom is not as trivial as its simple mapping description might suggest. The mapping (8.1.20) generates a nonvanishing fractal in the kicked hydrogen phase space whose analysis requires more refined tools than were introduced in Section 2.3.

8.2 Phase-space fractal and powerlaw decay

We investigate now whether the kicked hydrogen atom, in analogy with the microwave-driven hydrogen atom, is capable of generating a phase-space fractal. We focus on the rectangle $\mathcal{R} = [0, 2\pi] \times [0.9, 1.1]$ (compare with Fig. 8.1) and propagate the phase-space points in $\mathcal{R}$ for $\xi = 1$ with the mapping T defined in (8.1.20).

The black region in Fig. 8.2(a) shows the phase-space points in $\mathcal{R}$ that survive one kick, i.e. a single application of the mapping T. The black regions in Fig. 8.2(b) and (c) represent phase-space points that are not ionized after two and three kicks, respectively. It appears that the black regions in Fig. 8.2 indeed represent the first three stages in the construction of a fractal set $\Lambda_T^{(+)}$, i.e. the set of phase-space points that never ionize under the application of T. There are, however, two major differences with respect to the simple fractals encountered in Section 2.3. (i) Except for the first stage of the fractal generating process (see Fig. 8.2(a)) a whole *infinity* of subsets is deleted in every stage of the construction of $\Lambda_T^{(+)}$. This contrasts markedly with fractal generating processes of simple fractals such as Cantor's middle thirds fractal $\mathcal{C}$. In the case of

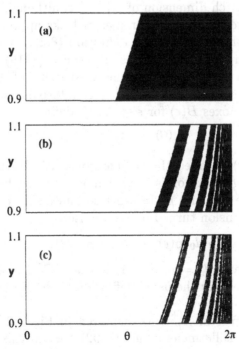

Fig. 8.2. Three stages in the construction of the one-dimensional kicked hydrogen fractal after (a) $N = 1$, (b) $N = 2$ and (c) $N = 3$ kicks.

the Cantor fractal an increasing but *finite* number of subsets is deleted
in every generation of the construction process. (ii) The fractal $\Lambda_T^{(+)}$ is
not a scaling fractal but changes its appearance slightly from generation
to generation. Since $\Lambda_T^{(+)}$ is not scaling, the methods for computing a
fractal dimension presented in Section 2.3 cannot be applied. We have to
generalize the concept of the fractal dimension. A useful approach is the
Hausdorff-Besicovitch dimension (see, e.g., Mandelbrot (1983)) defined
in the following way. The set is covered with boxes of side length ϵ. For
given ϵ one determines the number of boxes $B(\epsilon)$ necessary to cover the
set. Then, one studies the following function:

$$\mu^{(d)} = B(\epsilon)\,\epsilon^d. \qquad (8.2.1)$$

The Hausdorff-Besicovitch dimension is now defined as the critical value
d_0 at which $\mu^{(d)}$ jumps from 0 to ∞, i.e.

$$\mu^{(d)} = \begin{cases} 0, & \text{for } d > d_0 \\ \infty, & \text{for } d < d_0. \end{cases} \qquad (8.2.2)$$

The definition (8.2.2) of the Hausdorff-Besicovitch dimension can be applied to any set, fractal or nonfractal, scaling or nonscaling. Applied to
our fractal $\Lambda_T^{(+)}$, however, we obtain a paradox. It turns out that the
Hausdorff-Besicovitch dimension of $\Lambda_T^{(+)}$ is 2, although its area in phase
space converges to zero as the number of kicks approaches infinity. In
order to resolve this paradox, the definition (8.2.2) has to be extended
by including logarithmic corrections (Hausdorff (1919), Umberger *et al.*
(1986)). The idea is to retain the general structure of (8.2.1), but to ad-
mit a larger class of functions than ϵ^d to counterbalance the proliferation
of the number of boxes $B(\epsilon)$ for $\epsilon \to 0$. We define

$$\mu^{(G)} = B(\epsilon)\,G(\epsilon). \qquad (8.2.3)$$

The function $G(\epsilon)$ is called the "gauge function". In the sense of a test
function, the gauge function is in principle arbitrary, but not all choices
of G yield a finite measure μ. In order to find a suitable gauge function
the following expansion turned out to be useful:

$$G(\epsilon) = \exp\left[d_0 \ln(\epsilon) - d_1 \ln(\ln(1/\epsilon)) - \cdots\right]. \qquad (8.2.4)$$

For $d_0 = d \neq 0$ and $d_\nu = 0$, $\nu \geq 1$, we have $G(\epsilon) = \epsilon^d$. Thus, in this
special case, we recover the usual definition of the Hausdorff-Besicovitch
dimension.

The complete fractal $\Lambda_T^{(+)}$, as displayed in Fig. 8.2, was never completely analysed. Hillermeier *et al.* (1992) studied this problem for one
embedding dimension defined by considering a horizontal cut through the
fractal at $n = 1.1$. The reduced fractal, then, has a Hausdorff-Besicovitch

dimension of 1, while its measure is zero. Analysing this reduced version of $\Lambda_T^{(+)}$ with the concept of the uncertainty dimension discussed in Section 2.3, it turned out that a suitable gauge function for this simplified problem is given by

$$G(\epsilon) = \epsilon \exp\left[1.2\ln(\ln 1/\epsilon)\right]. \qquad (8.2.5)$$

The gauge function (8.2.5) contains a logarithmic term, $d_1 = -1.2$, which guarantees that, for the reduced fractal, $d_0 = 1$ and a zero measure are compatible with each other.

Compared with the fractal shown in Fig. 8.1, the fractal $\Lambda_T^{(+)}$ does not seem to contain any contiguous regions, for instance regular islands in phase space. Since $\Lambda_T^{(+)}$ is neither a scaling fractal – a class of fractals for which we expect exponential decay – nor a fractal with regions of finite measure – a class of fractals for which we expect algebraic decay – we do not know at this point what to expect for the decay properties of the kicked hydrogen atom. A numerical calculation helps to clarify the situation. We start once more with the regular grid of 137 500 phase-space points in $\mathcal{R}$ that was the starting situation for the computation of Fig. 8.1, and record the number of surviving points $P(N)$ as a function of the number of kicks N. The result is displayed in Fig. 8.3 in the form of a log-log plot. Clearly, the decay to the continuum is algebraic in time according to

$$P(N) = P_0/N^\alpha, \qquad (8.2.6)$$

where α turns out to be approximately 3.3. This result for the decay of the two-dimensional region $\mathcal{R}$ is consistent with the decay exponent $\alpha = 1.65$ computed by Hillermeier *et al.* (1992) for the decay of various one-dimensional horizontal cuts through the fractal $\Lambda_T^{(+)}$.

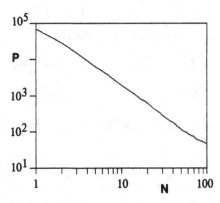

Fig. 8.3. Powerlaw decay of the phase-space probability for the kicked hydrogen atom.

In the following section we prove that, despite the observed algebraic decay, the fractal $\Lambda_T^{(+)}$ is indeed a "skinny" fractal, i.e. it does not contain any regions of finite measure.

8.3 Absence of elliptic points

In this section we provide an analytical proof for the absence of regular period-1 islands in the phase space of the positively kicked one-dimensional hydrogen atom. This proof may also serve as a template to prove the absence of elliptic points for other important atomic physics systems such as the stretched helium atom discussed in Chapter 10.

As discussed in Section 8.1 the phase-space dynamics of the kicked hydrogen atom is governed by the four-step mapping $T = T_4 \circ T_3 \circ T_2 \circ T_1$, where the partial mappings T_j, $j = 1, ..., 4$ are defined in (8.1.13) – (8.1.16), respectively. A period-1 point (θ, y) of T satisfies $T(\theta, y) = (\theta, y)$. This point is unstable if the trace of the Jacobian of T is larger than 2. Since T consists of four steps, the Jacobian J of T is the product of the four Jacobians J_j, $j = 1, ..., 4$ defined by

$$J_1 = \left(\frac{\partial x, p}{\partial \theta, I}\right), \quad J_2 = \left(\frac{\partial x', p'}{\partial x, p}\right), \quad J_3 = \left(\frac{\partial \theta', I'}{\partial x', p'}\right), \quad J_4 = \left(\frac{\partial \theta'', I''}{\partial \theta', I'}\right). \tag{8.3.1}$$

The Jacobian J_2 is trivial. We have $J_2 = 1$. The Jacobian J_3 can be expressed as the inverse of J_1 according to $J_3 = J_1^{-1}$, where J_1^{-1} is evaluated at the "primed" coordinates. Thus, we need to calculate only J_1 and J_4. They are given by

$$J_1 = \begin{pmatrix} y^3 P & 2X/y \\ -y^3/X^2 & -P/y \end{pmatrix} \tag{8.3.2}$$

and

$$J_4 = \begin{pmatrix} 1 & -6\pi\xi/y'^4 \\ 0 & 1 \end{pmatrix}. \tag{8.3.3}$$

We can now compute J explicitly. We obtain

$$J = \begin{pmatrix} \frac{r}{2}(1 + r^2 + I^2) + \frac{6\pi\xi y^3}{y'X^2} & -\frac{2X}{yy'} - \frac{3\pi\xi}{r^3 y'^4}(1 + r^2 + y^2) \\ -\frac{y^3 y'^3}{X^2} & \frac{1}{2r^3}(1 + r^2 + y^2) \end{pmatrix}, \tag{8.3.4}$$

where we define the auxiliary variable $r = y/y'$. From (8.3.4) we compute the trace of J. It is given by

$$\text{Tr}(J) = \frac{1}{2}\left[r^2 + \frac{1}{r^2}\right]\left[r + \frac{1}{r} + II'\right] + \frac{6\pi\xi r I^2}{x^2}. \tag{8.3.5}$$

At this point our restriction to the positively kicked hydrogen atom enters. If we confine our attention to $\xi > 0$ only, we can easily obtain a lower bound for Tr (J) by dropping the additive term proportional to ξ in (8.3.4). Since both y and y' are positive, we can also drop the yy' term and obtain

$$\text{Tr}\,(J) \;>\; \frac{1}{2}\left[r^2 + \frac{1}{r^2}\right]\left[r + \frac{1}{r}\right]. \tag{8.3.6}$$

Now we make use of the formula

$$z + 1/z \;\geq\; 2 \quad , \quad \forall z > 0, \tag{8.3.7}$$

to obtain the final estimate

$$\text{Tr}\,(J) \;>\; 2. \tag{8.3.8}$$

This estimate is independent of a particular location in phase space. This implies that *all* points of phase space, in particular the period-1 points, are linearly unstable. Thus we have proved that the positively kicked hydrogen atom does indeed not possess any first order elliptic islands in phase space. It is possible to extend this proof to period-N points and to show that all period-N points, N integer, are linearly unstable (Blümel (1993c)). This implies that the positively kicked hydrogen atom is completely chaotic. We emphasize that, as far as we know, the kicked hydrogen atom is the only model for a physically realizable system where a numerically motivated chaos conjecture was followed up by an analytical proof. In this sense the kicked hydrogen atom is a most remarkable system.

From the physical point of view the absence of stable islands means that all the phase-space probability eventually ionizes. Since all the atomic physics systems investigated to date possess a mixed phase space that shows regular islands embedded in a chaotic sea, the absence of stable islands in the kicked hydrogen atom is a very unique property.

This section marks the end of our overview of chaos in time dependent atomic physics systems. We will now turn to a discussion of time independent atomic physics systems that are chaotic in the classical description.

9

Chaotic scattering
with CsI molecules

In all of the atomic and molecular systems studied in the previous chapters the relevant dynamics was the bound-space dynamics with the continuum playing either no role at all (see, e.g., the kicked rotor and the driven CsI molecule), or only an auxiliary role for probing the bound-space dynamics with the help of the observed ionization signal (see, e.g., the driven surface state electrons and microwave-driven hydrogen atoms). In this chapter we focus on atomic and molecular scattering, i.e. on processes in which the continuum plays an essential role. This subject has recently attracted much attention as dynamical instabilities and chaos have been discovered in the simplest scattering systems. Complicated scattering in an atomic physics system was noticed as early as 1971 by Rankin and Miller in the theoretical description of a simple chemical reaction. In 1983 Gutzwiller observed complicated behaviour of the quantum phase shift in a schematic model of chaotic scattering. 1986 saw the publication of various important papers on chaotic scattering. Eckhardt and Jung (1986) reported on the occurrence of chaos in a model scattering system. Chaos was found by Davis and Gray (1986) in the classical dynamics of unimolecular reactions, and Noid *et al.* (1986) noticed fractal behaviour in the He $-$ I$_2$ scattering system. These papers were an important catalyst for the creation of a whole new field: *chaotic scattering*. These seminal papers were followed by an avalanche of research work on classical and quantum chaotic scattering in schematic models, as well as in atomic and molecular systems that can be studied in the laboratory.

A simple example of chaotic scattering is Box C. We have encountered this system already in Section 1.1. Many other simple scattering systems of this kind are known by now. The most illustrative chaotic scattering system is Eckardt's three-disk scattering system discussed in Section 2.4. The fundamental mechanism for chaotic scattering is the same in all these systems. Classical chaotic scattering occurs in phase-space regions characterized by the appearance of a fractal set of scattering singularities.

In a system that exhibits chaotic scattering the result of a scattering event cannot in general be predicted. In some regions of phase space the output variables of a chaotic scattering system depend sensitively and in a discontinuous way on the input variables. As a consequence the scattering cross-sections show structure on all scales This observation has an important consequence: as long as classical mechanics provides a good description of the system, the cross-sections cannot be resolved, no matter how carefully the system's parameters are controlled. This feature of a chaotic scattering system was already discussed in connection with the deflection function y^C of the chaotic box C in Section 1.1. Only within the framework of a quantum description of the scattering process do the cross-sections become smooth functions of the system parameters as Planck's constant $\hbar$ sets an ultimate scale. But even in the quantum description chaos leaves its mark. Scattering amplitudes and cross-sections are found to fluctuate in a characteristic way reminiscent of similar fluctuations that occur in nuclear physics in the context of neutron and heavy ion scattering. These fluctuations, called Ericson fluctuations in nuclear physics (Ericson (1960)), are currently much discussed in solid state physics (Baranger (1993a,b)). They also occur in atomic and molecular physics as we will see in this chapter and in Chapter 10. The subject of this chapter is the discussion of the scattering dynamics of the CsI molecule in an array of wires of mesoscopic size. This system was suggested recently (Blümel (1993b)) as a chaotic scattering system that can be investigated with the tools of atomic and molecular spectroscopy. Even if it should turn out that this system is too difficult to realize as an actual laboratory system, it may still serve as an excellent Gedanken experiment helpful for sharpening one's intuition in quantum chaotic scattering.

Both Eckhardt's three-disk system as well as Box C show that complicated systems are not needed to produce chaotic scattering. In order to appreciate what chaotic scattering means, we have to discuss the physical origin of scattering singularities introduced in Section 1.1. Depending on their origin we may classify them into "static" and "dynamic" scattering singularities. The physics of dynamic singularities is the subject of the following sections. Static singularities are a property of the scattering potential only and are easy to understand. Consider, for instance, the scattering of a particle from the repulsive potential $V(x) = V_0/[1 + (x/\sigma)^2]$, $(V_0 > 0)$ shown in Fig. 9.1. Here, V_0 is the strength of the potential and σ is the half-width of the potential at half maximum. The particle is

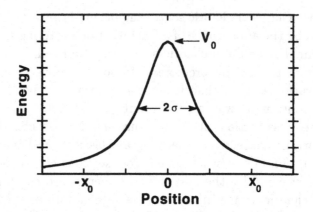

Fig. 9.1. A trivial scattering singularity. A particle launched with energy $E = V_0$ towards $x = 0$ gets "stuck" while approaching, but never quite reaching, the top of the potential barrier. (From Blümel (1993b).)

launched at $-x_0$ ($x_0 > 0$) toward the scattering potential. The particle's initial momentum is $P_0 > 0$. The scattering event is over as soon as the particle re-emerges at $x = \pm x_0$. The total scattering time from launch to re-emergence is denoted by T. Since T depends on P_0, and the total energy E is conserved, we may consider T as a function of E, i.e. $T = T(E)$. The time T is also referred to as the lifetime.

The total energy E of the particle is given by

$$E = \frac{P^2}{2m} + V(x), \qquad (9.0.1)$$

where P is the particle's momentum and m is its mass. Introducing the dimensionless position $\xi = x/\sigma$, the dimensionless energy $\epsilon = E/V_0$, the dimensionless time $\tau = t/t_0$, $t_0 = (m\sigma^2/V_0)^{1/2}$, and the dimensionless momentum $p = d\xi/d\tau = P/(mV_0)^{1/2}$, the total energy becomes

$$\epsilon = \frac{p^2}{2} + \frac{1}{1+\xi^2}. \qquad (9.0.2)$$

We distinguish three cases: (i) $\epsilon > 1$, (ii) $\epsilon < 1$ and (iii) $\epsilon = 1$.

For $\epsilon > 1$ the particle traverses the potential region without turning around and emerges as a "free" particle at $\xi_0 = x_0/\sigma$. For $\epsilon > 1$ and $\epsilon < 1$ the lifetimes can be computed analytically (Blümel (1993b)). We denote them by $T_>$ and $T_<$, respectively, and obtain

$$T_> = \int_{-\xi_0}^{\xi_0} \frac{d\tau}{d\xi}\, d\xi =$$

$$= \sqrt{\frac{2}{\epsilon}} \left\{ F(\alpha, q) - E(\alpha, q) + \xi_0 \left[\frac{1 + \xi_0^2}{b^2 + \xi_0^2} \right]^{1/2} \right\} \qquad (9.0.3)$$

and

$$T_< = \sqrt{\frac{2}{\epsilon}} \left\{ (F(\varphi, s) - E(\varphi, s))/\sqrt{\epsilon} + \frac{1}{\xi_0} \left[(\xi_0^2 + 1)(\xi_0^2 - \xi_t^2) \right]^{1/2} \right\},$$
$$(9.0.4)$$

where, $b = (1 - 1/\epsilon)^{1/2}$, $\alpha = \arctan(\xi_0/b)$, $q = 1/\sqrt{\epsilon}$, $\xi_t = ((1/\epsilon) - 1)^{1/2}$, $\varphi = \arccos(\xi_t/\xi_0)$, and $s = \sqrt{\epsilon}$, and F and E are the incomplete elliptic integrals of the first and second kind as defined, e.g., by Gradshteyn and Ryzhik (1994).

The case $\epsilon \to 1$ is the most interesting of the three. Evaluating (9.0.3) or (9.0.4) for $\epsilon \to 1$ we obtain

$$T \sim -\ln |\epsilon - 1|. \qquad (9.0.5)$$

This means that the lifetime of the particle diverges logarithmically for $\epsilon \to 1$. The singularity at $\epsilon = 1$ shows clearly in a plot of the lifetime T as a function of incident energy ϵ (see Fig. 9.2). For $\epsilon = 1$, it takes the particle an infinite amount of time to reach the point $\xi = 0$. This means that the particle gets "stuck" close to the top of the scattering potential and never re-emerges at $\xi = \pm \xi_0$. As a consequence, $\epsilon = 1$ is a scattering singularity, since the output function of the scattering (in this case the lifetime function T) cannot be defined. Since this type of scattering singularity is associated with the specific form of the scattering potential, and quite readily identified and predicted from a sketch of the potential, we call it a *static* scattering singularity. In contrast to the static singularities the *dynamic* singularities to be discussed below cannot, in general, be so easily predicted from a sketch of the potential alone. A detailed knowledge of the scattered particle's dynamics is necessary for a prediction of these singularities. Hence their name.

Static scattering singularities are well known in atomic and nuclear physics. A prominent example is the phenomenon of *orbiting*, which occurs whenever the effective scattering potential exhibits a local maximum (see, e.g., Nörenberg and Weidenmüller (1976)).

It is not difficult to design scattering potentials which exhibit more than one scattering singularity. With three local maxima the scattering potential shown in Fig. 9.3, e.g., exhibits three scattering singularities. But instead of increasing the complexity of the scattering potential there

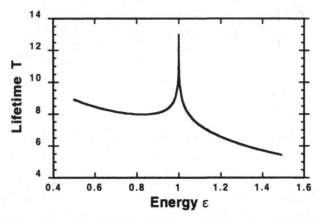

Fig. 9.2.   Lifetime function $T(\epsilon)$ for the scattering potential shown in Fig. 9.1. (From Blümel (1993b).)

is a much simpler way of producing an infinite number of scattering singularities. In fact an *uncountable* number of *dynamic* scattering singularities may be obtained if the simple scattering potential shown in Fig. 9.1 is suitably extended to two dimensions resulting in a system with two degrees of freedom. This "trick" is inspired by our experience with bounded chaotic systems, for instance the double pendulum discussed in Section 3.2. While the one-dimensional pendulum is trivially integrable, a two-dimensional version of the pendulum, the double pendulum, is chaotic. Thus, the *qualitative* change in the number of scattering singularities from *countable* to *uncountable* by a simple *quantitative* change of the dimension from one to two, although surprising at first, is completely in line with the general behaviour of Hamiltonian systems. The uncountable number

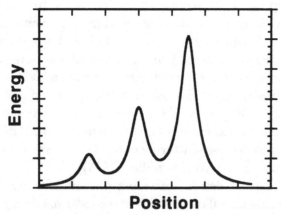

Fig. 9.3.   Sketch of a model potential which allows for three scattering singularities to occur. (From Blümel (1993b).)

of scattering singularities, in many cases arranged in complicated (fractal) sets, is the physical origin for the occurrence of chaotic scattering and the associated loss of predictability in a chaotic scattering system.

In the following section we discuss the classical and quantum dynamics of a CsI molecule scattered off a reaction zone consisting of an arrangement of inhomogenous fields. This system shows classical and quantum chaotic scattering. It can, at least in principle, be built as a laboratory experiment, which would enable the experimenter to check the theoretical predictions advanced in the following sections.

9.1 The CsI scattering system

The set-up of the proposed CsI chaotic scattering experiment is shown in Fig. 9.4. At time $t = 0$ the molecule is launched from $x = -x_0$ toward the electric field produced by two oppositely charged wires. The total energy of the molecule is denoted by E and its initial angular momentum is L_0. All dissipative effects, such as spontaneous decay and thermal noise, are neglected. Thus, this system is Hamiltonian and E is a conserved quantity. We treat the molecule in a two-degree-of-freedom approximation, i.e. we allow rotation to take place only in the $x - z$ plane and the centre of mass C of the molecule is restricted to move on the x axis. Thus, the angular momentum vector $\vec{L}$ is parallel to the y axis for all time. The rotation angle, defined as the angle between the vertical and the axis of the molecule, is denoted by θ. The two charged wires are arranged symmetrically with respect to the x axis, and are located a distance z_0 away from the x axis. In keeping with the model assumptions they are assumed to be parallel to the y axis. Assuming that the upper wire is positively charged, the field on the x axis ($z = 0$) is directed in the

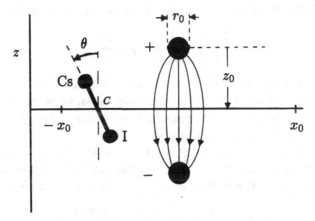

Fig. 9.4.   Sketch of the CsI scattering system. (From Blümel (1993b).)

negative z direction. It is assumed to be homogeneous over the extension of the molecule. Note that in this respect Fig. 9.4 is not drawn to scale. Therefore, since the molecule is a dipole, the z component of the force acting on the molecule vanishes if the molecule moves exactly on the x axis. The z component of the dipole force is small for small deviations of the molecule's centre of mass from the x axis. Therefore, given the relevant interaction times of the molecule in the interaction zone, the drift in the z direction is negligible and the molecule stays close to the x axis as it passes through the field. The rotation of the molecule is an additional factor which helps to reduce the time averaged dipole force, and thus the drift in z direction.

As discussed already in Section 5.4, the CsI molecule has a considerable electric dipole moment (see Table 5.1). Therefore, it can very effectively interact with the electric field of the two wires. In computing the interaction potential we first compute the electric field of a single straight wire of radius r_0 a distance r away from the wire. It is given by

$$E(r) = \frac{\alpha}{2\pi\epsilon_0 r}, \quad r \geq r_0. \tag{9.1.1}$$

Here, α is the charge of the wire per unit length and ϵ_0 is the electric permittivity of the vacuum ($\epsilon_0 = 8.85 \times 10^{-12} \, \text{C}^2/\text{N m}^2$). Superposing the fields of the two oppositely charged wires, we obtain the following expression for the electric field on the x axis

$$E(x, z = 0) = \frac{E_0}{1 + (x/z_0)^2}, \tag{9.1.2}$$

where

$$E_0 = \frac{\alpha}{\pi\epsilon_0 z_0}. \tag{9.1.3}$$

With (9.1.2) the Hamiltonian of the CsI molecule is given by

$$H = \frac{L^2}{2I} + \frac{P^2}{2m} + V_0 \frac{\cos(\theta)}{1 + (x/z_0)^2}, \tag{9.1.4}$$

where L is the angular momentum of the molecule, I is its moment of inertia, P is the momentum of the centre of mass, m is the total mass of the molecule, $V_0 = \mu E_0$ is the maximal interaction energy and μ is the electric dipole moment of the molecule.

Introducing the unit of length z_0, the unit of time $t_0 = (mz_0^2/V_0)^{1/2}$, the unit of momentum $\lambda = mz_0/t_0$, the unit of angular momentum $\gamma = t_0 V_0$ and V_0 as the unit of energy, the Hamiltonian (9.1.4) becomes

$$h = \frac{1}{2}\rho l^2 + \frac{1}{2}p^2 + \frac{\cos(\theta)}{1 + \xi^2}, \tag{9.1.5}$$

where ξ, p and l denote dimensionless position, momentum and angular momentum variables, respectively. Besides the total energy, the Hamiltonian (9.1.5) contains only one control parameter,

$$\rho = mz_0^2/I. \qquad (9.1.6)$$

The classical equations of motion of the CsI molecule can now be derived from (9.1.5). With $\tau = t/t_0$ they are given by

$$\frac{d\xi}{d\tau} = \frac{\partial h}{\partial p} = p, \qquad \frac{dp}{d\tau} = -\frac{\partial h}{\partial \xi} = \frac{2\xi \cos(\theta)}{[1+\xi^2]^2},$$

$$\frac{d\theta}{d\tau} = \frac{\partial h}{\partial l} = \rho l, \qquad \frac{dl}{d\tau} = -\frac{\partial h}{\partial \theta} = \frac{\sin(\theta)}{1+\xi^2}. \qquad (9.1.7)$$

The set of equations (9.1.7) is investigated in detail in the following section. It is shown that, despite its simple appearance, the solutions of the set (9.1.7) contain the full complexity of classical chaotic scattering, reminiscent of the complexity we encountered in Section 1.1 for the trajectories of box C.

9.2 Classical chaotic scattering

The purpose of this section is to show that the set of equations (9.1.7) has chaotic solutions. Analytical solutions of (9.1.7) are not known and one has to resort to numerical integration. Because of the simple structure of (9.1.7), special integration methods are not needed. Any standard algorithm for the solution of ordinary differential equations will do. We used a simple fourth order Runge-Kutta method with constant step size, as described, e.g., by Milne (1970). In order to prove the existence of complexity in the solutions of (9.1.7) we compute the lifetime function T as a function of the initial angle θ_0. For the total energy, the initial angular momentum and the rotational constant we choose $\epsilon = 0.5$, $l_0 = 0.3$ and $\rho = 1$, respectively. The lifetime T is defined as the time τ it takes the molecule to re-emerge at $\xi = \pm\xi_0 = \pm 4$. The result is shown in Fig. 9.5(a).

The graph of T is mostly smooth with two unresolved structures. The left hand structure is denoted by J. Fig. 9.5(b) shows a magnification of J. It consists of three sub-structures, which again appear to be unresolved. They are denoted by J_1, J_2 and J_3. A magnification of J_1 is shown in Fig. 9.5(c). Again, it consists of three sub-structures, and again they seem to be unresolved. They are denoted by J_{11}, J_{12} and J_{13}. Further magnifications show that there is a whole hierarchy of structures within structures. Every given structure $J_{k_1 k_2 ... k_n}$ consists of yet another

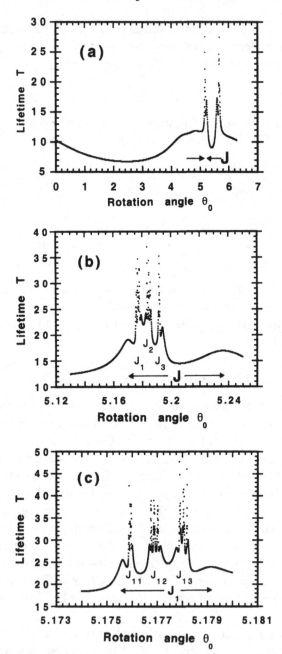

Fig. 9.5.   Lifetime function $T(\theta_0)$ for the CsI scattering model. Frames (b) and (c) show successive magnifications of $T(\theta_0)$ displayed in frame (a). (From Blümel (1993b).)

"generation" of three further sub-structures labelled $J_{k_1 k_2 ... k_n 1}$, $J_{k_1 k_2 ... k_n 2}$ and $J_{k_1 k_2 ... k_n 3}$. The indices $k_j \in \{1, 2, 3\}$ can be interpreted as the three symbols of a symbolic encoding (see Section 2.4) of the hierarchy of intervals $J_{k_1 ... k_j ...}$. There is no grammatical restriction on the permissible symbol combinations. All combinations are allowed. The most important implication of the structure of the sets $J_{k_1 ... k_j ...}$ is that $T(\theta_0)$ cannot be resolved in J. This unresolvability is the most interesting feature of the classical mechanics of the CsI scattering system. It has observable consequences even if the CsI system is treated quantum mechanically. This is shown in Section 9.3.

The results shown in Fig. 9.5 indicate that $T(\theta_0)$ is singular on a fractal set consisting of infinitely many dynamical singularities at which $T(\theta_0) \to \infty$. In contrast to the static scattering singularities the two-dimensional model discussed here produces *uncountably* many dynamic singularities. Thus, the transition from the one-dimensional situation to two dimensions results in a *qualitative* change of the scattering behaviour. At first glance this is astonishing given the fact that we did nothing but add another dimension to the problem. At most one would have expected an "unneccessary complication" of a basically well understood scattering problem. On second thought the qualitative change of the features of the system is consistent with the observation that, quite generally, dimensionality plays a decisive role for the qualitative behaviour of many physical systems.

The lifetimes $T(\theta_0)$ displayed in Fig. 9.5 were calculated for fixed l_0. In order to obtain a better impression where the fractal features of T are located in the θ_0, l_0 plane, $T(\theta_0, l_0)$ was calculated as a function of the two initial conditions θ_0 and l_0 for $\epsilon = 0.5$ and $\rho = 1$. The result is shown in Fig. 9.6 in the form of a grey-scale plot. The darker the shades in Fig. 9.6 the longer lived the molecule. Again there are apparently unresolved regions in Fig. 9.6(a). A magnification of the framed detail of Fig. 9.6(a) is shown in Fig. 9.6(b). Again there are apparently unresolved structures in Fig. 9.6(b). As before in the one-dimensional case one will never be able to resolve the two-dimensional features in $T(\theta_0, l_0)$ as more and more structure appears on smaller and smaller scales. Thus, $T(\theta_0, l_0)$ is a fractal function embedded in the two-dimensional $(\theta_0 - l_0)$ space.

Figs. 9.5 and 9.6 prove convincingly that the CsI system with one pair of wires is indeed a chaotic scattering system. The origin of complicated behaviour in the CsI scattering system, especially the origin of dynamical trapping, can be understood within the following physical picture. If the interaction between the molecule and the field is strong one can imagine a situation in which the initial conditions allow the molecule to

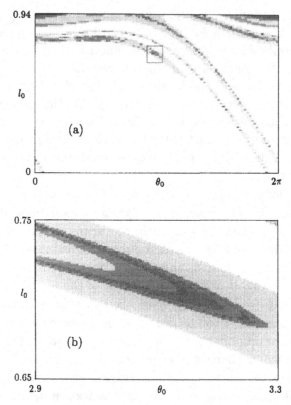

Fig. 9.6. Fractal lifetime function $T(\theta_0, l_0)$ for the CsI scattering model. The longer the lifetime of the molecule, the darker the shade of grey. The framed detail in (a) is magnified in (b). (From Blümel (1993b).)

enter the field region, but not to exit. This situation can arise since the molecule rotates inside the field, and one can imagine that for special initial conditions all of the molecule's rotational energy is transformed into potential energy shortly *before* the molecule tries to exit the potential region. Consequently, the molecule has to turn around. But it can happen that now the same situation occurs shortly before the molecule tries to exit again, and it is possible that this situation repeats itself over and over again. This is the physical reason for the occurrence of dynamical scattering singularities. Although the molecule has enough energy to exit the field region on the basis of energy conservation alone, the molecule always "manages" to balance its potential and rotational energies as a function of x in such a "clumsy" way that it is never allowed to exit the field region. While one can intuitively understand that such a situation may occur at countably many initial conditions, the surprise is that an uncountable number of initial conditions leads to this "pathological" behaviour.

It is intuitively clear that adding more wires to the interaction region increases the chances of the CsI molecule becoming dynamically trapped in the potential region. A dipole scattering model with three "wires" was studied by Blümel and Smilansky (1988). The model is defined by replacing the single Lorentzian form factor in the electric field (9.1.2) by a sum of three Gaussians. This model does not reproduce the Lorentzian shape of the electric field, especially not for large x, but is more convenient for numerical computations. It does not change the physics of the CsI model, since the origin of chaotic scattering is not to be sought in the tails of the potential, but in the interaction region where the field is strongest. In the model of the three "Gaussian wires" the field on the x axis is given by

$$E(x, z = 0) = E_0 \, f_\sigma(x), \qquad (9.2.1)$$

where

$$f_\sigma(x) = \sum_{j=1}^{3} \exp\left[-(x - x_j)^2/\sigma^2\right]. \qquad (9.2.2)$$

For this model the final angular momentum I_f of the classical CsI molecule was calculated for $\sigma = 1$. It is computed from (9.1.7) as a function of the initial angle θ_i and is defined by $I_f = I(T, \theta_i)$, where T is the lifetime of the scattering trajectory launched with initial angle θ_i at $x_i = -4$. The result is shown in Fig. 9.7. Fig. 9.7(a) shows that the angular momentum I_f exhibits a large set of scattering singularities. Since the singularities in Fig. 9.7 also correspond to singularities in T, Fig. 9.7 proves that more potential humps indeed lead to a larger set of chaotic scattering singularities. Although real wires produce a different kind of potential than the "Gaussian wires", the analogy is close enough that we may conclude that more wires in the CsI experiment will also lead to a larger set of scattering singularities. Magnifications of θ_i sub-intervals shown in Figs. 9.7(b) and (c) indicate that the set of scattering singularities is indeed a fractal set.

A situation where the CsI molecule gets dynamically stuck in the scattering potential, i.e. T or I_f exhibit a scattering singularity, can be interpreted in the quantum language as the formation of a long-lived scattering resonance. This interpretation reminds us strongly of the *compound nucleus*, a picture introduced by Niels Bohr (1936) in nuclear physics to account for the sometimes very long-lived intermediate scattering complexes that can occur in the scattering of a neutron off a heavy target

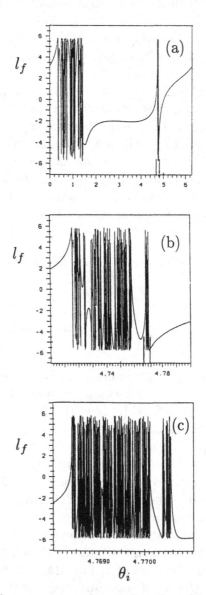

Fig. 9.7. Final angular momentum l_f as a function of initial rotation angle θ_i for the three-wire CsI scattering system. The vertical bars on the θ_i axis in frame (a) mark a θ_i interval shown magnified in frame (b). Frame (c) shows a magnified version of the interval marked in frame (b).

nucleus. We investigate this picture and its consequences in more detail in Section 9.4. It was found in many nuclear physics scattering experiments that the compound nucleus, once formed, selects its decay mode quasi-randomly from a large set of available exit channels. The lifetimes of individual scattering events are found to be exponentially distributed.

It is now very interesting to see whether our simple CsI scattering system follows the predictions of statistical theories of chemical and nuclear reactions. This indeed turns out to be the case, and, even more surprisingly, already on the classical level. Fig. 9.8 shows the distribution of classical lifetimes T of the intermediate dynamically trapped CsI scattering complex. This figure was generated by means of a Monte Carlo computation by launching thousands of scattering trajectories with initial angles θ_i equi-distributed in $0 < \theta_i < 2\pi$. The resulting lifetime distribution is approximately exponential and can be represented as

$$P(T) \sim e^{-\gamma T}, \tag{9.2.3}$$

where in analogy to compound nuclear scattering the inverse decay constant, $1/\gamma$, can be interpreted as the average lifetime of the intermediate scattering complex. Another interesting quantity to examine is the total accrued scattering angle Θ, i.e. the scattering angle not taken modulo 2π. The distribution of total angles is shown as the squares in Fig. 9.8. The distribution is exponential; arbitrarily large scattering angles can occur.

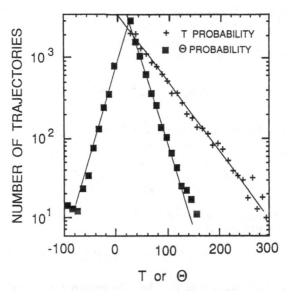

Fig. 9.8. Distribution of lifetimes (crosses) and scattering angles (squares) in the classically chaotic regime of the three-wire CsI scattering system. (Adapted from Blümel and Smilansky (1988).)

We have also examined the final angular momenta in regions with chaotic scattering. In close analogy with the random decay of a compound nucleus into its exit channels, we find that the final angular momenta of the CsI molecule, the classical analogue of "exit channels", are indeed evenly distributed in those regions of initial angles that correspond to chaotic scattering regions.

The most important question is now whether the existence of classical chaotic scattering leaves its fingerprints in the quantum description of the CsI system. In order to answer this question we discuss in the following section the quantum mechanics of a CsI molecule interacting with three pairs of wires. For technical reasons the correct three-wire potential is approximated by the three Gaussians defined in (9.2.2). This model shows clear manifestations of classically chaotic scattering in the scattering matrix and the scattering cross-sections of the CsI system.

9.3 Quantum chaotic scattering

The quantization of the dipole model defined in Sections 9.1 and 9.2 is not difficult. The Hamiltonian for the three-wire version of the CsI system is given by

$$\hat{H} = \frac{\hat{P}^2}{2m} + \frac{\hat{L}^2}{2I} + V_0\, f_\sigma(x)\, \cos(\theta), \qquad (9.3.1)$$

where $\hat{P} = -i\hbar\partial/\partial x$, $\hat{L} = -i\hbar\partial/\partial\theta$ and $f_\sigma(x)$ is defined in (9.2.2). The eigenfunctions $|n\rangle$ of $\hat{L}$ are given by

$$\hat{L}\,|n\rangle = \hbar\, n\, |n\rangle \qquad (9.3.2)$$

with

$$\langle\theta|n\rangle = \frac{1}{\sqrt{2\pi}}\, e^{in\theta}. \qquad (9.3.3)$$

The stationary scattering function for total energy E can be expanded according to

$$\langle x|\psi^{(n)}\rangle = \sum_{n'} \psi_{n'}^{(n)}(x)\,|n'\rangle, \qquad (9.3.4)$$

where the superscript n specifies the incoming channel. The wave function (9.3.4) is a solution of

$$\hat{H}\,|\psi^{(n)}\rangle = E\,|\psi^{(n)}\rangle. \qquad (9.3.5)$$

The boundary conditions satisfied by the expansion amplitudes $\psi_{n'}^{(n)}(x)$ are given by

$$\psi_{n'}^{(n)}(x) \sim \tau_{nn'}\, e^{ik_{n'}x}, \qquad x \to +\infty$$

$$\psi_{n'}^{(n)}(x) \sim e^{ik_n x} + \rho_{nn'} e^{-ik_{n'} x}, \quad x \to -\infty. \qquad (9.3.6)$$

The incident wave is prepared in the state $|n\rangle$ as a plane wave in x. The channel wave numbers are given by

$$k_n = \left\{ \frac{2m}{\hbar^2} \left[E - \hbar^2 n^2 / 2I \right] \right\}^{1/2}. \qquad (9.3.7)$$

We denote by N the number of open channels, i.e. N is the largest integer n for which $E \geq \hbar^2 n^2 / 2I$. The scattered wave is a superposition of angular momentum states with amplitudes $\psi_{n'}^{(n)}(x)$. The notation τ and ρ is used for transmission and reflection amplitudes, respectively. In the calculations discussed below typically 15 to 20 open and 2 to 3 closed channels were explicitly taken into account.

The Hamiltonian (9.3.1) contains the control parameter V_0. On the classical level it was found that for small values of V_0 there is no chaotic scattering. There exists a critical value $V_0^{(cr)}$ which marks the first appearance of a nontrivial set of scattering singularities. Well developed chaotic scattering characterized by a large set of scattering singularities is observed for $V_0 > V_0^{(cr)}$. Thus, the control parameter V_0 allows us to switch between regular and chaotic scattering, where "regular" scattering is characterized by the presence of at most a countable set of scattering sigularities.

Quantum mechanically, because of the uncertainty principle, it is not possible to specify θ_i if the initial state is an angular momentum eigenstate. Therefore we study the behaviour of the transmission probabilities $\sigma_{nn'} = |\tau_{nn'}|^2$ as a function of the energy. Here, $\tau_{nn'}$ is the transmission amplitude defined in (9.3.6). It is the amplitude of observing a CsI molecule to the right of the three-wire interaction zone with $n'\hbar$ units of angular momentum when it was prepared on the left of the three-wire interaction zone in a rotational state containing $n\hbar$ units of angular momentum.

The behaviour of $\sigma_{nn'}$ as a function of E differs markedly in the regions $V_0 < V_0^{(cr)}$ and $V_0 > V_0^{(cr)}$. For $V_0 < V_0^{(cr)}$, corresponding to regular scattering, we observe a smooth dependence of $\sigma_{nn'}$ on E. This is demonstrated in Fig. 9.9(b), which shows σ_{79} as a function of E in the interval $1.5 \leq E \leq 2.4$. For $V_0 > V_0^{(cr)}$, σ_{79} exhibits irregular fluctuations, as shown in Fig. 9.9(a). The fluctuations exhibited by the transmission probabilities of the CsI molecule in the classically chaotic regime are strongly reminiscent of a fluctuation phenomenon first predicted to exist by Ericson in 1960. These fluctuations have since been called *Ericson fluctuations* and are an important phenomenon in compound nuclear

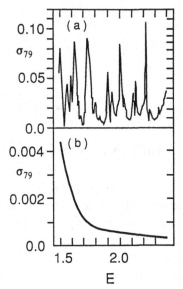

Fig. 9.9. Transmission probability σ_{79} of the CsI molecule as a function of the total energy E. (a) Irregular fluctuations in the classically chaotic regime ($V_0 > V_0^{(cr)}$). (b) Smooth behaviour in the classically regular regime ($V_0 < V_0^{(cr)}$). (Adapted from Blümel and Smilansky (1988).)

scattering. Ericson fluctuations were originally attributed to the complexity of a heavy nucleus consisting of many particles. It is therefore surprising to see Ericson fluctuations appear in our simple CsI scattering model, which is a system with only two degrees of freedom. Chaos is the reason for complexity in our system. In hindsight, therefore, it is no longer surprising to see Ericson fluctuations in the CsI system as well. The dipole model with only two degrees of freedom shows all the features of Bohr's compound nucleus. In Chapter 10 we predict that Ericson fluctuations appear even in the helium atom. Since Ericson fluctuations are such an important qualitative feature of chaotic scattering systems we discuss this topic in more detail in Section 9.4.

It is straightforward to calculate the scattering matrix $S_{nn'}$ for the CsI model. It is unitary and symmetric because $\hat{H}$ is time reversal invariant. Another feature of compound nuclear theory is the applicability of random matrix theory. Therefore, if the similarity between the CsI model and a compound nucleus is indeed as close as claimed above, we should expect that the S matrix for the CsI system shows features of a random matrix. In 1962 Dyson showed that the nearest neighbour spacings of the eigenangles of a large unitary and symmetric random matrix are approximately Wigner distributed. Therefore, on diagonalizing the S

matrix

$$S\,|\alpha_j\rangle \;=\; e^{i\alpha_j}\,|\alpha_j\rangle, \quad j = 1,...,N \tag{9.3.8}$$

we should find that the normalized spacings s are Wigner distributed, i.e.

$$P(s) \;=\; \frac{\pi}{2}\,s\,\exp[-\pi s^2/4]. \tag{9.3.9}$$

Fig. 9.10(a) shows that a distribution resembling the Wigner distribution is indeed obtained for the S matrix eigenangle spacings in the classically chaotic regime of the CsI model. For parameter values corresponding to regular scattering, $V_0 < V_0^{(cr)}$, large deviations from a Wignerian distribution are observed. This is shown in Fig. 9.10(b).

In summary, classically chaotic scattering manifests itself quantum mechanically in the following way:

(a) Cross-sections fluctuate strongly as a function of the energy.
(b) The statistics of eigenangles and matrix elements of the S matrix follow the predictions of random matrix theory.

While the investigation of the statistics of energy levels is a well established technique in atomic and molecular physics, the study of cross-section fluctuations is not a standard technique. The purpose of the following section is to discuss the subject of Ericson fluctuations in some

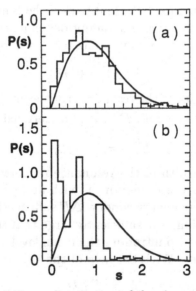

Fig. 9.10. Nearest neighbour distribution of the eigenangles of the S matrix in the chaotic regime (a) and the regular regime (b). (Adapted from Blümel and Smilansky (1988).)

more detail and to strengthen its ties to the field of quantum chaotic scattering.

9.4 Ericson fluctuations

In this section we define the concept of Ericson fluctuations and discuss their relevance in quantum chaotic scattering systems of atomic and molecular physics. Our discussion follows the excellent introduction to Ericson fluctuations in Ericson's own paper of 1963. A nuclear compound reaction is taken as an illustrative example.

Let us assume that we are dealing with a heavy nucleus at an excitation energy where the nucleus has access to many reaction channels. We denote the channels by the subscript α. According to a general result in scattering theory, the cross-section for inelastic scattering from channel α to channel α' at energy E can be computed from the scattering matrix S by

$$\sigma_{\alpha\alpha'} = \pi \left(\frac{\lambda_\alpha}{2\pi}\right)^2 |S_{\alpha\alpha'}(E)|^2. \qquad (9.4.1)$$

According to Feshbach (1962) the S matrix splits into two parts according to

$$S_{\alpha\alpha'} = S_{\alpha\alpha'}^{(p)} + \sum_k \frac{a_k}{E - E_k}. \qquad (9.4.2)$$

The first part of the S matrix in (9.4.2) is due to potential scattering, indicated by the superscript "p". This part of the S matrix is only weakly dependent on the energy, i.e. its energy dependence is "smooth". The second part of (9.4.2) is a sum over pole terms where

$$E_k = E_k^{(r)} + iE_k^{(i)} \qquad (9.4.3)$$

are the complex energies of the resonances of the compound nucleus. The imaginary part of the energies, $E_k^{(i)}$, is parametrized according to

$$E_k^{(i)} = -\Gamma_k/2, \qquad (9.4.4)$$

where $\Gamma_k > 0$ is the width of the resonance number k. In the following we approximate $\Gamma_k \approx \Gamma$ and identify Γ with an "effective width" of the compound nucleus in an energy window ΔE sufficiently small that $\Gamma_k \approx \Gamma$ is a good approximation, but sufficiently large that many resonances are contained in ΔE. We also introduce the mean level spacing D of the real parts of the resonances. With the help of the Ericson parameter

$$\Lambda = \Gamma / D \qquad (9.4.5)$$

we characterize the relationship between the mean spacing and the mean width of the resonances. We distinguish three different cases: (i) $\Lambda \ll 1$,

(ii) $\Lambda \sim 1$ and (iii) $\Lambda \gg 1$. Case (i) is the familiar "resonance" case. In this case the cross-section (9.4.1) shows a structure which consists of individual isolated resonances with a Lorentzian profile. Case (ii) is the case of overlapping resonances. Both cases are illustrated in Fig. 9.11(a), where we show an example of a cross-section with isolated and weakly overlapping resonances. Fig. 9.11(a) was generated artificially in the following way. First, we drew ten random numbers φ_i, $i = 1, 2, ..., 10$ from the interval $[0, 1]$. With the help of the φ_i we define the amplitudes $a_i = \exp(i\varphi_i)$. The real parts $E_k^{(r)}$ of the resonance energies were again drawn from a uniform distribution of random numbers in the unit interval. The locations of $E_k^{(r)}$ of the resulting ten resonances are marked by the vertical bars in Fig. 9.11(a). The imaginary parts of the resonances were set to $\Gamma = 0.02$. With $S_{\alpha\alpha'}^{(p)}$ assumed to be zero we define the cross-section according to

$$\sigma^{(r)}(E) = \frac{1}{4000} \left| \sum_{k=1}^{10} \frac{a_k}{E - E_k} \right|^2. \qquad (9.4.6)$$

The resulting cross-section is shown in Fig. 9.11(a). The distinguishing feature of Fig. 9.11(a) is that every peak or shoulder in the cross-section can be lined up uniquely with one of the bars in Fig. 9.11(a), i.e. every feature of the cross-section (9.4.6) can be assigned uniquely to one of the resonances E_k. Another typical feature is the characteristic Lorentzian shape of the isolated resonances in σ.

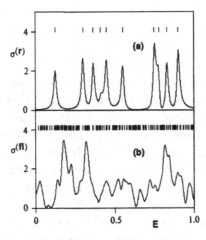

Fig. 9.11. Simulated cross-sections for (a) the regime of isolated and weakly overlapping resonances, $\Lambda \lesssim 1$, and (b) the Ericson fluctuation regime characterized by $\Lambda \gg 1$. $\Lambda = \Gamma/D$ is the Ericson parameter. The real parts (locations) of the resonances are indicated by the vertical bars.

We will now "synthesize" a cross-section corresponding to case (iii), the regime of *strongly* overlapping resonances. This time we choose 200 random numbers φ_i for generating the amplitudes a_i, and 200 random $E_k^{(r)}$ values, again equi-distributed in $[0,1]$, and indicated by the bars in Fig. 9.11(b). The resulting mean level spacing is $D = 1/200 = 5 \times 10^{-3}$. We choose $\Gamma = 0.03$ so that this time the resonance width is six times larger than the mean spacing. We define the cross-section according to

$$\sigma^{(fl)}(E) = \frac{1}{6 \times 10^4} \left| \sum_{k=1}^{200} \frac{a_k}{E - E_k} \right|^2. \tag{9.4.7}$$

The result is shown in Fig. 9.11(b). Although the widths of the resonances underlying Fig. 9.11(b) are much larger than their spacings, the resulting cross-section is not smooth. It shows a fluctuating pattern with roughly 30 distinguishable features. The features themselves look like "resonances" or "shoulders", but clearly cannot be assigned to any one resonance alone, as the vertical bars in Fig. 9.11(b) clearly show. Also, their shapes are not Lorentzian. Since the widths of the resonances are so large and their spacings are so small, many individual resonances contribute to any given feature in Fig. 9.11(b). According to (9.4.7) every one of the resonances contributes coherently with a random phase and a random amplitude resulting in the fluctuating pattern of the cross-section shown in Fig. 9.11(b). The coherent contribution of many resonances within an energy width of the order of Γ is the physical origin of the fluctuations.

Ericson's prediction (1960) of cross-section fluctuations in case (iii) was subsequently confirmed in many nuclear physics experiments. The important lesson one can learn from Ericson's discussion is that large damping, i.e. large Γ, does not automatically result in a smooth cross-section if the mean level spacing is much smaller than the widths of the participating resonances. In nuclear physics it turns out that this situation is the rule rather than the exception. In nuclear physics the small level spacing, i.e. the large level density, is brought about by the many-body aspect of a heavy compound nucleus, which, due to the presence of its many constituent nucleons, has many quantum states available. It was argued by Blümel and Smilansky in 1988 that a complicated many-body situation is not necessary to bring about the Ericson fluctuation phenomenon. This argument is based on our experience with low-dimensional chaotic systems which generate complexity dynamically. A case in point is the CsI model. It has only two degrees of freedom, but nevertheless, as demonstrated in Fig. 9.9(a), shows Ericson fluctuations. The fluctuating pattern in Fig. 9.9(a) can now be compared with the "ideal" case of Ericson fluctuations shown in Fig. 9.11(b). Apart from a trivial difference in the energy scale, there is a close resemblance in structure.

Ericson's central result (1960) is the extraction of the average width Γ from the fluctuating cross-sections. This is done in the following way. We define the energy auto-correlation function $C_{\alpha\alpha'}(\epsilon)$ of the cross-section $\sigma_{\alpha\alpha'}$ according to

$$C_{\alpha\alpha'}(\epsilon) = \langle[\sigma_{\alpha\alpha'}(E+\epsilon) - \bar{\sigma}_{\alpha\alpha'}][\sigma_{\alpha\alpha'}(E) - \bar{\sigma}_{\alpha\alpha'}]\rangle_E =$$

$$\langle\sigma_{\alpha\alpha'}(E+\epsilon)\sigma_{\alpha\alpha'}(E)\rangle_E - \bar{\sigma}^2_{\alpha\alpha'}, \qquad (9.4.8)$$

where the angles $\langle\rangle_E$ indicate averaging over an energy interval large compared with Γ, and $\bar{\sigma}_{\alpha\alpha'} = \langle\sigma_{\alpha\alpha'}\rangle_E$ is the energy averaged mean cross-section. Thus, (9.4.8) is essentially the energy auto-correlation function of the fluctuating part $\sigma^{(fl)}_{\alpha\alpha'}$ of the cross-section. Ericson was able to show that

$$C_{\alpha\alpha'}(\epsilon) \sim \frac{1}{1 + (\epsilon/\Gamma)^2}, \qquad (9.4.9)$$

i.e. the energy averaged cross-section auto-correlation function is essentially a Lorentzian with width Γ. Since $\hbar/\Gamma$ is the lifetime of the intermediate scattering complex, Ericson's method allows us to extract the lifetime of the compound nucleus from the fluctuating cross-sections. In the 1960s and 1970s this method was widely employed for the experimental determination of compound nuclear lifetimes. In the context of chaotic scattering, the result (9.4.9) offers the possibility of assigning a lifetime to the intermediate scattering complex that is generated dynamically in the case of chaotic scattering. There is more. In Section 9.2 we computed a classical lifetime of the intermediate scattering complex by evaluating the statistics of the lifetimes of classical scattering trajectories. This offers the exciting possibility of comparing the mean lifetime of classical trajectories, $\bar{T}$, with the quantum lifetime $\hbar/\Gamma$ of the intermediate scattering complex determined from the cross-section auto-correlation function (9.4.9). It was shown by Blümel and Smilansky (1988) that both quantities are indeed in good agreement with each other.

We show now that Ericson fluctuations also play a fundamental role in Gutzwiller's example of a quantum chaotic scattering system, the first of its kind reported in the literature. Gutzwiller (1983) considered the scattering of a particle on a surface of negative curvature. The fascinating feature of his model is that he was able to compute the scattering amplitude analytically. The wave numbers κ of the scattering resonances in his model are given by the zeros of $\zeta(1-2i\kappa)$, where $\zeta(z)$ is Riemann's ζ function. If a famous conjecture by Riemann is true, then the nontrivial zeros of $\zeta(z)$ are all lined up on the line $z = 1/2 + iy$, where y

is real. Let us assume that Riemann's conjecture is true, and label the zeros $z_j^{\pm} = 1/2 \pm i y_j$, $j = 1, 2, ...$, of ζ with $y_j > 0$. Then, the resonance wave numbers of Gutzwiller's model occur at

$$\kappa_j = y_j/2 - i/4. \tag{9.4.10}$$

The resonance energies are given by

$$E_j = \kappa_j^2 = \left(\frac{y_j^2}{4} - \frac{1}{16}\right) - i\frac{y_j}{4}. \tag{9.4.11}$$

It can be shown that, under the assumption that Riemann's conjecture holds, the staircase function of Riemann zeros is given to leading order by (Whittaker and Watson (1927))

$$\mathcal{N}(y) = \frac{y}{2\pi}\left[\log\left(\frac{y}{2\pi}\right) - 1\right]. \tag{9.4.12}$$

From (9.4.12) we compute the density of Riemann zeros:

$$D(y) = \frac{1}{2\pi}\log\left(\frac{y}{2\pi}\right). \tag{9.4.13}$$

From (9.4.11) we compute the density of resonances, whose inverse is the mean spacing s of resonances. The result is

$$s = y/2D. \tag{9.4.14}$$

According to (9.4.4) and (9.4.11) the width of the resonances is given by

$$\Gamma = y/2. \tag{9.4.15}$$

Thus, the Ericson parameter for Gutzwiller's model is given by

$$\Lambda = D(y) = \frac{1}{2\pi}\log\left(\frac{\kappa}{\pi}\right), \tag{9.4.16}$$

where we used the fact that according to (9.4.10) the real part of κ is $y/2$. Equation (9.4.16) shows that Gutzwiller's model possesses all three scattering regimes. The regime (ii) of overlapping resonances ($\Lambda = 1$) is reached for $\kappa = \kappa_1 = \pi \exp(2\pi)$. Regime (i) occurs for $\kappa \ll \kappa_1$, and regime (iii), the Ericson fluctuation regime, occurs for $\kappa \gg \kappa_1$. We note that because of the exponential dependence of κ on Λ, $\kappa = \exp(2\pi\Lambda)$, the fully developed Ericson regime ($\Lambda \gg 1$) is very hard to reach in Gutzwiller's model.

Although ubiquitous in nuclear physics, schematic scattering models and simple models of atoms and molecules, the topic of Ericson fluctuations in atomic physics has only recently attracted attention (Main and Wunner (1994)). A possible reason may be that in the Ericson regime there are no individual levels to resolve. Nevertheless, valuable information on the mean lifetime of resonances can be extracted from the quasi-randomly fluctuating cross-sections. On the basis of the discussion in this and the previous three sections, it appears that the CsI molecule is a promising candidate for the detection of Ericson fluctuations in molecular physics.

The experimental feasibility of a quantum chaotic scattering experiment with CsI molecules was investigated by Blümel (1993b). As a result of this investigation it appears that the direct experimental implementation of the single-wire system proposed in Fig. 9.4 is not impossible, but very hard to do. In particular, it turns out that the implementation of the CsI system requires mesoscopically small components on the nano scale. This is immediately clear since in order to generate chaotic scattering the interaction potential produced by the wires has to vary appreciably on the scale of the molecule itself, since otherwise the interaction of the molecule with the field is quasi-adiabatic, not leading to a fractal set of dynamic scattering singularities. Recent advances in sub-micron technology, however, might make the CsI experiment more realistic. Mesoscopic wires may be fabricated with the help of nano scale atomic lithography as developed by Timp *et al.* (1992) and McClelland *et al.* (1993). It was also shown experimentally that wires produced by the technique of laser focussed deposition of atoms are very close to parallel. Thus, the proper alignment of the wires in the CsI experiment is not a problem. In addition it was demonstrated in Section 9.2 that using more wires to produce the interaction potential actually enhances the chaotic scattering. Therefore it seems plausible that some of the technical difficulties may be overcome by using an interaction zone consisting of three or more pairs of wires. More theoretical work is necessary to fine-tune the parameters in order to make the CsI scattering system a realistic proposal for experimental investigation.

10

The helium atom

The stability of the solar system is one of the most important unsettled questions of classical mechanics. Even a simplified version of the solar system, the three-body problem, presents a formidable challenge. An important breakthrough occurred when Poincaré, with some assistance from his Swedish colleague Fragmen, proved in 1892 that, apart from some notable exceptions, the three-body problem does not possess a complete set of integrals of the motion. Thus, in modern parlance, the three-body problem is chaotic.

The helium atom is an atomic physics example of a three-body problem. On the basis of Poincaré's result we have to expect that the helium atom is classically chaotic. Richter and Wintgen (1990b) showed that this is indeed the case: the helium atom exhibits a mixed phase space with intermingled regular and chaotic regions (see also Wintgen *et al.* (1993)). Thus, conceptually, the helium atom is a close relative of the double pendulum studied in Section 3.2. Given the classical chaoticity of the helium atom we are confronted with an important question: How does chaos manifest itself in the helium atom?

In order to provide clues for an answer to this question we study in this chapter a one-dimensional version of the helium atom, the "stretched helium atom" (Watanabe (1987), Blümel and Reinhardt (1992)). This model, although only a "caricature" of the three-dimensional helium atom, is realistic enough to capture some of the most important physical features of the helium atom. For instance, the one-dimensional model exhibits the phenomenon of "overlapping series" (see Sections 10.1, 10.4, 10.5), the key mechanism for the emergence of complicated quantum behaviour in the helium atom. But since the series overlap only at sufficiently high excitation energy, we do not expect to see pronounced effects of classical chaos in the low-lying bound states and resonances of the helium atom. This is consistent with experiment, which shows regular sequences of resonances and smooth scattering cross-sections for the low-lying resonances of helium (see, e.g., Domke *et al.* (1995)). The same observation holds for H^-, another classically chaotic three-body system (see, e.g., Harris *et al.* (1990)).

But this regularity is not expected to extend to all energy regimes. For instance, at sufficiently high excitation energy, Blümel and Reinhardt (1992) predict the occurrence of Ericson fluctuations (see Sections 9.4 and 10.5) of the scattering cross-section of the $He^+ + e^-$ system. Furthermore, we predict a chaotic "gas of resonances" in the complex energy plane of helium. Thus, the gas of resonances as well as the fluctuating scattering cross-sections can be used as fingerprints of chaos in the helium atom. These topics are studied in detail in Section 10.5.

The focus of this chapter is not on the accurate computation of helium energy levels, but on the development of a global understanding of the different dynamical regimes of the helium atom, and on identifying them in the helium spectrum.

This chapter is organized as follows. In Section 10.1 we present a brief review of the history of the helium problem. We also discuss some elementary properties of the helium spectrum which are important in connection with the prediction of new quantum dynamical regimes. In Section 10.2 we define our model, the stretched one-dimensional helium atom. In Section 10.3 we present a detailed analysis of the classical dynamics of the one-dimensional helium atom. The quantum analysis of the one-dimensional model is presented in Section 10.4. In Section 10.4.1 we introduce the method of complex rotation and use it for the numerical computation of resonances in the complex energy plane. In Section 10.4.2 we construct a simple analytical fit formula which reproduces the real parts of the energies of low-lying bound states and resonances astonishingly well. In Section 10.4.3 we use Gutzwiller's trace formula to extract periodic orbit information from the helium spectrum computed in Section 10.4.1. The quantum manifestations of chaos in the helium spectrum are discussed in Section 10.5. There, we classify the helium spectrum into a repetitive sequence of three qualitatively different regimes: Ericson, Wigner and Rydberg. Both the Ericson regime and the Wigner regime are manifestations of chaos in the helium atom. The Ericson regime is characterized by fluctuating cross-sections in analogy to those encountered in Chapter 9. It may be detected by investigating the scattering cross-section of $He^+ + e^-$ scattering. Resonances in the Wigner regime are characterized by small widths so that bound state random matrix theory may be approximately applied. In the Wigner regime the underlying classical chaos in the helium atom manifests itself in level repulsion with a Wignerian nearest neighbour spacing distribution.

10.1 Brief history of the helium problem

Bohr's papers "On the constitution of atoms and molecules" (1913a,b) mark the birth of modern theoretical atomic physics. On the basis of his quantum postulates, Bohr derived a formula for the spectrum of the hydrogen atom, which by 1913 standards was in perfect agreement with experimental results. Encouraged by his success, Bohr immediately extended his theory to many-electron atoms, in particular to the helium atom. But, although of "the same order of magnitude" (Bohr (1913b)) his theoretical result for the helium ionization potential (27 eV) was several eV off the value of 20.5 eV measured by Franck and Hertz (1913) and interpreted as the helium ionization potential. Although it soon turned out that the 1913 Franck-Hertz experiment determined the transition to the first excited helium state rather than the helium single-electron ionization potential, Bohr's result was still not in satisfactory agreement with the experimental ionization potential (Mayer-Kuckuk (1977) cites 24.6 eV). Many researchers, including Bohr himself, tried to improve the calculations within the framework of the "old" pre 1925/26 quantum theory. The central idea in all these investigations is to identify two-electron periodic orbits and then quantize their actions on the basis of the Bohr-Sommerfeld quantization rules. Notable contributions to Bohr's theory of helium include the work of Landé (1919), Franck and Reiche (1920) and Langmuir (1921). None of these authors was able to improve substantially on Bohr's early and inaccurate result. The state of affairs was reviewed in 1922 by Van Vleck, who painted a gloomy picture of the theory of two-electron atoms within the framework of the old quantum theory. According to Van Vleck the negative results obtained for the helium atom indicate a fundamental flaw in the structure of the old quantum theory. This opinion was shared by Born and Heisenberg (1923), who reviewed and re-analysed all known helium work reported up to 1923 including a paper by Kramers (1923). Born and Heisenberg found none of the theoretical work on the helium atom in satisfactory agreement with experiment. They suggested abandoning the old quantum theory and to start looking for a new quantum theory by means of introducing "new hypotheses".

Born and Heisenberg's intuition turned out to be right on the mark when about two years later the "new" quantum theory was proposed by Heisenberg (1925) and Schrödinger (1926). Heisenberg (1926) was the first to apply the new theory to the helium atom. He obtained excellent qualitative agreement with experiments. In 1927 Kellner worked out the first quantitative approximation to the helium ground state energy. In 1928 and 1929 Kellner's result was significantly improved upon by

Hylleraas. These results showed that the new quantum theory is a reliable basis for the computation of the spectra of many-electron systems.

With the quantum theory finally on firm ground, the race was on for a more and more accurate computation of the helium ground state energy and its excited states and resonances. Early high accuracy computations of the ground state energies of the iso-electronic sequence of two-electron atoms and ions were performed by Pekeris (1958, 1959). He used the WEIZAC, one of the world's first all-purpose large-scale electronic computing machines. Additional high accuracy results on the helium ground state were reported by Kinoshita (1959).

A new field of research opened with the experimental work of Madden and Codling (1963), who proved the existence of autoionizing states embedded in the continuum. High accuracy calculations of resonances were subsequently reported by many researchers. We mention, e.g., the work by Ho (1981), Bhatia and Temkin (1984), Lindroth (1994), Müller *et al.* (1994), and Bürgers (1995).

A "reactionary" movement started with the work of Leopold and Percival (1980). Using modern semiclassical techniques these authors were able to show that the "old" quantum mechanics was not so bad after all. Improving the old theory with the help of Maslov indices and variational techniques, Leopold and Percival showed that the old quantum theory yields results for the ground state and excited states of helium that are within the experimental accuracy achieved by the 1920s. Thus, Leopold and Percival turned the "failure" of the old quantum theory into a success, since the accuracy of the semiclassical theory improves with increasing quantum numbers and turns out to be a very useful tool for the computation of highly excited states.

The most recent advance in the theory of the helium atom was the discovery of its classically chaotic nature. In connection with modern semiclassical techniques, such as Gutzwiller's periodic orbit theory and cycle expansion techniques, it was possible to obtain substantial new insight into the structure of doubly excited states of two-electron atoms and ions. This new direction in the application of chaos in atomic physics was initiated by Ezra *et al.* (1991), Kim and Ezra (1991), Richter (1991), and Blümel and Reinhardt (1992). The discussion of the manifestations of chaos in the helium atom is the focus of this chapter.

The chaotic aspects of the helium atom are not important for the low-lying bound states and resonances of the helium atom. High-lying doubly excited states of the helium atom, however, strongly feel the underlying chaos of the helium atom and give rise to new dynamical regimes. Since accurate numerical calculations in these energy regimes are not yet possible, we restrict our attention in the following sections mainly to a qualitative discussion of these new regimes.

A valuable guide in all our investigations is the independent particle model introduced by Bohr in 1913. In this model, we assume that the first electron of the helium atom is tightly bound in a state with principal quantum number N, whereas the second electron is a little further away from the helium nucleus. It is in a state with principal quantum number $N' > N$, and "sees" a nuclear charge screened by the first electron by one unit. The independent particle model completely neglects $e^- - e^-$ correlations, but qualitatively reproduces some of the most important features of the helium spectrum. For $N' \gg N$ the model is even a good quantitative approximation to the helium levels. The energy spectrum of the independent particle model is given by

$$E_{N;N'} = -54.4\,\text{eV}/N^2 - 13.6\,\text{eV}/N'^2. \qquad (10.1.1)$$

For fixed N, the second electron describes a Rydberg series of states converging to an ionization threshold at

$$E_N^* = -54.4\,\text{eV}/N^2. \qquad (10.1.2)$$

The spectrum (10.1.1) defines a doubly infinite sequence of states. They can be organized into series labelled by the first quantum number N. The states within a series are labelled by N'. The first 12 series are shown in Fig. 10.1. For energies larger than E_N^*, electron number 2 ionizes. Thus, every Rydberg series in Fig. 10.1 possesses an ionization continuum, which is also sketched.

Fig. 10.1 indicates that the low-lying Rydberg series behave in an orderly way since the only complication is the overlap of the discrete states predicted by the independent particle model with the ionization continua of lower Rydberg series. This overlap turns the "states" into auto-ionizing resonances. The low-lying resonances can be treated efficiently by quantum defect theory (see, e.g., Seaton (1983)). But Fig. 10.1(b) shows that the resonance series start to overlap from $N = 9$ on. While the overlapping of a few series for moderately high $N > 9$ can still be treated by quantum defect theory (Bürgers (1995), Domke *et al.* (1995)), it can be shown easily on the basis of (10.1.1) that for sufficiently high N arbitrarily many resonance series can overlap. In this case a description on the basis of quantum defect theory, while still possible in principle, becomes progressively more tedious. This observation indicates that a new physical picture is required. It is based on the theory of chaos and is described in the following sections.

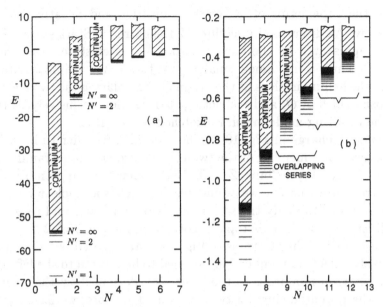

Fig. 10.1. Energy spectrum predicted by the independent particle model. (From Blümel and Reinhardt (1992).)

10.2 The one-dimensional model

Because of the $e^- - e^-$ repulsion, the lowest energy configuration of two electrons in the helium atom is expected to be a stretched configuration where the two electrons reside on opposite sides of the nucleus. This does not mean that meta-stable configurations cannot exist where the two electrons reside on the same side of the nucleus. Such "frozen planet" configurations were recently predicted by Richter and Wintgen (1990a). "Exotic" configurations of this type have already found important applications (Richter *et al.* (1991)).

We model stretched helium configurations using two-electron product states built from the extremal Stark states $|S_n^{(\leftrightarrow)}\rangle$ defined in (7.2.7). As shown in Fig. 7.4 these states are quasi-one-dimensional states highly elongated in the z direction. But a trivial rotation of the coordinate system results in a coordinate system Σ in which the states $|S_n^{(\rightarrow)}\rangle$ and $|S_n^{(\leftarrow)}\rangle$ point into the positive and negative x directions, respectively. On the basis of the extremal Stark states $|S_n^{(\leftrightarrow)}\rangle$, we construct stretched helium states according to

$$|\Psi_{nn'}^{(\pm)}\rangle = \frac{1}{\sqrt{2}}\left[|S_n^{(\rightarrow)}(1)\rangle|S_{n'}^{(\leftarrow)}(2)\rangle \pm |S_n^{(\rightarrow)}(2)\rangle|S_{n'}^{(\leftarrow)}(1)\rangle\right], \quad (10.2.1)$$

where the superscript of Ψ refers to positive and negative parities, respectively. As discussed in Section 7.2 the extremal Stark states $|S_n^{(\leftrightarrow)}\rangle$ are well represented by the one-dimensional surface states discussed in Chapter 6. Thus, a one-dimensional model of stretched helium is obtained by fixing the helium nucleus at the origin O of Σ, placing the two electrons on opposite sides of O, and restricting their motion exclusively to the x axis. The resulting model atom is sketched in Fig. 10.2.

The nuclear charge in Fig. 10.2 is not specified in order to be able to describe any other member of the two-electron iso-electronic family, such as H^-, Li^+, etc. For helium, $Z = 2$. Even when focussing on a specific two-electron atom or ion we would like to keep the nuclear charge Z variable in order to study the sensitivity of bound states and resonances to small changes of Z. This topic is covered in Section 10.5.2.

The positions of the electrons in Fig. 10.2 are denoted by $-x_1$ and x_2, respectively. Electron number 1 is assumed to be located to the left of the nucleus, electron number 2 to the right. We do not allow the electrons to cross the potential singularity at $x = 0$. Therefore, we have $x_1 > 0$, $x_2 > 0$ for all times. In analogy to the surface state electrons discussed in Chapter 6 we assume perfectly elastic reflection at $x = 0$. Thus, in the absence of the $e^- - e^-$ interaction, and because $x_1, x_2 > 0$, both electrons are governed by the same single-particle Hamiltonian

$$h(x, p; Z) = \begin{cases} \infty & \text{for } x \leq 0 \\ \dfrac{p^2}{2} - \dfrac{Z}{x} & \text{for } x > 0. \end{cases} \tag{10.2.2}$$

Except for the explicit occurrence of Z in (10.2.2) the Hamiltonian h is essentially the same as the SSE Hamiltonian (6.1.9). The complete

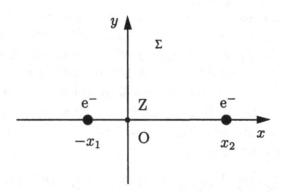

Fig. 10.2.   The one-dimensional helium atom. (Adapted from Blümel and Reinhardt (1992).)

Hamiltonian, taking the mutual $e^- - e^-$ repulsion into account, reads

$$H' = h(x_1, p_1; Z) + h(x_2, p_2; Z) + \frac{1}{x_1 + x_2}. \tag{10.2.3}$$

For the remainder of this chapter we work with Z-scaled coordinates $x \to x/Z$ and $p \to pZ$. This transformation is canonical and yields

$$H = \frac{p_1^2}{2} + \frac{p_2^2}{2} + V(\epsilon; x_1, x_2), \tag{10.2.4}$$

where

$$V(\epsilon; x_1, x_2) = -\frac{1}{x_1} - \frac{1}{x_2} + \frac{\epsilon}{x_1 + x_2}. \tag{10.2.5}$$

The scaled Hamiltonian H is obtained from H' via

$$H = H'/Z^2. \tag{10.2.6}$$

The scaled $e^- - e^-$ interaction strength ϵ is given by

$$\epsilon = 1/Z. \tag{10.2.7}$$

The Hamiltonian (10.2.4), the "stretched helium" Hamiltonian, is the basis for all our investigations. It was first investigated by Watanabe (1987). The Hamiltonian (10.2.4) is more realistic than other models of one-dimensional helium, for instance the one proposed by Lapidus (1975), where the smooth potential (10.2.5) is replaced by short-range δ-function interactions. The advantage of these models is that the δ-function potentials are easily treated quantum mechanically. The Hartree-Fock equations for δ-function models, e.g., are analytically solvable (Nogami *et al.* (1976)). The disadvantage is that because of the singular nature of the δ-function interactions, these models do not have a classical analogue.

One of the best reasons for studying the one-dimensional model of helium is the search for the quantum mechanical manifestations of classical chaos in the helium atom. Since (10.2.4) contains much of the essential physics of the three-dimensional helium atom, it is a natural starting point for quantum chaos investigations.

Although already investigated in great detail by many researchers, the physics content of (10.2.4) is far from exhausted. Many open questions concerned with both the classical and the quantum mechanics of (10.2.4) remain unanswered to date. Some of these questions are discussed in detail in later sections. Others are proposed for future research. In the following section we start the investigation of (10.2.4) with a discussion of its classical mechanics.

10.3 Classical dynamics

In this section we interpret the Hamiltonian (10.2.4) as a classical Hamiltonian function of the conjugate variables (x_1, p_1) and (x_2, p_2). Transforming to action and angle variables with the help of the canonical transformation (6.1.18), the Hamiltonian (10.2.4) becomes:

$$H = E = -\frac{1}{2I_1^2} - \frac{1}{2I_2^2} + \frac{\epsilon}{2[I_1^2 \sin^2(\eta_1) + I_2^2 \sin^2(\eta_2)]} \qquad (10.3.1)$$

with

$$\theta_i = 2\eta_i - \sin(2\eta_i); \quad i = 1, 2. \qquad (10.3.2)$$

The equations of motion for I_i and θ_i, $i = 1, 2$, are given by:

$$\dot{I}_i = \frac{\epsilon}{4} \frac{I_i^2 \cot(\eta_i)}{A(I_1, \eta_1; I_2, \eta_2)}, \quad \dot{\theta}_i = \frac{1}{I_i^3} - \frac{\epsilon I_i \sin^2(\eta_i)}{A(I_1, \eta_1; I_2, \eta_2)}, \qquad (10.3.3)$$

where

$$A(I_1, \eta_1; I_2, \eta_2) = \left[I_1^2 \sin^2(\eta_1) + I_2^2 \sin^2(\eta_2) \right]^2. \qquad (10.3.4)$$

In order to be useful, (10.3.3) has to be regularized. Defining a new time τ according to

$$\frac{dt}{d\tau} = \sin^2(\eta_1) \sin^2(\eta_2) \qquad (10.3.5)$$

and choosing η_i instead of θ_i as the angle variables, we obtain the following set of equations of motion:

$$\frac{dI_1}{d\tau} = \frac{\epsilon I_1^2}{4A(I_1, \eta_1; I_2 \eta_2)} \cos(\eta_1) \sin(\eta_1) \sin^2(\eta_2), \qquad (10.3.6a)$$

$$\frac{dI_2}{d\tau} = \frac{\epsilon I_2^2}{4A(I_1, \eta_1; I_2, \eta_2)} \cos(\eta_2) \sin(\eta_2) \sin^2(\eta_1), \qquad (10.3.6b)$$

$$\frac{d\eta_1}{d\tau} = \frac{1}{4} \left\{ \frac{1}{I_1^3} - \frac{\epsilon I_1 \sin^2(\eta_1)}{A(I_1, \eta_1; I_2, \eta_2)} \right\} \sin^2(\eta_2), \qquad (10.3.6c)$$

$$\frac{d\eta_2}{d\tau} = \frac{1}{4} \left\{ \frac{1}{I_2^3} - \frac{\epsilon I_2 \sin^2(\eta_2)}{A(I_1, \eta_1; I_2, \eta_2)} \right\} \sin^2(\eta_1). \qquad (10.3.6d)$$

For all practical purposes the set (10.3.6) is regularized. But there remains a problem when both η_1 and η_2 simultaneously tend to zero modulo π. This means that x_1 and $x_2 \to 0$ simultaneously. We call this situation the "triple collision case".

The classical Hamiltonian (10.3.1) has an important scaling property. Consider a solution $(I_1(t), I_2(t), \theta_1(t), \theta_2(t))$ of the equations of motion

(10.3.6) at fixed energy E. A new solution $(I_1'(t'), I_2'(t'), \theta_1'(t'), \theta_2'(t'))$ at energy E' is generated if we scale the quantities in (10.3.6) according to

$$I_1' = I_1 \sqrt{E/E'}, \quad I_2' = I_2 \sqrt{E/E'}, \quad t' = (E/E')^{3/2} t, \quad (10.3.7)$$

and leave the angle variables unchanged. At fixed energy E the action along a path Γ in the $(I_1, \theta_1, I_2, \theta_2)$ phase space is given by

$$S_\Gamma(E) = \int_\Gamma [I_1 \, d\theta_1 + I_2 \, d\theta_2] = \sqrt{E'/E} \, S_{\tilde{\Gamma}}(E'), \quad (10.3.8)$$

where $\tilde{\Gamma}$ is the scaled path at energy E'. Because of (10.3.7) the time $T_\Gamma(E)$ it takes to traverse the path Γ at energy E scales according to

$$T_\Gamma(E) = (E'/E)^{3/2} T_{\tilde{\Gamma}}(E'). \quad (10.3.9)$$

For $E < 0$ we choose the representative reference energy $E' = -1$ and obtain:

$$I_i(E) = |E|^{-1/2} I_i(E = -1), \quad S(E) = |E|^{-1/2} S(E = -1),$$

$$T(E) = |E|^{-3/2} T(E = -1). \quad (10.3.10)$$

Closed orbits are of special interest for the semiclassical quantization of the helium atom. For a closed orbit Γ we have

$$T_\Gamma(E) = \frac{\partial S(E)}{\partial E} = \frac{1}{2} S_{\tilde{\Gamma}}(E = -1). \quad (10.3.11)$$

We have used the set of equations (10.3.6) to search for regular islands in the phase space of the stretched helium model. We did not find any. We suspect that in analogy to the kicked hydrogen atom discussed in Chapter 8 there really are no regular islands. Therefore, the stretched helium atom may well turn out to be another physical example of a strongly chaotic system. In the absence of any regular islands, this system is auto-ionizing for any generic choice of initial conditions. Only phase-space points that are members of (unstable) periodic orbits do not ionize.

In Chapter 8 we were able to present an analytical proof for the absence of regular islands. An analogous proof for the stretched helium atom is currently not known. Therefore, although backed by strong numerical evidence, it is currently not "safe" to claim that the stretched helium atom is a strongly chaotic system. Similar conjectures for a related two-dimensional Hamiltonian system with an "$x^2 y^2$ potential" turned out to be wrong when Dahlqvist and Russberg (1990) found tiny stable islands in its phase space. We think that the degree of chaoticity of the stretched helium model is an important unsolved problem.

10.3.1 Periodic orbits

Currently available numerical results indicate that the one-dimensional helium atom is completely chaotic. The best-known semiclassical quantization procedure for completely chaotic systems is Gutzwiller's trace formula (see Section 4.1.3), which is based on classical periodic orbits. Therefore we search for simple periodic orbits of the one-dimensional helium atom. Since a two-electron orbit is periodic if the orbits $(n_1(t)$, $\theta_1(t))$ and $(n_2(t)$, $\theta_2(t))$ of the first and second electron have a common period, the periodic orbits of the one-dimensional model can be labelled with two integers, m and n, which count the 2π-multiplicity of the angle variables θ_1 and θ_2 after completion of the orbit. Therefore, if for some periodic orbit

$$\theta_1(T) = 2m\pi + \theta_1(0), \quad \theta_2(T) = 2n\pi + \theta_2(0), \qquad (10.3.12)$$

we call this periodic orbit an $m : n$ periodic orbit. Some simple periodic orbits at energy $E = -1$ and $\epsilon = 1/2$ (helium) are shown in Figs. 10.3 and 10.4. Fig. 10.3 shows four $m : 1$ - type periodic orbits. Fig. 10.4 shows four examples of $m : n$ - type periodic orbits.

There is one immediate problem with the "$m : n$" classification scheme. It concerns the possibility of retrograde motion, i.e. the possibility that one of the angles θ_1 or θ_2, or both, do not increase monotonically while traversing the periodic orbit. In this case, two different periodic orbits may end up with the same "$m : n$" label. It is currently not clear whether this situation can occur. A careful study has to be based on the equations of motion (10.3.3).

Besides the "$m : n$" classification there is another way to code periodic trajectories. Figs. 10.3 and 10.4 show that the periodic trajectories can be characterized by their bounces with the x_1 and x_2 axes. If we code a bounce with the x_1 axis by the letter "1" and a bounce with the x_2 axis by the letter "2", we can characterize periodic orbits by periodic symbol sequences over the alphabet consisting of the two letters "1" and "2". Thus, the periodic orbits shown in Figs. 10.3(a) – (d) can be characterized by the strings $\overline{12}$, $\overline{122}$, $\overline{1222}$ and $\overline{12222222222}$, respectively. As introduced in Section 2.4 the overbar indicates that the string under the bar has to be repeated indefinitely. The periodic orbits shown in Fig. 10.4(a) – (d) are represented by the strings $\overline{22121}$, $\overline{2221221}$, $\overline{2212121}$, and $\overline{22222111}$, respectively. A "reduced code", based on the symbols "+" and "−" was introduced by Ezra *et al.* (1991). It is constructed from the code based on the symbols "1" and "2" according to the following rule: scan a given string in the "1,2" representation. Replace every symbol

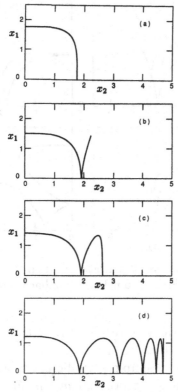

Fig. 10.3. Examples of $m:1$ - type periodic orbits in the $x_1 - x_2$ plane. (a) 1:1, (b) 2:1, (c) 3:1, (d) 10:1. (From Blümel and Reinhardt (1992).)

combination "12" or "21" by the symbol "$-$", and replace every combination "11" or "22" by "$+$". Thus, two consecutive bounces on the same axis are denoted by "$+$". Two consecutive bounces on different axes are denoted by "$-$". In the "$\pm$" classification scheme the periodic orbits shown in Fig. 10.3(a) – (d) are coded by "$-$", "$+--$", "$++--$" and "$+++++++++--$", respectively. For long periodic orbits, such as the 10:1 orbit shown in Fig. 10.3. (d), the code consists of strings of many repeated symbols of the same kind. In order to have a more compact notation, we use subscripts to denote the number of repeated symbols. This notation is constructed in analogy to chemical formulae. In this "chemical" shorthand notation the codes for the periodic orbits in Fig. 10.3 are given by "$-_1$" $\equiv$ "$-$", "$+_1-_2$" $\equiv$ "$+-_2$", "$+_2-_2$", and "$+_9-_2$", respectively. Using the shorthand notation the reduced code for the periodic orbits shown in Fig. 10.4 are given by "$+-_4$", "$+_2-_2+-_2$", "$+-_6$", and "$+_4-+_2-$", respectively. The "$\pm$" code is not only useful for labelling trajectories, it also performs automatic symmetry reduction of

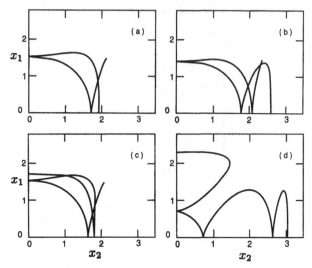

Fig. 10.4.   Some $m : 2$ and $m : 3$ periodic orbits in the $x_1 - x_2$ plane. (a) 3:2, (b) 5:2, (c) 4:3, (d) 5:3. (From Blümel and Reinhardt (1992).)

periodic orbits symmetric with respect to the line $x_1 = x_2$. This feature is discussed in detail by Richter (1991).

The properties of a few simple periodic orbits are displayed in Table 10.1.   Column 1 indicates the orbit in $m : n$ notation, column 2 is its reduced symbolic code, column 3 lists the action $S_m = \oint p_1 \, dx_1$ of the first electron and column 4 lists the action $S_n = \oint p_2 \, dx_2$ of the second electron.   Column 5 displays the total action S of the orbit and column 6 lists the scaled traversal time τ of the orbit.   The natural traversal times T of the orbits are not listed since they are obtained trivially from (10.3.11) as $T = S/2$.

The collection of orbits and their properties listed in Table 10.1 turn out to be useful in Section 10.4.3, where we extract periodic orbit information from the exact quantum spectrum of the one-dimensional model. The properties of the periodic orbits corroborate the claim that the one-dimensional helium atom is completely chaotic: All periodic orbits found so far are unstable with a positive Lyapunov exponent.

10.3.2 Helium: a chaotic scattering system

The stretched helium model can be interpreted as a chaotic scattering system.   In order to establish the connection, we first consider the potential (10.2.5) of the Hamiltonian (10.2.4).   Some equi-potential lines for $V(\epsilon; x_1, x_2)$ are shown in Fig. 10.5 for $\epsilon = 1/2$.   This potential is very similar to the potential of "elbow" scattering systems investigated

Table 10.1. Properties of a few simple periodic orbits.
(Adapted from Blümel and Reinhardt (1992).)

$m:n$	code	S_m	S_n	S	τ
1:1	−	5.75	5.75	11.49	14.44
2:1	$+-_2$	10.59	6.36	16.95	21.78
3:1	$+_2-_2$	15.31	6.81	22.12	28.56
4:1	$+_3-_2$	19.96	7.19	27.15	35.07
5:1	$+_4-_2$	24.59	7.51	32.10	41.40
6:1	$+_5-_2$	29.18	7.80	36.98	47.60
7:1	$+_6-_2$	33.76	8.05	41.82	53.71
8:1	$+_7-_2$	38.33	8.29	46.62	59.74
9:1	$+_8-_2$	42.89	8.50	51.39	65.71
10:1	$+_9-_2$	47.44	8.71	56.14	71.64
3:2	$+-_4$	16.36	12.14	28.50	36.87
5:2	$+_2-_2+-_2$	25.90	13.18	39.09	50.61
4:3	$+-_6$	22.12	17.89	40.01	51.59
5:3	$+_4-+_2-$	26.09	17.55	43.64	61.78

analytically, numerically and experimentally by Doron *et al.* (1990) in connection with chaotic scattering.

The potential shown in Fig. 10.5 clearly exhibits two asymptotic entry (exit) channels. The channel $x_1 \to \infty$ is labelled C_1, the channel $x_2 \to \infty$ is labelled C_2. Scattering potentials of the type shown in Fig. 10.5 also

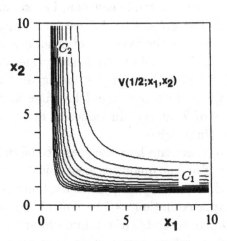

Fig. 10.5. Equi-potential lines for the potential $V(\epsilon = 1/2; x_1, x_2)$ for $V = -1.5$ to $V = -0.5$ in steps of $\Delta V = 0.1$. (Adapted from Blümel and Reinhardt (1995).)

occur in the theoretical description of chemical reactions. Since complicated dynamics and chaotic scattering were reported to occur in these
systems it is natural to expect chaotic scattering to occur in the potential $V(\epsilon; x_1, x_2)$. In order to prove this point we launched 800 scattering
trajectories at $I_1^{(j)} = 1$, $\theta_1^{(j)} = j \cdot (2\pi/801), j = 1, 2, ..., 800$, $x_2 = 10$ and
$p_2 = -0.5$. According to these specifications, the initial condition corre-
sponds to electron number 1 residing in a bounded bouncing state with
action $I_1 = 1$ while the second electron enters through channel C_2 headed
for a collision with electron number 1. According to the initial conditions the total energy of the scattering process is negative. Therefore,
two-electron break-up is impossible and there remain three possible final
states of the scattering process. (1) Electron number 1 remains bounded
in a bouncing state with action I_f while electron number 2 leaves through
either of the channels, C_1 or C_2, respectively. (2) Electron number 2 gets
caught while electron number 1 is kicked out of the potential and leaves
through C_1 or C_2. This is an exchange reaction. (3) No period of waiting
will reveal a "final state". The two electrons are dynamically trapped in
the potential, although according to their energy one of the electrons is
free to leave. As discussed in Chapter 9 we call this situation a dynamic
scattering singularity. In order to reveal whether scattering singularities
exist for the one-dimensional helium model, we calculated the time T it
takes to reach the "asymptotic" regions C_1 and C_2, which, for practical
reasons, we define to be $x_1 \geq 10$ or $x_2 \geq 10$. The result is shown in
Fig. 10.6. The delay time T varies smoothly with the initial phase θ_1 for
$0 < \theta_1 < 4.5$. At around $\theta_1 \approx 4.55$ we see a cusp-like behaviour which
is due to the occurrence of a triple collision, i.e. x_1 and x_2 both tend to
zero simultaneously. For $4.7 < \theta_1 < 5.7$ we see complicated dependence
of T on θ_1, which indicates the presence of chaotic scattering. Moreover,
the delay time T is unbounded at particular values of θ_1. For this reason we had to cut the delay time shown in Fig. 10.6 at some level. For
graphical reasons we chose to cut the delay time at the level $T = 200$.
The unboundedness of T at certain values of θ_1 proves the presence of
dynamic scattering singularities.

It is interesting to speculate how the presence of classical chaotic scattering manifests itself on the quantum mechanical level. In Chapter 9 we
saw that classical chaotic scattering yields a quantum scattering matrix,
which to a good approximation can be considered to be a representative of Dyson's circular ensemble of random unitary symmetric matrices.
Since in the stretched helium system time reversal symmetry holds, we
expect to see the same behaviour for the scattering matrix describing
electron scattering off the one-dimensional He$^+$ ion. The situation is

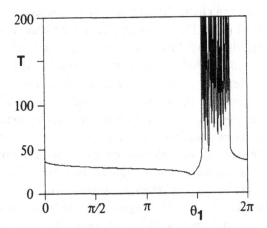

Fig. 10.6. Distribution of lifetimes for a chaotic scattering experiment with the one-dimensional model. Irregular behaviour and dynamic scattering singularities in the interval $4.7 < \theta_1 < 5.7$ indicate the presence of chaotic scattering in the stretched helium atom. (Adapted from Blümel and Reinhardt (1995).)

somewhat complicated by the presence of thresholds in the helium spectrum. Since close to thresholds we expect to see regular progressions of Rydberg states, we expect good agreement with (at least some of) the predictions of random matrix theory only in between and far away from thresholds. According to the arguments presented in Chapter 9 we also expect to find an energy regime which exhibits Ericson fluctuations. This idea is pursued further in Section 10.5.1.

10.4 Quantum analysis

The quantized version of the Hamiltonian (10.2.4) is given by

$$\hat{H} = \hat{H}_0(x_1) + \hat{H}_0(x_2) + \frac{\epsilon}{x_1 + x_2}, \qquad (10.4.1)$$

where $\hat{H}_0$ is the unperturbed SSE Hamiltonian defined in (6.1.22). The Hamiltonian $\hat{H}$ defined in (10.4.1) is invariant under the exchange transformation $x_1 \leftrightarrow x_2$. This means that parity is a good quantum number. Thus the eigenstates $|\psi(x_1, x_2)\rangle$ of $\hat{H}$,

$$\hat{H}\,|\psi(x_1, x_2)\rangle = E\,|\psi(x_1, x_2)\rangle, \qquad (10.4.2)$$

can be labelled according to parity. Introducing the parity quantum number σ, which can take the values ± 1, even and odd eigenstates are defined according to:

$$|\psi^{(\sigma)}(x_2, x_1)\rangle = \sigma|\psi^{(\sigma)}(x_1, x_2)\rangle. \qquad (10.4.3)$$

In Section 10.4.1 we use the technique of complex rotation to compute the spectrum (10.4.2) of (10.4.1). In Section 10.4.2 we construct a useful analytical energy formula. This formula is used to assign, wherever possible, quantum numbers to the helium resonances computed in Section 10.4.1. It is also used in Section 10.5.1 to estimate the onset of the regime of overlapping resonances in the one-dimensional helium atom. Resonance overlap is the first step on the way to Ericson fluctuations, the regime of strongly overlapping resonances. In analogy to the discussion of Ericson fluctuations in Chapter 9, we call the regime of strongly overlapping resonances the Ericson regime. It also marks the onset of quantum chaotic scattering in the $He^+ + e^-$ scattering system. In section 10.4.3 we use recently developed techniques (Wintgen (1987)) to extract classical periodic orbit information from the quantum spectrum computed in Section 10.4.1.

10.4.1 Complex rotation

The method of complex rotation, introduced by Aguilar and Combes (1971) and Balslev and Combes (1971) is reviewed, e.g., by Junker (1982), Reinhardt (1982), Ho (1983) and Buchleitner *et al.* (1994). It is a convenient and powerful tool for the calculation of positions and widths of bound states and resonances. Introducing the complex coordinates

$$x \to e^{i\theta} x, \quad p \to e^{-i\theta} p, \tag{10.4.4}$$

where θ is a real angle, the Hamiltonian (10.4.1) becomes:

$$\hat{H} = e^{-2i\theta} \left(\frac{\hat{p}_1^2}{2} + \frac{\hat{p}_2^2}{2} \right) - e^{-i\theta} \left(\frac{1}{x_1} + \frac{1}{x_2} \right) + \frac{\epsilon e^{-i\theta}}{x_1 + x_2}. \tag{10.4.5}$$

The Hamiltonian (10.4.5) is symmetric but no longer Hermitian. A property of (10.4.5) is that it rotates the continuum cuts of the two-electron problem by an angle 2θ into the lower complex energy plane, thus exposing the resonances. This property of (10.4.5) is illustrated in Fig. 10.7. The idea of the complex rotation method is to diagonalize the complex Hamiltonian (10.4.5) in a complete basis of square normalizable states. This technique is effective because, as a result of complex rotation, resonances are represented by square integrable wave functions. Most important for our purposes is the fact that a single diagonalization of (10.4.5) yields both the resonance positions (the real parts of the eigenvalues of (10.4.5)) and the resonance widths (the imaginary parts of the eigenvalues of (10.4.5)).

We diagonalize the Hamiltonian (10.4.5) in the space of two-particle states

$$\psi_{mn}^{(\sigma)}(x_1, x_2; \lambda) = N_{mn}^{(\sigma)} \left\{ \psi_m(x_1; \lambda)\, \psi_n(x_2; \lambda) + \sigma\, \psi_m(x_2; \lambda)\, \psi_n(x_1; \lambda) \right\},$$
(10.4.6)

where the normalization constants $N_{mn}^{(\sigma)}$ in (10.4.6) are chosen such that the states (10.4.6) are normalized. The single-particle states in (10.4.6) are obtained from the SSE Sturmian states defined in (6.2.3) for $\alpha = 2/\lambda$, where λ is a scale parameter. They are given explicitly by

$$\psi_n(x; \lambda) = \left[\frac{\lambda}{n(n+1)} \right]^{1/2} (\lambda x)\, L_{n-1}^{(2)}(\lambda x)\, \exp(-\lambda x/2).$$
(10.4.7)

The scale parameter λ in (10.4.7) is used to adjust the spatial range of the basis functions to the spatial extension of the wave functions of the model atom.

We diagonalized the Hamiltonian (10.4.5) for $\epsilon = 1/2$ (helium) for positive parity ($\sigma = +1$) using the first 39 single-particle states (10.4.7). For such a large basis the resonances should be only weakly dependent on the rotation angle θ. Therefore, we fixed θ at $\theta = 10°$. Optimizing λ, however, is essential. The scale parameter λ was adjusted such that the computed resonance energies are approximately stationary with respect to variations of λ. The resulting energies and widths of some low-lying states and resonances are presented graphically in the form of resonance charts in Fig. 10.8. The representation of the data in Fig. 10.8 makes use of the fact that the low energy states of the helium atom can be classified into series characterized by the ionization threshold index m.

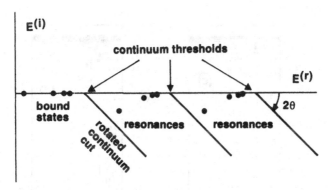

Fig. 10.7. Bound states and resonances in the complex plane with axes $E^{(r)} = \Re e(E)$ and $E^{(i)} = \Im m(E)$. Complex scaling with rotation angle θ rotates the continuum cuts into the negative imaginary energy plane (angle 2θ) thus exposing the resonances. (Adapted from Blümel and Reinhardt (1992).)

This feature reminds us of the independent particle model discussed in Section 10.1 (see Fig. 10.1). Thus the real parts $E_{mn}^{(r)}$ and the imaginary parts $E_{mn}^{(i)}$ of the resonances can be classified according to the threshold index m and the series index n corresponding to a given threshold m. Typically we computed about 13 states for every given threshold index m, with m ranging from 1 to 10. While the assignment of a state to a given m sequence is unique for $m \leq 5$, the assignment becomes more and more difficult for higher m. This progressive loss of our ability to assign quantum numbers with increasing m is a definite manifestation of the chaotic properties of the one-dimensional helium atom.

The dots shown in the ten panels of Fig. 10.8 indicate the positions of bound states and resonances in the complex energy plane. The horizontal

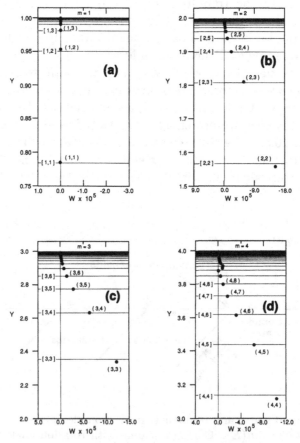

Fig. 10.8.   Positive parity bound states and resonances of the one-dimensional model for $\epsilon = 1/2$ and $m = 1, ..., 10$. (From Blümel and Reinhardt (1992).)

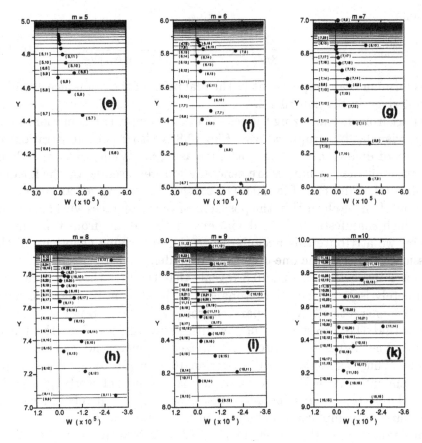

Fig. 10.8   Continued

axis of each panel corresponds to the widths of the resonances, the vertical axis to the real parts of their energies. Every one of the panels corresponds to a specific m threshold. Whenever graphically possible we have labelled the resonances with their m and n quantum numbers. For graphical reasons we do not use the real part of the energies $E^{(r)}$ themselves, but represent these real parts by the scaled energy

$$Y = \frac{1}{\sqrt{-2E^{(r)}}}. \qquad (10.4.8)$$

For $n \to \infty$ the real parts of the resonances of an m series converge toward the ionization threshold energy $E_m = -1/2m^2$. Therefore, as shown in Fig. 10.8, the scaled energy of the resonances converges toward $Y = m$ for highly excited members of the m series. The fine horizontal lines in the panels correspond to the energy levels predicted by the analytical formula

$$E_{mn}^{(r)} = -\frac{1}{2m^2} - \frac{\zeta^2}{8n^2}, \quad \zeta = 1 + \frac{\sqrt{e}}{(1+n/m)^{3/2}}. \qquad (10.4.9)$$

This formula is an empirical fit to the real parts of the resonance energies. It is derived in Section 10.4.2.

Apart from summarizing the results of our computer calculations in an easily visualizable graphical way, Fig. 10.8 is also a good illustration for the loss of quantum numbers with increasing m. While up to $m = 3$ the real parts of the resonances are generally well represented by the formula (10.4.9) (the "dots" in Fig. 10.8 fall on the "lines") we see an increasing amount of "irregularity" in the positions of the resonances with increasing m, which manifests itself in deviations of the "dots" from the "lines". Again, we attribute this irregularity to the underlying chaoticity of the classical version of the one-dimensional model.

Closer inspection of Fig. 10.8 reveals that the first three m series are completely isolated, i.e. the resonances with different m quantum numbers are energetically well separated from each other. But Fig. 10.8 (d) shows that the lowest state of the fifth series intrudes into the fourth series. The higher series are characterized by more and more intruder states. Thus, as conjectured in Section 10.1 on the basis of the simple independent-particle $He^+ + e^-$ model, it is the onset of overlapping series with the accompanying intruder states that, on the quantum level, are the source of complicated behaviour in the helium spectrum.

In connection with the intruder states an interesting detail emerges. According to Fig. 10.8(d) it looks as if the (4,10) resonance was pushed toward the real axis by the intruder (5,5). According to Figs. 10.8(e) and (f), e.g., this happens for the higher m sequences as well. Sometimes the resonance is so close to the real axis that one is tempted to call it a bound state embedded in the continuum. Investigations of the resonances of the three-dimensional helium atom show the same behaviour (Bürgers and Wintgen (1994)). It is possible that a resonance close to the real axis can be made to lie exactly on axis, i.e. can be forced to be bound, by a particular choice of ϵ.

In the following section we show how we arrived at the concise analytical expression (10.4.9) as a good fit to the real parts of the resonance energies.

10.4.2 The helium spectrum: an empirical formula

The purpose of this section is to derive an analytical formula for the positions of bound states and resonances of the one-dimensional model. The derivation of the energy formula proceeds in two steps. First, we use a perturbative argument to motivate the functional form of the analytical formula. Its final form is then obtained by graphically analysing the resonance data computed in Section 10.4.1.

For a perturbative calculation of the spectrum of (10.4.1) we assume that electron 1 is in the single-particle SSE state $|m\rangle$ and electron 2 is in a highly excited state, much more weakly bound than electron 1. In this situation electron 2 spends most of its time in regions with $x_2 \gg x_1$. Therefore, we assume that the eigenfunctions of (10.4.1) are approximately of the form

$$\Phi_{mn}(x_1, x_2) = \varphi_m(x_1) \, \beta_{mn}(x_2), \qquad (10.4.10)$$

where φ_m is defined in (6.1.24). Since the purpose of the following perturbative arguments is merely to suggest the scaling behaviour of our empirical energy formula, we neglect to symmetrize (antisymmetrize) (10.4.10). Using (10.4.10) in (10.4.2) we obtain

$$\left\{ -\frac{1}{2}\frac{d^2}{dx_2^2} - \frac{1}{x_2} + \epsilon \int_0^\infty \frac{|\varphi_m^2(x_1)|^2}{x_1 + x_2} dx_1 + E_m \right\} \beta_{mn}(x_2) =$$

$$E_{mn} \, \beta_{mn}(x_2), \qquad (10.4.11)$$

where $E_m = -1/2m^2$. This equation can be interpreted as a mean field equation for electron number 2, which moves in the combined potentials $-1/x$, generated by the nuclear charge, and the mean field potential

$$v_m(x) = \int_0^\infty \frac{|\varphi_m^2(x_1)|^2}{x_1 + x} dx_1, \qquad (10.4.12)$$

generated by electron 1. Asymptotically the mean field potential behaves like $v_m(x) \sim 1/x$. This suggests the definition of the reduced potential

$$\tilde{v}_m(x) = v_m(x) - \frac{1}{x}. \qquad (10.4.13)$$

Together with

$$\tilde{E}_{mn} = E_{mn} - E_m, \qquad (10.4.14)$$

the mean field equation (10.4.11) becomes

$$\left\{ -\frac{1}{2}\frac{d^2}{dx^2} - \frac{1-\epsilon}{x} + \epsilon\tilde{v}_m(x) \right\} \beta_{mn}(x) = \tilde{E}_{mn} \, \beta_{mn}(x). \qquad (10.4.15)$$

Neglecting $\tilde{v}$ in (10.4.15), the mean field equation is solved by the zeroth order wave functions

$$\beta_{mn}(x) = \varphi_n(x; Z^*) =$$

$$(Z^*)^{1/2} \, n^{-3/2} \, (2Z^* x/n) \, L_{n-1}^{(1)}(2Z^* x/n) \, e^{-Z^* x/n}, \qquad (10.4.16)$$

where

$$Z^* = 1 - \epsilon. \tag{10.4.17}$$

The energy eigenvalue is given by

$$\tilde{E}_{mn} = -\frac{Z^{*2}}{2n^2}. \tag{10.4.18}$$

The approximate spectrum of (10.4.1) thus becomes

$$E_{mn} = -\frac{1}{2m^2} - \frac{(1-\epsilon)^2}{2n^2}. \tag{10.4.19}$$

The energy shift ΔE_{mn} in first order perturbation theory is given by the expectation value of $\tilde{v}$ with the zeroth order wave functions (10.4.16). An exact calculation of the expectation value is difficult. But since we are only interested in the scaling of ΔE_{mn} in m and n for large m, n, we evaluate it semiclassically. Semiclassically, φ_m is given by

$$\varphi_m(\theta) = \frac{1}{\sqrt{2\pi}} e^{im\theta} \quad ; \quad 0 \le \theta < 2\pi, \tag{10.4.20}$$

and

$$x = 2\frac{n^2}{Z^*} \sin^2(\eta). \tag{10.4.21}$$

Since semiclassically $|\varphi_m|^2 = 1$, we obtain

$$\Delta E_{mn} \sim \int_0^{2\pi} d\theta \int_0^{2\pi} d\theta' \frac{1}{n^2 \sin^2(\eta)/Z^* + m^2 \sin^2(\eta')} \sim \frac{1}{n^2} \tilde{f}(n/m). \tag{10.4.22}$$

For $\epsilon = 1/2$ and including the energy correction (10.4.22) the eigenvalues E_{mn} in (10.4.11) can now be written as

$$E_{mn} = -\frac{1}{2m^2} - \frac{\zeta^2}{8n^2}. \tag{10.4.23}$$

This is exactly the energy formula (10.4.9) used in Section 10.4.1. Here ζ^2 is of the form $1 + f(n/m)$, where the function f, a small correction, scales in the variable n/m. Since f is assumed to be small, we plotted $\zeta - 1$ versus n/m using all the data displayed in Fig. 10.8. We obtained a straight line, slightly bent at $n/m \approx 1$, with an asymptotic slope of approximately -1.5 for $n/m \gg 1$. This indicates that $\zeta - 1$ approximately scales in the variable $C + n/m$. For $C = 1$ we obtain the result shown in Fig. 10.9. The data points, representing all the resonances computed in Section 10.4.1, approximately collapse on a straight line with slope ≈ -1.5. Therefore we obtain an excellent fit to the data by

$$\zeta - 1 = \frac{\alpha}{(1 + n/m)^{1.5}}. \tag{10.4.24}$$

The constant α is obtained directly from Fig. 10.9. It turns out to be close to $\sqrt{e}$. Thus we arrive at the formula (10.4.9). This fit formula approximates our numerically computed resonance data for all n and m to better than 3%. For $n \geq m + 1$ the accuracy is better than 1%.

Based on the above perturbation arguments the formula (10.4.9) may be improved by plotting $\zeta^2 - 1$ versus $C + n/m$. We would like to improve (10.4.9) since it allows us to compute the average level density for the one-dimensional helium atom. The average level density is an important quantity to know. It is used, e.g., in Section 10.4.3 to decompose the level density into its mean and fluctuating parts. The fluctuating part of the level density is then related to classical periodic orbits on the basis of Gutzwiller's trace formula discussed in Section 4.1.3. The mean level density is also an essential ingredient for the prediction of the Ericson regime. Therefore, formula (10.4.9) plays an important role in Section 10.5.1 as well.

10.4.3 Periodic orbits and scaled energy spectroscopy

According to Gutzwiller's result (4.1.72) the density of states can be computed as a generalized Fourier sum which contains only classical periodic orbit information. Therefore, using a suitable projection technique it should be possible to "invert" the transform and extract classical periodic orbit information from the level density. The inversion can be achieved using *scaled energy spectroscopy*, a technique first introduced by

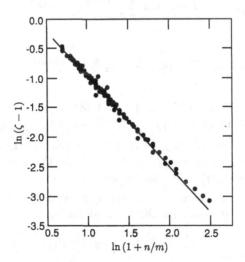

Fig. 10.9. Plot of $\ln(\zeta - 1)$ versus $\ln(1 + n/m)$ for the real parts of bound states and resonances displayed in Fig. 10.8. The resonances (full circles) lie close to a straight line with slope -1.5. (From Blümel and Reinhardt (1992).)

Wintgen (1987). This section consists of three parts. In part (a) we derive the trace formula for the one-dimensional helium atom, a system with an "odd-even" symmetry. In part (b) we use the classical scaling properties of the one-dimensional helium atom to apply the scaled energy technique. In part (c) we generalize the technique to apply to autonomous systems without scaling symmetries.

(a) *The trace formula in the presence of symmetries*

In order to account for the symmetries of our model system, we have to construct the properly symmetrized and antisymmetrized Green's functions and use them in the trace formula to obtain the semiclassical energies for positive and negative parity states, respectively. The symmetric and antisymmetric energy dependent Green's functions $G_\pm$ are given by:

$$G_\pm(x_1''x_2'', x_1'x_2'; E) = \frac{1}{2}\left\{G(x_1''x_2'', x_1'x_2'; E) \pm G(x_2''x_1'', x_1'x_2'; E)\right\}.$$
(10.4.25)

We use (10.4.25) to obtain the level densities of positive and negative parity states, $\rho_+(E)$ and $\rho_-(E)$, respectively, according to

$$\rho_\pm(E) = -\frac{1}{\pi}\Im m \int dx_1 dx_2\, G_\pm(x_1 x_2, x_1, x_2; E) =$$

$$\bar\rho_\pm(E) + \tilde\rho^{(1)} \pm \tilde\rho^{(2)},$$
(10.4.26)

where

$$\tilde\rho^{(1)}(E) = \frac{1}{4\pi^2\hbar^2}\Im m \int dx_1 dx_2 \sum_{c.tr.} D^{1/2} e^{\frac{i}{\hbar}\oint_{x_1 x_2}^{x_1 x_2}\vec{p}d\vec{x} - i\times\text{phases}}$$
(10.4.27)

and

$$\tilde\rho^{(2)}(E) = \sum_{c.tr.} D^{1/2} e^{\frac{i}{\hbar}\int_{x_1 x_2}^{x_2 x_1}\vec{p}d\vec{x} - i\times\text{phases}}.$$
(10.4.28)

The constant D in (10.4.27) and (10.4.28) is the Van Vleck determinant (Gutzwiller (1990)). Since $\rho(E) = \rho_+(E) + \rho_-(E)$, $\tilde\rho^{(2)}$ does not contribute to ρ. Therefore $\tilde\rho^{(1)}$ has to be exactly 1/2 of the familiar Gutzwiller formula. Since $\rho_+(E) \approx \rho_-(E)$ in the semiclassical regime, the factor 1/2 is exactly what we expect.

Evaluating $\tilde\rho^{(2)}$ in stationary phase approximation we get something new. The stationary phase condition is given by

$$p_1'' = p_2', \qquad p_2'' = p_1',$$
(10.4.29)

where p_1 and p_2 are the momenta conjugate to x_1 and x_2. According to (10.4.29) the trajectories contributing to $\tilde\rho^{(2)}$ leave at (x_1, x_2) with momentum (p_1, p_2) and arrive at the mirror reflected point (x_2, x_1) with mirror reflected momentum (p_2, p_1). In other words, contributions to $\tilde{\rho}^{(2)}$ come from periodic orbits which are invariant under mirror reflection at $x_1 = x_2$. An example is the orbit displayed in Fig. 10.3(a). The condition (10.4.29) is met after traversal of half the periodic orbit. According to Figs. 10.3 and 10.4 there are basically three different classes of periodic orbits: (i) the symmetric stretch orbit, (ii) symmetric orbits and (iii) antisymmetric orbits. They contribute three different types of terms to the trace formula for the positive and negative parity level densities whose appearance resembles the appearance of the general trace formula (4.1.72). In the following $\tilde{\rho}_\pm$ denotes the fluctuating part of the level densities, i.e. $\tilde{\rho}_\pm = \rho_\pm - \bar{\rho}_\pm$, and $\Delta\tilde{\rho}_\pm$ denotes the contribution of a specific type of orbit to the Gutzwiller sum.

(i) For the symmetric stretch orbit the momentum and its mirror image coincide. Therefore, it does not contribute to $\tilde{\rho}_-$ since in this case $\tilde{\rho}^{(1)}$ and $\tilde{\rho}^{(2)}$ cancel. The symmetric stretch orbit contributes with double weight to $\tilde{\rho}_+(E)$,

$$\Delta\tilde{\rho}_+(E) = \frac{1}{2\pi} T(E) \sum_{k=1}^{\infty} \frac{\cos[k(S - \frac{\alpha\pi}{2} - \beta\pi)]}{\sinh(k\lambda/2)}, \qquad (10.4.30)$$

provided the Lyapunov exponent $\lambda < \infty$. It has been argued by Ezra *et al.* (1991) that λ is in fact infinite. Therefore, according to (4.1.72), the symmetric stretch orbit does not contribute to $\tilde{\rho}_+$ either. However, it is possible that the symmetric stretch orbit contributes a "diffraction term", analogous to similar situations in ray-optics, where classical rays bounce off corners.

(ii) The contribution of a symmetric orbit to the fluctuating part of the level density is given by

$$\Delta\tilde{\rho}_\pm(E) = \frac{T(E)}{4\pi} \sum_{k=1}^{\infty} \left\{ \frac{\cos[k(S(E) - \alpha\pi/2 - \beta\pi)]}{\sinh(k\lambda/2)} \right.$$

$$\left. \pm \left(1 - \frac{1}{2}\delta_{k1}\right) \frac{\cos[(k - 1/2)(S(E) - \alpha\pi/2 - \beta\pi)]}{\sinh((k - 1/2)\lambda/2)} \right\}. \qquad (10.4.31)$$

(iii) Asymmetric orbits always come in pairs. Summing only over pairs of asymmetric orbits, we obtain

$$\Delta\tilde{\rho}_\pm(E) = \frac{1}{2\pi} T(E) \sum_{k=1}^{\infty} \frac{\cos[k(S - \alpha\pi/2 - \beta\pi)]}{\sinh(k\lambda/2)}. \qquad (10.4.32)$$

We are now ready for a discussion of scaled energy spectroscopy.

(b) *Scaled energy spectroscopy*

This technique uses scaling properties of the classical Hamiltonian in order to extract classical periodic orbit information from the fluctuating part of the level density. We illustrate the technique with the help of the positive parity states computed in Section 10.4.1. In order to compute $\tilde{\rho}_+$ we need the average level density. We solve the empirical formula (10.4.9) in the form

$$E = -\frac{1}{2m^2} - \frac{\zeta^2}{8n^2} \qquad (10.4.33)$$

for n, differentiate the result, and obtain

$$\bar{\rho}_+(E) = \sum_{m=1}^{\infty} \frac{dn(E;m)}{dE}. \qquad (10.4.34)$$

The structure of (10.4.30) – (10.4.32) suggests a Fourier transform of $\tilde{\rho}_+$ according to

$$R(S) = \int_{-\infty}^{0} \tilde{\rho}_+(E) \exp[-iS/ \mid E \mid^{1/2}] dE. \qquad (10.4.35)$$

Since the round-trip times of the classical periodic orbits scale according to $1/\mid E \mid^{3/2}$, a change of variables to $1/\sqrt{\mid E \mid}$ shows that the Fourier transform is sharply peaked at the scaled actions $S = S_j^{(0)} = S_j(E = -1)$ of the periodic orbits. We performed this transform on the basis of all the data obtained in Section 10.4.1. The real part of the resulting R-transform is shown in Fig. 10.10.

All the peaks in the real part of the R-transform can be assigned to classical periodic orbits (see labels in Fig. 10.10). As predicted by (10.4.31) the first peak in Fig. 10.10 occurs at half the action of the 1:1 periodic orbit. A peak corresponding to the symmetric stretch orbit is missing. This was first noticed by Ezra *et al.* (1991).

The symmetric stretch orbit is not the only periodic orbit emerging from the point $x_1 = x_2 = 0$. There is a whole family of other periodic orbits which emerge from, and end in, the origin. These are the "triple collision" orbits briefly discussed in Section 10.3. The triple collision orbits are also characterized by large Lyapunov exponents, and, just like the symmetric stretch orbit, should be absent in the R-transform. Indeed, we do not see any indication of these orbits in Fig. 10.10. But as discussed in (i) above, the question of whether the signatures of triple collision orbits are totally absent from $R(S)$ is not yet decided. Triple collision

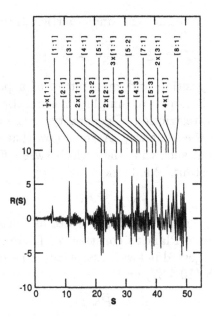

Fig. 10.10. Real part of the generalized Fourier transform of the fluctuating part of the positive parity level density of the one-dimensional helium model. (From Blümel and Reinhardt (1992).)

orbits, especially the symmetric stretch orbit, may leave their trace in $R(S)$ as "diffraction peaks", much smaller than the contributions from the orbits listed in Fig. 10.10, but nevertheless important for the fine details of $R(S)$. Many more resonances have to be included in the evaluation of $R(S)$ in order to settle this question conclusively.

Fig. 10.10 proves that a close connection exists between the classical mechanics and the quantum mechanics of the simple one-dimensional two-electron model. On the basis of the evidence provided by Fig. 10.10, there is no doubt that classical periodic orbits determine the structure of the level density in an essential way. The key element for establishing the one-to-one correspondence between the peaks in R and the actions of periodic orbits is the scaling relations (10.3.10). Similar relations hold for the "real" helium atom. Therefore, it should be possible to establish the same correspondence for the three-dimensional helium atom. First steps in this direction were taken by Ezra *et al.* (1991) and Richter (1991).

(c) *Generalization of scaled energy spectroscopy*

It is not necessary for the classical Hamiltonian of a system to possess intrinsic scaling symmetries in order to apply the technique of scaled energy spectroscopy. Consider the single-particle Hamiltonian

$$H = E = \frac{\vec{p}^2}{2} + \lambda V(\vec{x}), \qquad (10.4.36)$$

where $\vec{x}$, $\vec{p}$ are the position and momentum of a particle, and λ is a control parameter. In order to simplify the notation we assume that $E > 0$. This is not a restriction of generality as the case $E < 0$ can be treated in essentially the same way. Assume now that (10.4.36) is studied as a function of E, but such that the ratio between E and the potential strength remains constant. We define

$$\lambda = r E. \qquad (10.4.37)$$

The condition of constant r can be interpreted as a study of the quantum problem as a function of Planck's constant $\hbar$. Therefore, the generalized Fourier transform suggested below is essentially a Fourier transform with respect to $1/\hbar$. With (10.4.37) and

$$\vec{p} = \vec{P}\sqrt{E}, \qquad (10.4.38)$$

we obtain

$$H' = 1 = \frac{\vec{P}^2}{2} + r V(\vec{x}), \qquad (10.4.39)$$

where $H' = H/E$. Thus, the dynamics at E is now seen to be equivalent with the dynamics at $E = 1$ upon a trivial rescaling of the momentum $\vec{p}$. In particular, the shapes of periodic orbits remain the same upon this rescaling, as well as their Lyapunov exponents.

We are now interested in the quantum mechanics of (10.4.36). According to (4.1.72) the semiclassical density of states is given by

$$\rho(E) = \rho_0(E) + \sum_{nl} A_{nl} T_l \exp\left\{in\left[S_l(E)/\hbar + \alpha_l\right]\right\}, \qquad (10.4.40)$$

where l labels the primitive periodic orbits, n is the repetition index, A_{nl} are complex amplitudes, T_l is the round-trip time of the orbit number l, $S_l(E)$ is its action at energy E and α_l is the phase associated with orbit number l. The action S_l is given by

$$S_l(E) = \oint \vec{p}\, d\vec{x} = \hbar k \sigma_l. \qquad (10.4.41)$$

In (10.4.41) we used

$$k = \frac{1}{\hbar^2}\sqrt{2mE} \qquad (10.4.42)$$

and defined the scaled action

$$\sigma_l = \oint \sqrt{1 - r V(\vec{x})}\, \vec{p}^{\,0}\, d\vec{x}, \qquad (10.4.43)$$

where $\vec{p}^{\,0}$ is a unit vector in the $\vec{p}$ direction. The round-trip time is given by

$$T_l \;=\; \frac{dS_l(E)}{dE} \;=\; \hbar \sigma_l \frac{dk}{dE}. \qquad (10.4.44)$$

In (10.4.43) the scaled energy condition is assumed, i.e. the ratio of the total potential strength and the energy are constant. Therefore, using (10.4.43) in Gutzwiller's trace formula does not predict the usual level density $\rho(E)$, but the level density under the condition of a constant potential to energy ratio. We denote this modified level density by $d(E)$. It is given by

$$d(E) \;=\; \sum_j \delta(E - E_j), \qquad (10.4.45)$$

where E_j are the scaled energy levels. The semiclassical prediction of d is now given by

$$d(E) \;=\; d_0(E) \;+\; \hbar \frac{dk}{dE} \sum_{nl} A_{nl}\, \sigma_l \exp\Big[in(k\sigma_l + \alpha_l)\Big]. \qquad (10.4.46)$$

The structure of (10.4.46) suggests a Fourier transform in k. The Fourier transform of $d(E)$ is given by

$$F(\sigma) \;=\; \int d(E)\, \exp[-ik\sigma]\, dE \;=\; \sum_j \exp[-ik_j\sigma], \qquad (10.4.47)$$

where (10.4.42) and (10.4.45) were used to compute the right hand side of (10.4.47). The Fourier transform F can be evaluated as soon as the scaled energy levels E_j are known. The levels E_j may originate from a theoretical computation, as shown below, or directly from experiment by measuring energy levels, keeping the ratio between the potential strength and the energy constant. The Fourier transform (10.4.47) can also be evaluated using (10.4.46). The result is

$$F_0(\sigma) \;+\; \sum_{nl} A_{nl} \exp[in\alpha_l]\, \hbar \sigma_l \int \frac{dk}{dE} \exp[ink(\sigma_l - \sigma)]\, dE, \qquad (10.4.48)$$

where

$$F_0(\sigma) \;=\; \int d_0(E)\, \exp[-ik\sigma]\, dE \qquad (10.4.49)$$

is the Fourier transform of the smooth part of the level density. It does not contain any oscillations and is responsible for a peak in the Fourier transform F at small σ. This peak is not interesting. Much more interesting is the second term in (10.4.48). It is a sum of δ functions at integer multiples of the scaled actions σ_l of the periodic orbits l. Thus, the final result for the Fourier transform of the scaled level density is

$$F(\sigma) \;=\; F_0(\sigma) \;+\; 2\pi\hbar \sum_{nl} A_{nl} \exp[in\alpha_l]\,\sigma_l\,\delta[n(\sigma_l - \sigma)]. \qquad (10.4.50)$$

The relation (10.4.50) establishes a direct connection between a quantum spectrum characterized by discrete energy levels and the actions of classical periodic orbits. We now show how to actually compute the scaled energy levels E_j.

Defining $\hat{T} = \hat{p}^2/2$, the Hamiltonian (10.4.36) with (10.4.37) defines the following eigenvalue problem:

$$\hat{H}\,|\psi\rangle \;=\; \left[\hat{T} + rE\hat{V}\right]|\psi\rangle \;=\; E\,|\psi\rangle. \qquad (10.4.51)$$

This is not an ordinary eigenvalue problem, since the potential term contains the eigenvalue explicitly. We rewrite this eigenvalue problem in the following form:

$$\hat{T}\,|\psi\rangle \;=\; E\,\hat{W}\,|\psi\rangle, \qquad (10.4.52)$$

where

$$\hat{W} \;=\; 1 - r\hat{V}. \qquad (10.4.53)$$

The operator $\hat{W}$ is Hermitian. Therefore, it can be diagonalized and spectrally decomposed according to

$$\hat{W} \;=\; \hat{U}\,\hat{w}\,\hat{U}^\dagger, \quad \hat{U}^\dagger\hat{U} \;=\; 1, \qquad (10.4.54)$$

where $\hat{U}$ is unitary and $\hat{w}$ is diagonal. We assume now that $\hat{w}$ is positive and define

$$\hat{M} \;=\; \frac{1}{\sqrt{\hat{w}}}\,\hat{U}^\dagger\,\hat{T}\,\hat{U}\,\frac{1}{\sqrt{\hat{w}}}. \qquad (10.4.55)$$

The transformation

$$|\varphi\rangle \;=\; \sqrt{\hat{w}}\,\hat{U}^\dagger\,|\psi\rangle \qquad (10.4.56)$$

turns the eigenvalue problem (10.4.52) into the ordinary eigenvalue problem

$$\hat{M}\,|\varphi\rangle \;=\; E\,|\varphi\rangle. \qquad (10.4.57)$$

Given the potential $\hat{V}$ the eigenvalue equation (10.4.57) can be solved using standard eigenvalue techniques. This means that the spectrum of (10.4.51) is known and thus so is the generalized density (10.4.45).

The most successful application of scaled energy spectroscopy to date is in the case of the hydrogen atom in a strong magnetic field (Wintgen (1987), Friedrich and Wintgen (1989)). Experimental techniques for directly measuring the scaled density $d(E)$ were developed, e.g., in Welge's group in Bielefeld (Main *et al.* (1992)). In this way the experimental spectra can be Fourier transformed directly and show prominent peaks that can be assigned uniquely to periodic orbits of the classical diamagnetic hydrogen atom.

10.5 Quantum signatures of chaos

In Section 10.3 we established that the one-dimensional helium atom is classically chaotic. In Section 10.4.1 we computed its quantum spectrum. We extracted periodic orbit information from the spectrum in Section 10.4.3. But so far the main question has not been addressed: How does chaos manifest itself in the helium atom? Although this question is still the subject of ongoing research, some preliminary answers are provided in this section.

First we would like to note a nearly trivial, but far-reaching, observation. In the previous section we established that a Fourier transform of $\rho(E)$ reveals the full spectrum of classical periodic orbits together with all their properties, such as actions, round-trip times and Lyapunov exponents. Thus, $\rho(E)$ "knows" about the full complexity of a classically chaotic problem, in this case the classical version of the stretched helium problem. This one-to-one correspondence can hold only if $\rho(E)$ contains the same complexity as found in the underlying classically chaotic problem. In order to code the classical chaos into $\rho(E)$, $\rho(E)$ must be an exceedingly complicated function. An example of similar complexity is Riemann's ζ-function. In fact, it is known that given a function $f(x)$ in $[x_1, x_2]$ there always exists an interval $[y_1, y_2]$ in which the ζ-function mimics $f(x)$ to any desired accuracy. Thus, the ζ-function contains the functional behaviour of any "imaginable" function and is thus in a very real sense "chaotic". We encountered the ζ-function in Section 9.4 as the essential part of the quantum phase shift of Gutzwiller's scattering model (1983). The correspondence between classical chaos and smooth, but "information loaded", functions such as Riemann's ζ-function was first noted by Gutzwiller (1983). Thus, following Gutzwiller's idea of the manifestation of chaos in exceedingly complex behaviour of smooth functions occurring in the quantum context, we expect $\rho(E)$ to be a function as complex as Riemann's ζ-function. The complexity, i.e. the unlimited information content of $\rho(E)$, is certainly a manifestation of classical chaos in the quantum helium atom.

One of the most basic features of the helium spectrum is its organization into an infinite sequence of ionization thresholds. This feature is not the result of intricate computations. It is already apparent on the level of the independent particle model of the helium atom (see Section 10.1). All predictions on the quantum manifestations of chaos have to take this basic structure into account. Our picture of the manifestations of chaos in the helium spectrum is firmly based on this natural partition of the helium spectrum. Fig. 10.11 shows a schematic sketch of helium resonances in the semiclassical regime in the energy regime of three consecutive thresholds with energies E_{M-1}, E_M and E_{M+1}, where $E_m = -1/2m^2$. Let us focus on the resonances in the energy interval $[E_{M-1}, E_M]$. We can identify a regular sequence of resonances close to E_M. This is the familiar sequence of Rydberg levels, well described in the independent particle model. In Z-scaled units the real parts of their energies are approximately given by $E_{Mn} = E_M - 1/8n^2$, $n >> M$, the Z-scaled equivalent of (10.1.1). The Rydberg levels are resonances with very small imaginary parts. Their defining characteristic is that both the real and the imaginary parts of the resonance energies are regular. Because of this regularity we call the energy regime immediately below the threshold energy E_M the *Rydberg regime*. The regularity in the Rydberg regime is broken only sporadically by intruder states. There is always an energy interval sufficiently close to E_M that still contains an infinity of Rydberg levels, but no intruder states.

At somewhat lower energies, but still within the interval $[E_{M-1}, E_M]$, we encounter a second class of resonances. Fig. 10.11 shows that their imaginary parts are still small, but their real parts show a considerable degree of irregularity. The real parts of these levels are no longer well

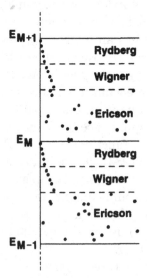

Fig. 10.11. Qualitative sketch (because of graphical reasons not to scale) of the repetitive sequence of three different dynamical regimes in the helium atom: Ericson, Wigner and Rydberg. The Ericson and the Wigner regimes are manifestations of chaos in the helium atom.

described by an independent particle model, and no simple energy formula exists. These are the states that grow out of the regular sequence of Rydberg states, but start to "feel" the underlying chaos of the classical helium atom. Since these resonances are still close to the real energy axis, they can be approximately described by real energy levels. We expect that the nearest neighbour spacings of the real parts of the energies of these resonances are Wignerian distributed. The energy regime supporting this class of states is located between the threshold energies E_{M-1} and E_M. It is referred to as the *Wigner regime*.

In the lower portion of the energy interval $[E_{M-1}, E_M]$ we expect the resonances to develop large imaginary parts. This expectation is based on the behaviour of the resonances shown in Fig. 10.8. It is possible that in the semiclassical regime the imaginary parts of the resonance energies are larger than the spacings of the real parts of the resonance energies. In this regime the resonances resemble a disordered gas of particles. We explore the properties of this (strongly interacting) "gas of resonances" in more detail in Section 10.5.2. Because of the large imaginary parts of these resonances we can no longer represent these resonances by their real parts alone, and the situation is truly two-dimensional. Because of the underlying classical chaoticity, we still wish to model these resonances by the eigenvalues of an appropriate random matrix ensemble. The proper random matrix ensemble is no longer the ensemble of Hermitian random matrices appropriate only for purely real energies. The proper random matrix ensemble now is Ginibre's ensemble of complex random matrices whose eigenvalues explore the complex plane. Instead of computing nearest neighbour spacings on the real energy axis we now characterize the "spacings" of the resonances by their Euclidian distances in the complex plane. We expect these distances to be distributed according to Ginibre's statistics of the eigenvalues of complex matrices (Ginibre (1965)). Thus, this energy regime may be called the Ginibre regime.

We refer to the Ginibre regime as the *Ericson regime*. This name is appropriate since resonances in the Ginibre regime are characterized by large imaginary parts. As discussed in Chapter 9 this is precisely the defining characteristic of an Ericson regime where the widths of the resonances (imaginary parts) are larger than their spacings (real parts). Interpreting the helium atom as a chaotic scattering system, as we did in Section 10.3.2, we expect scattering cross-sections in this regime to exhibit irregular fluctuations (Ericson fluctuations). The cross-section fluctuations in the Ericson regime are a quantum manifestation of chaos in the helium atom. The scattering cross-sections are expected to behave in a very similar way to the scattering cross-sections studied in Chapter 9 in connection with the chaotic scattering of CsI molecules.

Since our choice of the threshold index M is arbitrary, we expect that the three different regimes, Ericson, Wigner and Rydberg, are characteristic for the energy region between any two thresholds. Therefore, we predict a repetitive infinite sequence of Ericson, Wigner and Rydberg regimes as a function of increasing energy (Blümel and Reinhardt (1995)). The Wigner and the Ericson regimes are direct manifestations of chaos in the helium atom. The repetitive sequence of these regimes is illustrated in Fig. 10.11, which shows that the distribution of resonances in the energy interval $[E_M, E_{M+1}]$ qualitatively resembles the distribution in the interval $[E_{M-1}, E_M]$. Defining Γ as the mean width of the resonances and s as their mean spacing, the three regimes are defined by Ericson: $\Gamma \gg s$, Wigner: $\Gamma \lesssim s$, and Rydberg: $\Gamma \ll s$.

There is no problem with the existence of the Rydberg regime. Its existence is guaranteed on the basis of the elementary independent particle model, an excellent approximation of the helium states in the Rydberg regime where $n \gg m$.

Although not yet firmly established with the help of computer calculations, there are strong numerical indications that the Ericson regime actually exists. Numerical and analytical arguments for the existence of the Ericson regime are presented in Section 10.5.1.

While the existence of the Rydberg and the Ericson regimes can be considered as established, the Wigner regime is harder to come by. We have not yet found a good analytical argument either for its existence, or for estimating the energy that separates the Rydberg and the Wigner regimes from each other. We are, however, able to present firm numerical evidence for the existence of the Wigner regime. Our first task is to identify a promising energy region for the investigation of the Wigner regime. In this task we are restricted by our technical possibilities, i.e. we have to identify a Wigner regime for a threshold index not exceeding $M = 15$. The energy region below the first ionization threshold is not suitable. Here we have a regular Rydberg series of bounded states. Even the states above the first threshold, but below the second, are very regular. The same is expected for energies up to the fifth threshold: regular sequences of "states" (autoionizing resonances) are only sporadically broken by "intruder states" from a sequence above. For resonances above the fifth threshold, however, intruders are more frequent, and we expect to see a cross-over to the Wigner regime.

As discussed in Section 10.4.1 it is possible to calculate the resonances of the one-dimensional helium atom with considerable accuracy using the method of complex scaling in a complete L^2 basis. For the demonstration of the existence of the Wigner regime, however, we adopted a cheaper, only slightly less accurate, approach, namely diagonalization of (10.4.1) in a set of two-particle product states constructed from the single-particle hydrogen states (6.1.24). Since the states (6.1.24) span only the negative energy space of the one-dimensional hydrogen atom, a two-particle basis constructed from (6.1.24) is not complete. Nevertheless, comparing the results of this approach with the results of an L^2 calculation in a complete basis, the relative accuracy of the real parts of the resonances is typically better than 1% .

Within this approach, we computed several hundred levels of the one-dimensional helium atom. The resulting levels are shown in the form of a level diagram in Fig. 10.12. Again we use the Y variable defined in (10.4.8) to represent the energy. Fig. 10.12 shows an interesting metamorphosis of the qualitative appearance of the energy levels with increasing energy. Fig. 10.12(a) shows that up to the fifth threshold the spectrum can be naturally characterized by sequences of levels that converge to the respective ionization thresholds in the spirit of the independent particle model. Fig. 10.12(b) shows that from the fifth threshold on the energy levels start to fill in more evenly the space between ionization thresholds, a tendency which is even more pronounced in Fig. 10.12(c). The sequence of frames Fig. 10.12(a) – (c) shows that the regularity, together with our ability to classify the helium levels into regular sequences converging to thresholds, is progressively lost with higher excitation energy. We interpret this loss of regularity as a manifestation of chaos in the helium atom.

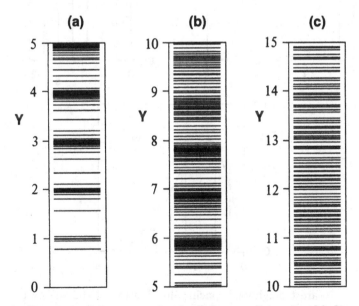

Fig. 10.12. Diagram of the energy levels of the one-dimensional helium atom in the three energy regimes (a) $m = 1$ to $m = 5$, (b) $m = 6$ to $m = 10$, (c) $m = 11$ to $m = 15$. (Adapted from Blümel and Reinhardt (1995).)

The metamorphosis of the qualitative appearance of energy levels with
higher energy is quantitatively supported by Figs. 10.13(a) – (c), which
show the probability distribution $P(s)$ of nearest neighbour spacings of
the energy levels shown in Figs. 10.12(a) – (c), respectively. Fig. 10.13(a)
shows that the probability distribution of the first group of levels is
sharply peaked at $s = 0$. This is understandable since, as shown in
Fig. 10.12(a), the first group of levels consists mainly of sequences of
states that converge in a completely orderly way to their respective ionization thresholds. The situation looks very different for the spacing
distribution of levels from the sixth to the tenth threshold shown in
Fig. 10.13(b). The distribution $P(s)$ starts to move towards Wignerian
statistics. Fig. 10.13(c) represents the level distribution for thresholds
from $m = 11$ to $m = 15$. Here, the distribution is already very close to
Wignerian.

How does chaos manifest itself in the helium atom? This question,
raised at the beginning of this section, can now be answered in the fol-

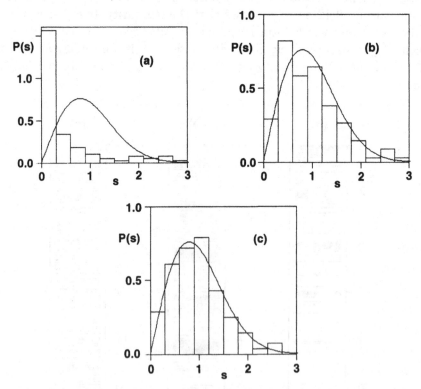

Fig. 10.13.   Nearest neighbour spacing distribution of the energy levels of the
one-dimensional helium atom in the three energy regimes of Fig. 10.12(a) – (c),
respectively. Histograms: numerical data. Smooth lines: Wignerian distribution.
(Adapted from Blümel and Reinhardt (1995).)

lowing way. Both the Wigner regime and the Ericson regime are quantum manifestations of chaos in the helium atom. The Wigner regime is firmly established on the basis of Fig. 10.13(c). Numerical and analytical arguments for the existence of the Ericson regime are advanced in the following section.

10.5.1 The Ericson regime

In Chapter 9 we saw that the regime of strongly overlapping resonances is especially interesting in connection with the manifestations of chaos in atomic physics. In Section 9.4 we defined the Ericson regime via the

$$\text{Ericson condition:} \quad \Lambda = \Gamma/D \gg 1, \qquad (10.5.1)$$

where Γ is the mean resonance width and D is the mean spacing of the resonances. We convinced ourselves of the existence of an Ericson regime in the stretched helium atom by computing ~ 600 resonances of the one-dimensional helium atom in the energy regime up to the $m = 30$ threshold. In these computations we see explicity the regime $\Gamma/D \sim 1$, which may be called the regime of *weakly* overlapping resonances. Numerically, the resonances with the largest widths scale according to $\Gamma \sim \mid E \mid^{\alpha}$, where $\alpha < 1$. The spacings D of the resonances are smaller than the spacings of thresholds. In fact, since the thresholds are accumulation points of the spectrum, there is always at least one resonance between consecutive thresholds. Therefore we also have $D < \mid E \mid^{3/2}$, and consequently $\Gamma/D > \mid E \mid^{\alpha-3/2} > \mid E \mid^{-1/2} \to \infty$ for $E \to 0$. Thus, there is no doubt that the Ericson condition is fulfilled if the energy approaches the two-electron break-up threshold at $E = 0$.

In order to corroborate the numerical results with analytical estimates of the onset of the Ericson regime we need estimates of the mean level density $\bar{\rho} = 1/D$ and the average resonance width Γ. The following analytical arguments are based on the assumption that subsets of classical phase space decay exponentially to the continuum. This assumption may not be true as indicated by the powerlaw results obtained in Chapter 8. Nevertheless, the argument is illustrative, and may be modified eventually to take the true decay mechanism of the helium phase space into account.

Since we expect the onset of the Ericson regime in the helium atom at high values of the actions of the two electrons, we make use of classical scaling arguments. First, we compute the scaling behaviour of the average resonance widths.

Assume that a classical phase-space volume at energy $E = -1$ decays according to

$$P_B(\tilde{t}) = \exp(-\gamma\tilde{t}), \qquad (10.5.2)$$

where $\tilde{t}$ is the scaled time at energy $E = -1$. Since according to (10.3.7) time scales according to $\tilde{t} = t \mid E \mid^{3/2}$, we obtain at energy E

$$P_B(t) = \exp(-\Gamma t), \tag{10.5.3}$$

where

$$\Gamma = \gamma \mid E \mid^{3/2}. \tag{10.5.4}$$

We assume now that the classical scaling law (10.5.4) also describes the energy scaling of the average width of the helium resonances. Inspection of the resonance widths computed in Section 10.4.1 yields $\gamma \approx 0.005$. Therefore,

$$\Gamma \approx 0.005 \mid E \mid^{3/2}. \tag{10.5.5}$$

Compared with the numerically obtained scaling of the resonance widths the energy exponent in (10.5.5) is somewhat on the "high side", but not too far from the numerical result. The deviations are probably due to the assumption of exponential decay.

Next we compute the mean level density. We use the empirical formula (10.4.9) as an estimate of the helium energy levels. Also, since we expect the Ericson regime to occur at $n \approx m$, we replace the fraction n/m in (10.4.9) by 1 and obtain in the interesting energy regime

$$E_{mn} = -\frac{1}{2m^2} - \frac{1}{2n^2}. \tag{10.5.6}$$

The mean level density is given by

$$\bar{\rho} = \int_{m_<}^{m_>} dm \int dn \; \delta(E - E_{mn}), \tag{10.5.7}$$

where, because of the expected high density of states, we have replaced summation over discrete states (resonances) by integration. Not all m series have discrete states that overlap with E. This is the reason for the occurrence of the integration limits $m_<$ and $m_>$ in (10.5.7). The upper limit $m_>$ is determined by the condition that the lowest state $(m, n = m)$ of (10.5.6) still reaches E. We obtain

$$m_> = 1/\sqrt{\mid E \mid}. \tag{10.5.8}$$

Since the most interesting states for the Ericson regime are characterized by $m \approx n$, we demand that $m_<$ is such that E is located exactly in the middle of the discrete states of the $m_<$ series. This results in

$$m_< = \frac{\sqrt{3}}{2\sqrt{\mid E \mid}}. \tag{10.5.9}$$

The limits known, we perform the integrals in (10.5.7). The result is

$$\bar{\rho} = \frac{1}{4\sqrt{2}\,|\,E\,|^2}. \qquad (10.5.10)$$

As a first result we obtain that for $|\,E\,| \to 0$ the mean spacing $D \sim |\,E\,|^2$ shrinks faster than the mean resonance width $\Gamma \sim |\,E\,|^{3/2}$. This proves that for $E \to 0$ there is always an Ericson regime in the helium atom. Using (10.5.5) and (10.5.10) in (10.5.1) we obtain as an estimate for the onset of Ericson fluctuations in the one-dimensional helium atom

$$m > 34. \qquad (10.5.11)$$

This estimate agrees with the numerically determined onset of the weakly overlapping regime at $m \approx 30$. It also agrees qualitatively with an independently derived earlier estimate ($m > 26$) by Blümel and Reinhardt (1992). The onset of the Ericson regime also marks the onset of quantum chaotic scattering, which, according to (10.5.11), is expected to occur for $m > 34$. Once the Ericson condition (10.5.1) is fulfilled, the scattering cross-sections are no longer dominated by isolated resonances, but instead exhibit a fluctuation pattern very similar to the fluctuating cross-sections studied in Chapter 9.

10.5.2 Resonance dynamics

In Section 4.1.2 we studied the dynamics of levels for Hermitian Hamiltonians. The levels were represented as fictitious particles restricted to move on a line, the one-dimensional real energy axis. The sequence of frames shown in Fig. 10.8 inspired the idea to study the motion of resonances in the plane. In analogy to the real case studied in Section 4.1.2 we expect that the resonances shown in Fig. 10.8 move in the complex energy plane as a function of the fictitious time $\epsilon = 1/Z$, the strength of the perturbation in (10.4.1). Thus we are led from level dynamics to *resonance dynamics*. The dynamic evolution of the resonances in ϵ corresponds physically to a continuous switch from one two-electron system to another, from U^{90+}, say, to H^-. Since the uranium ion U^{90+} corresponds to $\epsilon = 1/92$, and H^- corresponds to $\epsilon = 1$, the fictitious time ϵ ranges from 1/92 to 1. The available "time" interval $\Delta\epsilon \approx 1$ seems to be rather short. However, $\Delta\epsilon$ has to be compared with the typical collision time τ of two resonances in the complex plane. At high excitation energies in the Rydberg and planetary atom regime (Percival (1977)), the ratio $\Delta\epsilon/\tau$ is very large. In fact, it tends to infinity as m, n with $n \approx m$ approach infinity.

The equations of motion for the energy levels derived in Section 4.1.2 can easily be generalized to the case of motion in the complex plane if we choose the complex Hamiltonian (10.4.5) as our starting point. In contrast to (4.1.48) the eigenvalue equation

$$\hat{H}(\epsilon) \mid n(\epsilon)\rangle \; = \; E_n(\epsilon) \mid n(\epsilon)\rangle \qquad (10.5.12)$$

now yields complex eigenvalues E_n. But this is not an essential handicap, and all the formal derivations presented in Section 4.1.2 go through if we define the inner product (Elander and Brändas (1989))

$$\langle j|k\rangle \; = \; \sum_n x_n^{(j)} x_n^{(k)} \; = \; 0, \quad j \neq k, \qquad (10.5.13)$$

where $x_n^{(j)}$ are the components of the eigenvectors of the Hamiltonian matrix H_{nm} expressed in a convenient real Sturmian basis

$$\sum_m H_{nm} x_m^{(j)} \; = \; E_j x_n^{(j)}. \qquad (10.5.14)$$

Note that the "scalar product" (10.5.13) does not involve complex conjugation and is therefore not positive definite. The final set of differential equations describing the motion of the helium resonances in the complex energy plane is identical with the set (4.1.57) with the only difference that the quantities appearing in (4.1.57) are now complex.

For Hermitian Hamiltonians Dyson interpreted the set of energy levels E_n as an ensemble of interacting particles moving on a line. Inspired by the formal existence of dynamical equations given by the complex version of (4.1.57), we suspect that the helium resonances behave like an interacting gas of particles moving in two dimensions, the complex energy plane. In analogy to the one-dimensional gas in the Hermitian case, even a "temperature" may be assigned to the interacting two-dimensional gas of resonances. In order to prove that the idea of a "gas" is indeed fruitful, we study the properties of the helium resonance "gas" at "time" $\epsilon = 0.5$. In this spirit we interpret the frames of Fig. 10.8 as "snapshots" of the gas of resonances in the complex energy plane at "time" $\epsilon = 0.5$.

Analysing the behaviour of the resonances as a function of increasing energy Fig. 10.8 shows that there is no two-dimensional gaseous behaviour for $m \leq 6$. In this regime the resonances are seen to follow regular patterns, only sporadically perturbed by intruder states from higher m sequences. Only for $m \geq 7$ can we identify irregular behaviour that resembles what one would intuitively interpret as a "gas phase". In the previous section we estimated that the Ericson regime occurs at $m \gg 30$. We suspect that it is in the Ericson regime where we can talk about a fully developed gas phase. Checking this conjecture is currently not possible given that numerical diagonalization schemes cannot yet reliably reach the $m \gg 30$ regime. Fortunately the resonances with $m \geq 7$ already

exhibit some of the features we expect from a well developed gas of resonances. Studying this class of resonances already allows us to illustrate
the main features of the gas. We restrict ourselves to the energy regime
corresponding to the thresholds $7 \leq m \leq 10$ and compute 26 resonances
for $\epsilon = 0.5$ (helium) using the complex rotation method.

The resonances at $\epsilon = 0.5$ are shown as the full circles in Fig. 10.14.
In order to determine their local velocities at $\epsilon = 0.5$, we also computed
the location of these resonances in the complex plane at $\epsilon = 0.51$. The
resulting locations at $\epsilon = 0.51$ are shown as the open circles in Fig. 10.14.
As expected the resonances move in the complex plane as a function
of ϵ. Moreover, the movement is not only in one direction. Especially
close to the real energy axis, the resonances move in different directions.
This is already close to what we expect from a gas of particles. A finite
difference approximation to the local velocity of the resonances at $\epsilon = 0.5$
is computed according to

$$v_\alpha^{(r)} = \left[E_\alpha^{(r)}(\epsilon = 0.51) - E_\alpha^{(r)}(\epsilon = 0.50) \right] / \Delta\epsilon,$$

$$v_\alpha^{(i)} = \left[E_\alpha^{(i)}(\epsilon = 0.51) - E_\alpha^{(i)}(\epsilon = 0.50) \right] / \Delta\epsilon, \qquad (10.5.15)$$

where $\Delta\epsilon = 0.01$ and $\alpha = 1, ..., 26$. A more quantitative indication
for a gaseous phase of the helium resonances is a broad distribution of
resonance velocities, ideally Maxwellian distributed. In order to check

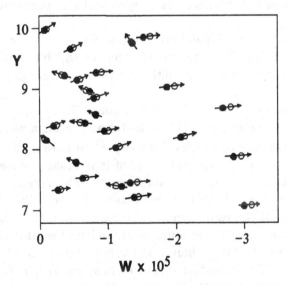

Fig. 10.14. Gas of resonances in the complex energy plane. Shown are snapshots
of the gas particles at $\epsilon = 0.5$ (bullets) and $\epsilon = 0.51$ (circles). The direction of
the resonance velocities is indicated by arrows.

whether our "gas" of 26 particles exhibits the expected broad distribution, we display the probability distributions $P(v^{(r)})$ and $P(v^{(i)})$ in the form of two histograms in Fig. 10.15. A Gaussian, the Maxwellian distribution in two dimensions, with the same mean value and standard deviation as calculated from the data points, is also shown in Fig. 10.15 for comparison. Both $P(v^{(r)})$ and $P(v^{(i)})$ are peaked at nonzero values. This shows that the resonance velocities contain a "collective" component. The $v^{(r)}$ data peak at $\bar{v}^{(r)} = 2.15 \times 10^{-3}$. This corresponds to a collective "drift" whose origin and magnitude can be understood qualitatively. Using the independent particle model for a rough estimate of the energy levels we have

$$E_{mn}^{(r)} \approx -\frac{1}{2m^2} - \frac{(1-\epsilon)^2}{2n^2} \qquad (10.5.16)$$

and therefore

$$v_{mn}^{(r)} = \frac{dE_{mn}^{(r)}}{d\epsilon} \approx \frac{1}{2n^2}. \qquad (10.5.17)$$

The mean n value, $\bar{n}$, of all the resonances considered is $\bar{n} \approx 15$. Therefore, $\bar{v}^{(r)} \approx 2.2 \times 10^{-3}$, which agrees well with the observed collective drift.

The nonzero mean of the imaginary part of the resonance velocities corresponds to a collective expansion into the negative imaginary energy plane. In view of the "barrier" at $\Im m(E) = 0$ this behaviour is not too surprising.

The widths of the two Gaussians shown in Fig. 10.15 correspond to two definite "temperatures" of the position and the width degrees of freedom of the helium resonances. Currently the physical meaning of these two temperatures is not clear.

We show now that the gas-like behaviour at large m is very different from the orderly behaviour of bound states and resonances for the low m sequences. According to (10.5.17) we expect to obtain $v_n^{(r)} = 1/2n^2$. It turns out that for $m = 1$ the velocities of individual n states follow this rule nearly exactly, and not just on average. The regular behaviour of velocities of the $m = 1$ bound states is therefore very different from that of the velocities of the resonances in the region $m = 8, ..., 10$. The latter velocities contain a stochastic component, and are best described by their ensemble properties, rather than as individual states. Additional computations show that the velocities of the low-m series with $m \le 6$ also follow the $1/2n^2$ rule nearly exactly. Consequently, there is a definite qualitative difference between the behaviour of low-lying and high-lying states. The low-lying states behave in an orderly fashion, whereas, because of the

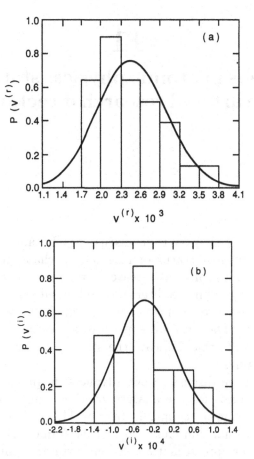

Fig. 10.15. Probability distribution of (a) real parts and (b) imaginary parts of the velocities of the helium resonances in the complex energy plane. (From Blümel and Reinhardt (1992).)

underlying chaoticity of the helium atom, the high-lying states resemble the stochastic jitter of a gas of interacting particles.

Thus far, quantum chaos theory has mainly concentrated on the statistical description of real energy levels and their spacings (see Section 4.1.1). The complex helium resonances with their two-dimensional character require a different approach. It may be based on Ginibre's ensemble of complex random matrices (Ginibre (1965)). The investigation of the statistical properties of the helium resonances in the complex plane is certainly a good subject for future research.

11

Chaos in atomic physics: state of the art and research directions

The purpose of this chapter is to discuss briefly, and as far as we are aware of it, the present status of research on chaos in atomic physics including trends and promising research directions. Given the enormous and rapidly growing volume of literature published every year, we cannot provide within the scope of this chapter a complete overview of existing published results. The best we can do is to select – in our opinion – representative results that indicate the status and trends in the field of chaos in atomic physics.

In Section 11.1 we discuss recent advances in quantum chaology, i.e. the semiclassical basis for the analysis of atomic and molecular spectra in the classically chaotic regime. In Section 11.2 we discuss some recent results in type II quantum chaos within the framework of the dynamic Born-Oppenheimer approximation. Recent experimental and theoretical results of the hydrogen atom in strong microwave and magnetic fields are presented in Sections 11.3 and 11.4, respectively. We conclude this chapter with a brief review of the current status of research on chaos in the helium atom.

11.1 Quantum chaology

Quantized chaos, or quantum chaology (see Section 4.1), is about understanding the quantum spectra and wave functions of classically chaotic systems. The semiclassical method is one of the sharpest tools of quantum chaology. As discussed in Section 4.1.3 the central problem of computing the semiclassical spectrum of a classically chaotic system was solved by Gutzwiller more than 20 years ago. His trace formula (4.1.72) is the basis for all semiclassical work on the quantization of chaotic systems.

There are currently two main streams of quantum chaology: the fundamental and the applied. Fundamental research in quantum chaology addresses the question of the analytical structure and the convergence properties of (4.1.72) and its derivatives. It was a major breakthrough in fundamental quantum chaology when Berry in 1986 noticed the similarities between Gutzwiller's trace formula and certain representations of Riemann's ζ-function. Henceforth the ζ-function served as a model for studying the analytical properties of semiclassical trace formulae (Berry (1986), Keating (1993), Bogomolny and Leboeuf (1994)). We expect that research on Riemann's ζ-function will continue to be a central theme in fundamental quantum chaology.

But there is more to the ζ-function than learning about proper resummations and how to construct faster convergent trace formulae. It has been argued that there may be a completely chaotic bounded dynamical system whose (real and discrete) spectrum is identical to the imaginary parts of the nontrivial zeros of Riemann's ζ-function. In our opinion the search for this elusive dynamical system is the "holy grail" of fundamental quantum chaology. The existence of this dynamical system would not only offer deep insights into the analytical connection between classical chaos and quantum energy levels, but would also prove the Riemann conjecture, i.e. $\zeta(1/2 + iz) = 0$ in $|\Im m(z)| < 1/2$ for $\Re e(z) = 0$ only.

On the applied side of quantum chaology we find serious efforts to forge the semiclassical method into a handy tool for easy use in connection with arbitrary classically chaotic systems. Quite frankly, the current status of semiclassical methods is such that they are immensely helpful in the interpretation of quantum spectra and wave functions, but are only of limited power when it comes to accurately predicting the quantum spectrum of a classically chaotic system. In this case numerical methods geared toward a direct numerical solution of the Schrödinger equation are easier to handle, more transparent, more accurate and cheaper than any known semiclassical method. It should be the declared aim of "applied semiclassics" to provide methods as handy and universal as the currently employed numerical schemes to solve the spectral problem of classically chaotic quantum systems.

Another topic of current interest concerns quantum chaology in higher dimensions. The semiclassical method is expected to be accurate only to order $\hbar^2$ independent of dimension. The level density of a D-dimensional system scales like $\hbar^{-D}$, which implies a level spacing $\sim \hbar^D$. Therefore, if we use conventional semiclassical methods to predict energy levels in higher dimensions, we have to be prepared for a relative error in the level spacing that scales according to $\hbar^{2-D}$. Thus, in the limit $\hbar \to 0$, the error in the spacings does not vanish for $D \geq 2$. According to this argument the semiclassical method is not accurate enough to predict individual levels in $D > 2$ dimensions. The case $D = 2$ is "marginal". Based on current results in two-dimensional systems, it appears that conventional semiclassical methods are powerful enough to control the error in the $\hbar \to$

0 limit. Results on higher-dimensional systems are still scarce. The three-dimensional Sinai billiard, e.g., was quantized only recently (Primack and Smilansky (1995)). Thus, in our opinion, quantum mechanical and semiclassical work on higher-dimensional systems is a promising avenue for further research.

There are some recent developments concerning the connection between chaos and the three canonical matrix ensembles, i.e. the Gaussian orthogonal (GOE), the unitary (GUE) and the symplectic (GSE) ensembles (see Section 4.1.1). Bogomolny gave an example of a completely chaotic system that does not exhibit GOE nearest neighbour spacing statistics. Systems of this type are called "arithmetic". They are reviewed by Bogomolny *et al.* (1995). The existence of nongeneric systems means that, although the relation "chaos ↔ GOE/GUE/GSE" may still be universal, one has to be prepared to find exceptions here and there. The reason for the deviation from GOE statistics in arithmetic systems is the existence of an exponentially large number of periodic orbits with exactly degenerate actions. Systems such as the arithmetic ones provide a number theoretic handle on the length spectrum of classical periodic orbits. They are very convenient as far as exact mathematical investigations are concerned, but they are not "typical" systems, for which GOE statistics is expected. On the other hand, the spectrum generated by the zeros of Riemann's ζ-function behaves very accurately according to GUE, even as far as n-point correlation functions are concerned. This result is backed numerically (Odlyzko (1989)) as well as analytically (see, e.g., Keating (1993)).

We still lack a proper understanding of how classical chaos induces the universal fluctuations in the energy levels that are so accurately described by the statistics of random matrix ensembles. Berry (1985) was the first to prove semiclassically some important results on the connection between chaos and energy level correlation functions of classically chaotic systems. The fundamental question, however, of how classical chaos generates quantum spectra amenable to a random matrix description is still not completely solved in all its details.

Additional topics of current interest are diffraction and refraction corrections of semiclassical procedures. Diffraction in classically chaotic systems was studied by Vattay *et al.* (1994). Diffraction is important whenever the wavelength is larger than a local classical curvature in the system. Refraction is important whenever the potential of a system changes on a length scale much smaller than the wavelength. Additionally, the step in the potential has to extend in the "transverse" direction over a classical length much longer than the wavelength. The refraction correction of the mean level density was computed analytically by Prange *et al.* (1996).

We conclude this section by mentioning another promising research area: chaos assisted tunnelling. This process was suggested by Bohigas *et al.* in 1993. It consists of the tunnelling of waves between two regular islands in phase space separated by a chaotic sea. Chaos manifests itself in this process by increasing the tunnel splitting of energy levels by several orders of mangitude.

11.2 Type II quantum chaos

As discussed in Section 4.2 type II systems (Blümel and Esser (1995)) show exponential sensitivity in the quantum subsystem. This allows the possibility of investigating the characteristics of true quantum chaos (type III quantum chaos) using the quantum subsystems of type II systems as a model. Apart from these new possibilities for fundamental research, type II systems have already found an important application. The mixed classical/quantum description, the basis of type II chaos, is a natural starting point for the investigation of the physics of dimers. In these systems chaos may result when electronic and vibronic degrees of freedom are coupled (Hennig and Esser (1992), Esser and Schanz (1995)).

It is often pointed out that quantum chaos can easily be generalized to wave chaos. This generalization is fine as far as type I systems are concerned. When studying spectra or the wave manifestations of ray dynamics it makes little difference whether the focus of our investigation is a quantum system or any other wave system such as an acoustic, electrodynamic or optical system. But in the case of type II systems a new quality emerges. For classical waves the existence of a classical boundary is an integral part of the theory. Thus, for an acoustic system, for instance, the type II case is *real*, i.e. no approximation at all. This example proves that *genuine* wave chaos exists in nature. Another example of type II wave chaos is provided by classical electrodynamics (Blümel (1995c)). In classical electrodynamics the Maxwell field may be coupled to classical conductors which move according to classical equations of motion under the forces generated by the radiation pressure of the electromagnetic field. Thus, within the framework of classical electrodynamics the separation into a wave system and a dynamic boundary is quite real. Consequently, in the right dynamical regime, the Maxwell field is chaotic.

The above arguments show that type II wave chaos is a genuine wave phenomenon in classical wave systems. In the context of quantum mechanics, however, type II quantum chaos is only an approximation. This is because "classical walls" or dynamic boundaries do not exist in quantum mechanics. The dynamical degrees of freedom of the walls, or boundaries, have to be quantized too, resulting in a higher-dimensional, but purely quantum, system, usually of type I. This fact leads us to a promising research direction: What happens if type II chaos is quantized? Do we see traces of type II chaos in the spectrum of the corresponding fully quantized system?

11.3 Microwave ionization

Bayfield and Koch (1974) provided the first experimental results on a manifestly quantum, but classically chaotic, system: hydrogen Rydberg atoms in a strong microwave field. Both pioneers, Bayfield at Pittsburgh and Koch at Stony Brook, continue to contribute actively to the investigation of time dependent chaos in Rydberg atoms.

While for a long time microwave ionization experiments addressed the linear polarization (LP) case only, experimental results on elliptic polarization (EP) are now available from Stony Brook (Koch and van Leeuwen (1995), Bellermann et al. (1996)). According to a widely used rule-of-thumb, EP ionization thresholds are expected to be higher than LP ionization thresholds. Bellermann et al. have shown that this is not generally the case. Bellermann et al. also provide experimental evidence for the importance of classical phase-space structures in the EP case. The EP case adds a new dimension to the microwave ionization problem. It provides an additional testing ground for the manifestations of chaos in atomic physics.

The latest work from Pittsburgh (Bayfield et al. (1996)) addresses a relatively new phenomenon, *phase-space metamorphoses*, a mechanism first suggested theoretically by Lai et al. (1992b). The essence of phase-space metamorphoses is the creation and destruction of phase-space features, such as island chains, as the microwave field strength changes. The Pittsburgh experiments with their half-sine pulse shapes are ideally suited for the investigation of phase-space metamorphoses. Bayfield et al. (1996) have found evidence for phase-space metamorphoses in the behaviour of the microwave ionization signal which shows characteristic ionization steps.

The Pittsburgh experiments differ from the Stony Brook experiments by the presence of a small static electric field which produces and stabilizes the extremal quasi-one-dimensional Stark states used in the Pittsburgh apparatus. The quasi-one-dimensional character of the Rydberg states used in the Pittsburgh experiments makes a direct comparision with inexpensive one-dimensional calculations possible.

The best way, at least in principle, of implementing a one-dimensional atomic system is still the use of surface state electrons. Indeed, research on the SSE system continues at a steady pace. New experimental and theoretical results on surface state electrons are reported, e.g., by Saville et al. (1993) and Saville and Goodkind (1994). This work focusses on the behaviour of the low-lying SSE states and concentrates mainly on the tunnelling regime. As discussed in Chapter 6, no chaos pictures are necessary in this regime to understand the physical processes involved. Nevertheless, we think that the work by Saville *et al.* is important to get a better grip on the delicate SSE system, and ultimately to explore the high-n semiclassical regime.

In Chapter 6 we discussed several severe obstacles in the way of achieving the high-n semiclassical regime in the SSE system. Here we point out yet another one: many-body effects. These are important whenever the SSE starting wave function is not exactly prepared as a two-dimensional sheet, or quantum coherence is lost in the direction parallel to the helium surface. On the basis of a classical picture it seems intuitively clear that the higher the principal quantum number, i.e. the larger the mean distance from the helium surface, the more important the electron-electron interactions within the sheet. This problem can be avoided only by reducing the density of the surface state electrons, which unfortunately results in a reduced ionization signal. On the other hand, this apparent "handicap" may be turned into an advantage if one wants to study the SSE system in the context of charged (one-component) plasmas.

There are also important advances concerning the theoretical description of Rydberg atoms in strong radiation fields. Buchleitner *et al.* (1995) report on fully fledged three-dimensional computations of the microwave ionization problem. They use the method of complex rotation discussed in Section 10.4.1, adapted to the computation of the resonances of the Floquet operator. The computed ionization probabilities are in good overall agreement with existing experimental data.

One of the central results in microwave excitation is the existence of an Anderson localization regime for driving frequencies larger than the Kepler frequency of the unperturbed Rydberg atom. The localization theory in microwave excitation is due to Casati *et al.* (1987) and was briefly discussed in Section 7.4. The localization theory was recently subjected to intense experimental scrutiny by Koch *et al.* (1992). It was found lacking in its ability to account for experimental details. We think that the basic ideas of the localization theory are correct. But it is also true that the localization theory is not, and may not have been intended to be, a theory which describes experimental detail. The localization theory introduced a new global way of thinking about microwave ionization at high frequencies. In this spirit, the localization theory predicted successfully the existence of quite a general effect in high frequency excitation of Rydberg atoms: the suppression of diffusion excitation with a corresponding increase of ionization thresholds. We think that the existence of dynamical localization in high frequency driven Rydberg atoms is a remarkable effect. We suggest that the influence of symmetry breaking on the localization length of microwave-driven hydrogen atoms should be investigated.

11.4 Hydrogen in a strong magnetic field

The hydrogen atom in a strong magnetic field is known as the "diamagnetic Kepler problem" (DKP). As a function of increasing magnetic field strength the hydrogen atom passes through essentially three dynamical regimes, the linear Zeeman regime (I), the quadratic Zeeman regime (II) and the Landau regime (III). Regimes I and III are the "ordered phases". They are dynamically simple, since the hydrogen atom is approximately separable in both regimes and perturbation theory works very well. Regime II is the most interesting of the three. The dynamics in regime II is chaotic, which is reflected in GOE statistics of nearest neighbour spacings of the energy levels of the atom (see Fig. 4.5). After more than two decades of intensive research, the DKP is currently the most intensively studied autonomous classically chaotic atomic physics system.

Regime II became of experimental relevance when in 1969 Garton and Tomkins discovered regular oscillations, the "quasi-Landau resonances" (Edmonds (1970)), in the absorption cross-section of alkaline earth elements in strong magnetic fields. Edmonds was one of the first to appreciate the importance of the problem. Based on his war-time experience with magnetrons, he explained the quasi-Landau resonances semiclassically on the basis of simple *classical* orbits (Edmonds (1970)). He expanded on these results in a series of papers in collaboration with Pullen. But Edmonds' and Pullen's ideas were far ahead of their time. Although at this time Gutzwiller's periodic orbit theory was already published, the idea of the relevance of classical mechanics in atomic physics was not appreciated, and the Edmonds-Pullen papers were never published. This episode illustrates that only about 20 years ago the idea of explaining an atomic physics effect on the basis of classical mechanics was still "revolutionary", if not a "cardinal sin". In 1973 Edmonds proved to be a visionary in another respect. He fully realized the importance of the DKP for (i) solid state physics, (ii) astrophysics and (iii) Rydberg atoms. An illustrative example for Edmonds' remark (i) are recent experimental results in solid state physics published by von Klarenbosch *et al.* in 1990. The importance of the DKP for astrophysics and the field of Rydberg atoms will become clear below.

In regime II the Lorentz force F_L acting on the hydrogen's electron has to be on the same order of magnitude as the binding Coulomb force F_C. Otherwise the electron's dynamics is either Coulomb- or magnetic-field dominated resulting in the regular regimes I and III, respectively. The required magnetic field B necessary to achieve $F_L \approx F_C$ can be worked out immediately on the basis of Bohr's model of the hydrogen atom. The radius of the nth Bohr orbit is given by

$$r_n = 4\pi\epsilon_0 \frac{n^2\hbar^2}{me^2}, \tag{11.4.1}$$

where m is the reduced mass of the electron-proton system. Classically the electron's velocity in the nth Bohr orbit is given by

$$v_n = \left[\frac{e^2}{4\pi\epsilon_0 m r_n}\right]^{1/2}. \tag{11.4.2}$$

Equating the Lorentz force

$$F_L = e v_n B \tag{11.4.3}$$

and the Coulomb force

$$F_C = \frac{e^2}{4\pi\epsilon_0} \frac{1}{r_n^2}, \tag{11.4.4}$$

we obtain the following estimate for the critical magnetic field required for the onset of chaos:

$$B_n = \frac{\alpha^2 m^2 c^2}{e\hbar} \frac{1}{n^3} = \frac{B^*}{n^3}, \tag{11.4.5}$$

where α is the fine structure constant, and

$$B^* \approx 2.35 \times 10^5 \, \text{T}. \tag{11.4.6}$$

The estimate (11.4.5) together with the numerical value (11.4.6) of B^* show clearly that for a hydrogen atom in its ground state ($n = 1$) one does not have to worry about chaos since the generation of static magnetic fields on the order of B^* is well beyond the possibilities of present day technology. However, fields of this order of magnitude may occur on white dwarf stars and on neutron stars (Garstang (1977), Wunner (1989)). Thus Edmonds' remark (ii) was right on the mark when in 1973 he noted the importance of the DKP for astrophysics.

But there is another way to reach regime II in the magnetized hydrogen atom. Writing (11.4.5) as

$$B_n = \frac{B^*}{30^3} \left(\frac{30}{n}\right)^3 \approx 8.5 \, \text{T} \times \left(\frac{30}{n}\right)^3, \tag{11.4.7}$$

we see that in the case of Rydberg atoms (see Edmonds' remark (iii) above) regime II can be reached even for relatively modest fields provided the hydrogen atoms are prepared with principal quantum numbers on the order of $n = 30$ or larger. This possibility, well within the reach of present day experiments, has boosted the hydrogen atom in a magnetic field from an elementary textbook example (regimes I and III) to the forefront of current scientific research.

The status of the DKP up to 1989 has been reviewed by Friedrich and Wintgen (1989). In the meantime the field evolved essentially along the lines delineated by Friedrich and Wintgen. With the further evolution and refinement of experimental techniques, a closer look at the underlying periodic orbit structure of the DKP is currently a central theme. A paper by Main *et al.* (1994) illustrates this point. This paper presents an excellent summary of experimental and theoretical results on the connection of the Fourier transform of photo-absorption data and periodic orbits using the technique of scaled energy spectroscopy, introduced in Section 4.1.3. It is also a good example of state-of-the-art spectroscopy of the hydrogen atom in a strong magnetic field (see also Main *et al.* (1992)).

The DKP Hamiltonian in scaled cylindrical coordinates is given by (Friedrich (1990))

$$H = \frac{p_\rho^2}{2} + \frac{p_z^2}{2} + \frac{L_z}{2\rho^2} - \frac{1}{\sqrt{\rho^2 + z^2}} + \frac{1}{8}\rho^2. \tag{11.4.8}$$

The quantized version of (11.4.8) is the basis for numerical state-of-the-art computations on super-computers. Over the past ten years these computations have evolved into an astonishingly accurate and reliable tool for the interpretation of experimental spectra. An example of the accuracy that can be achieved routinely nowadays is given by Iu *et al.* (1991). The computations by Iu *et al.*, converged *ab initio* solutions of (11.4.8) with no adjustable parameters, are able to reproduce existing experimental results astonishingly well.

In 1992 Wintgen and Friedrich reviewed the status of semiclassical methods in connection with the DKP. Traditional semiclassical methods are not applicable in the vicinity of bifurcations in phase space. An improved semiclassical theory, applicable at bifurcation points, is discussed and experimentally tested by Courtney *et al.* (1995).

Novel resonance effects occur in nonhydrogenic atoms in a strong magnetic field. As noted by Dando *et al.* (1995) these effects cannot be explained on the basis of hydrogenic periodic orbit theory. Their relation to "ghost orbits" is discussed by Hüpper *et al.* (1995).

In 1993 Howard and Farrelly realized that the Hamiltonian (11.4.8) is a special case of the class of Hamiltonians

$$H = \frac{p_\rho^2}{2} + \frac{p_z^2}{2} + \frac{\nu^2}{2\rho^2} + \frac{\sigma}{\sqrt{\rho^2 + z^2}} + \frac{1}{2}(\rho^2 + \lambda^2 z^2), \tag{11.4.9}$$

where λ and σ can assume any real value. Upon rescaling, however, only $\sigma = 0, \pm 1$ are dynamically different. The Hamiltonian (11.4.9) is related to the Hamiltonian of the generalized van der Waals problem introduced by Alhassid *et al.* (1987). It encompasses a wide variety of systems. For $\lambda = 0, \sigma = -1$, it is equivalent (up to scaling) to the DKP Hamiltonian (11.4.8). For $\lambda = 1, \sigma = -1$, it describes the spherical quadratic Zeeman effect studied in some detail by Silva and Canuto in 1984. For $\sigma = 0$ it describes a trivial separable harmonic oscillator. For $\sigma = 1$ (11.4.9) describes two charged particles in a Penning trap or in the pseudo-potential of the Paul trap (Farrelly and Howard (1994)). In 1993 several constants of the motion were derived by Howard and Farrelly for the Hamiltonian (11.4.9) whose quantum signatures were obtained as deviations from GOE statistics by Moore and Blümel (1993). Since the constants exist for selected values of $\lambda \neq 0$ ($\lambda = 1/2, 1, 2$), it would be interesting to investigate the signatures of these constants in the case of the DKP. In order to generate a nonvanishing λ, one might think of adding a quadrupole electric field to the DKP. This may be a promising direction for future research.

In order to strengthen the link between classical chaos and the applicability of random matrix theory it is desirable to change the symmetry class of the DKP. In order to do this, we have to break the time reversal symmetry. One may argue that the application of the magnetic field already has this effect. But this is not so. As discussed by Haake (1991) the magnetized hydrogen atom possesses several generalized antiunitary symmetries that act like the time reversal symmetry and are not destroyed by the application of a homogeneous magnetic field. However, these symmetries can be destroyed by the application of a magnetic field that is strongly inhomogeneous over the extension of the Rydberg atom. This sounds like a formidable experimental task. But due to the relatively large size of a Rydberg atom it is not impossible. Once all the antiunitary symmetries in the diamagnetic hydrogen atom are destroyed, we expect its energy levels to exhibit GUE nearest neighbour spacing statistics (see (4.1.45)). We think that this line of research may turn out to be particularly promising.

11.5 The helium atom

The helium atom is a classically chaotic system. In Chapter 10 we saw that chaos chooses the way of overlapping resonance series to make its presence felt. The existence of overlapping series in the helium atom was confirmed experimentally by Domke *et al.* (1995). Using synchrotron radiation, helium resonances up to the $N = 9$ threshold were investigated. At the same time good agreement with theoretical complex rotation computations was obtained. In the energy region from the $N = 7$ to the $N = 9$ threshold the excitation functions measured by Domke *et al.* show resonance structures that are just beginning to overlap. Thus the experimental results confirm that the Ericson regime, the regime of strongly overlapping resonances, occurs for $N \gg 10$.

Apart from using the complex rotation method discussed in Section 10.4.1 the existence of the Ericson regime may also be proved by setting up a traditional coupled channels scattering calculation. The task is to compute the S matrix for electron scattering off He^+. Choosing $E < 0$, two-particle break-up is not possible, which considerably facilitates these calculations. For $E < 0$ the scattering channels can be labelled by the quantum numbers of He^+. This is a set of single-particle hydrogenic quantum numbers which characterize the bound state in which the He^+ electron resides prior to and after the impact with the second electron. In the Ericson regime the energy dependent partial cross-sections computed from the full quantum S matrix are expected to fluctuate in analogy to the cross-sections of the CsI molecule discussed in Chapter 9.

The availability of highly accurate experimental data on doubly excited helium is a welcome challenge for computational physics. The challenge was met by Wintgen and Delande (1993), Bürgers and Wintgen (1994), Bürgers (1995) and Bürgers *et al.* (1995).

While all efforts are welcome that address the difficult problem of solving the Schrödinger equation for a real-life three-body problem, it is very hard to learn about new dynamical regimes. Because of this reason many researchers are still interested in simple one-dimensional models of the helium atom. Apart from the collinear models discussed in Section 10.2 there exists another one-dimensional model of the helium atom introduced by Handke *et al.* in 1993: s-wave helium. In contrast to the collinear models which represent electron motion in states that are strongly correlated in angle, the electrons of s-wave helium occupy spherical shells with angular momentum zero. Thus, s-wave helium does not have any built-in angular correlations and is therefore complementary to the collinear models. It is interesting to note that s-wave helium appears to have the same periodic orbit structure as the collinear model with both electrons on different sides. It may be possible to make this correspondence mathematically exact. In this case the conjectured complete chaoticity of the collinear model implies the complete chaoticity of the s-wave model. Indeed, Handke *et al.* (1993) do not see any evidence of regular islands in the s-wave model and conjecture that the s-wave model is fully chaotic. Since the s-wave model has certain mathematical advantages over the collinear model, it may be possible to prove the chaoticity of the s-wave model first, and then obtain the chaoticity of the collinear model as a corollary, provided the symbolic dynamics of both systems is the same.

In connection with the conjectured complete chaoticity of the collinear model we point out that so far no proof exists for the completeness of the binary code for enumerating its periodic orbits studied in Section 10.3.1. The decay properties of phase-space regions, exponential or algebraic, are also not known. If the model should turn out to be completely chaotic and the decay is algebraic, this would be another interesting example of a system with algebraic decay in the absence of elliptic islands, a system analogous to the kicked hydrogen atom discussed in Chapter 8. In contrast to the collinear model the decay fractal of s-wave helium has already been investigated by Handke in 1994. No indications of regular islands were found.

The quantum mechanics of the s-wave model of the helium atom was investigated by Draeger *et al.* (1994). It turns out that for certain classes of states the energy levels of s-wave helium are very close to the energy levels of the real helium atom.

The collinear models are also useful close to the two-electron break-up threshold. In 1994 Rost was able to obtain the correct Wannier exponent by a semiclassical treatment of electron impact ionization of hydrogen, another important quantum problem which involves nonintegrable three-body dynamics (see also Rost (1995)).

In conclusion it appears that the basic facts about the classical/quantum correspondence of classically chaotic atoms and related atomic systems is by now well understood, at least on the conceptual phenomenological level. For the near future we see the main research direction in strengthening the classical-quantum correspondence using the ever more powerful semiclassical methods that are currently being developed and will soon be a part of the tool-kit of modern atomic physics.

References

Abramowitz, M. and Stegun, I.A. (1964). *Handbook of Mathematical Functions* (National Bureau of Standards, Washington, D.C.).

Aguilar, J. and Combes, J.M. (1971). A class of analytic perturbations of one-body Schrödinger Hamiltonians, *Comm. Math. Phys.* **22**, 269–279.

Alhassid, Y., Hinds, E.A. and Meschede, D. (1987). Dynamical symmetries of the perturbed hydrogen atom: the van der Waals interaction, *Phys. Rev. Lett.* **59**, 1545–1548.

Altshuler, B.L., Lee, P.A. and Webb, R.A. (1991). *Mesoscopic Phenomena in Solids* (North-Holland, New York).

Anderson, H.L. (1989). *A Physicist's Desk Reference* (American Institute of Physics, New York).

Anderson, M.H., Ensher, J.R., Matthews, M.R., Wieman, C.E. and Cornell, E.A. (1995). Observation of Bose-Einstein condensation in a dilute atomic vapor, *Science* **269**, 198–201.

Anderson, P.W. (1958). Absence of diffusion in certain random lattices, *Phys. Rev.* **109**, 1492–1505.

Arnold, V.I (1989). *Mathematical Methods of Classical Mechanics* (Springer, New York).

Arnol'd, V.I. and Avez, A. (1968). *Ergodic Problems of Classical Mechanics* (W. A. Benjamin, New York).

Artuso, R., Aurell, E. and Cvitanović, P. (1990a). Recycling of strange sets: I. Cycle expansions, *Nonlinearity* **3**, 325–359.

Artuso, R., Aurell, E. and Cvitanović, P. (1990b). Recycling of strange sets: II. Applications, *Nonlinearity* **3**, 361–386.

Badii, R. and Meier, P.F. (1987). Comment on "Chaotic Rabi oscillations under quasiperiodic perturbations", *Phys. Rev. Lett.* **58**, 1045.

Baker, G.L. and Gollub, J.P. (1990). *Chaotic Dynamics* (Cambridge University Press, Cambridge).

Baldo, M., Lanza, E.G. and Rapisarda, A. (1993). Chaotic scattering in heavy-ion reactions, *Chaos* **3**, 691–706.

Balslev, E. and Combes, J.M. (1971). Spectral properties of many-body Schrödinger operators with dilatation-analytic interactions, *Comm. Math. Phys.* **22**, 280–294.

Baranger, H.U., Jalabert, R.A. and Stone, A.D. (1993a). Quantum chaotic scattering effects in semiconductor microstructures, *Chaos* **3**, 665–682.

Baranger, H.U., Jalabert, R.A. and Stone, A.D. (1993b). Transmission through ballistic cavities: chaos and quantum interference, in *Quantum Dynamics of Chaotic Systems*, eds. J.-M. Yuan, D. H. Feng and G. M. Zaslavsky (Gordon and Breach, Amsterdam).

Barnsley, M. (1988). *Fractals Everywhere* (Academic Press, Boston).

Bayfield, J.E. and Koch, P.M. (1974). Multiphoton ionization of highly excited hydrogen atoms, *Phys. Rev. Lett.* **33**, 258–261.

Bayfield, J.E. and Sokol, D.W. (1988). Excited atoms in strong microwaves: classical resonances and localization in experimental final-state distributions, *Phys. Rev. Lett.* **61**, 2007–2010.

Bayfield, J.E., Casati, G., Guarneri, I. and Sokol, D.W. (1989). Localization of classically chaotic diffusion for hydrogen atoms in microwave fields, *Phys. Rev. Lett.* **63**, 364–367.

Bayfield, J.E., Luie, S.-Y., Perotti, L.C. and Skrzypkowski, M.P. (1996). Ionization steps and phase-space metamorphoses in the pulsed microwave ionization of highly excited hydrogen atoms, *Phys. Rev.* **A53**, R12–R15.

Beddington, J.R., Free, C.A. and Lawton, J.H. (1975). Dynamic complexity in predator-prey models framed in difference equations, *Nature* **255**, 58–60.

Bellermann, M.R.W., Koch, P.M., Mariani, D.R. and Richards, D. (1996). Polarization independence of microwave "ionization" thresholds of excited hydrogen atoms near the principal resonance, *Phys. Rev. Lett.* **76**, 892–895.

Berry, M.V. (1985). Semiclassical theory of spectral rigidity, *Proc. R. Soc. Lond.* **A400**, 229–251.

Berry, M.V. (1986). Riemann's zeta function: a model for quantum chaos?, in *Quantum Chaos and Statistical Nuclear Physics*, eds. T.H. Seligman and H. Nishioka (Springer, Berlin).

Berry, M.V. (1989). Quantum chaology, not quantum chaos, *Physica Scripta* **40**, 335–336.

Berry, M.V. (1992). True quantum chaos? An instructive example, *New Trends in Nuclear Collective Dynamics, Springer Proceedings in Physics 58*, eds. Y. Abe, H. Horiuchi and K. Matsuyanagi (Springer, Berlin).

Berry, M.V., Balazs, N.L., Tabor, M. and Voros, A. (1979). Quantum maps, *Ann. Phys. (N.Y.)* **122**, 26–63.

Bhatia, A.K. and Temkin, A. (1984). Line-shape parameters for ^{1}P Feshbach resonances in He and Li$^+$, *Phys. Rev.* **A29**, 1895–1900.

Bleher, S., Grebogi, C. and Ott, E. (1990). Bifurcation to chaotic scattering, *Physica* **D46**, 87–121.

Blümel, R. (1993a). On the integrability of the two-ion Paul trap in the pseudo potential approximation, *Phys. Lett.* **A174**, 174–175.

Blümel, R. (1993b). Quantum chaotic scattering with CsI molecules, *Chaos* **3**, 683–690.

Blümel, R. (1993c). Exotic fractals and atomic decay, in *Quantum Chaos*, eds. G. Casati, I. Guarneri and U. Smilansky, (North-Holland, Amsterdam).

Blümel, R. (1994a). Exponential sensitivity and chaos in quantum systems, *Phys. Rev. Lett.* **73**, 428–431.

Blümel, R. (1994b). Microwave ionization of hydrogen Rydberg atoms: Resonance analysis and critical fields, *Phys. Rev.* **A49**, 4787–4793.

Blümel, R. (1995a). Reply comment to "Exponential sensitivity and chaos in quantum systems", *Phys. Rev. Lett.* **75**, 582.

Blümel, R. (1995b). Dynamic Kingdon trap, *Phys. Rev.* **A51**, R30–R33.

Blümel, R. (1995c). Genuine electromagnetic wave chaos, *Phys. Rev.* **E51**, 5520–5523.

Blümel, R. and Esser, B. (1994). Quantum chaos in the Born-Oppenheimer approximation, *Phys. Rev. Lett.* **72**, 3658–3661.

Blümel, R. and Esser, B. (1995). Type II quantum chaos, *Zeit. Physik* **B98**, 119–131.

Blümel, R. and Reinhardt, W.P. (1992). Where is the chaos in two-electron atoms?, in *Quantum Non-Integrability*, eds. D.H. Feng and J.-M. Yuan (World Scientific, Singapore).

Blümel, R. and Reinhardt, W.P. (1995). Stretched helium: a model for quantum chaos in two-electron atoms, in *Quantum Chaos*, eds. G. Casati and B.V. Chirikov (Cambridge University Press, Cambridge).

Blümel, R. and Smilansky, U. (1984). Suppression of classical stochasticity by quantum-mechanical effects in the dynamics of periodically perturbed surface-state electrons, *Phys. Rev.* **A30**, 1040–1051.

Blümel, R. and Smilansky, U. (1987). Microwave ionization of highly excited hydrogen atoms, *Z. Phys.* **D6**, 83–105.

Blümel, R. and Smilansky, U. (1988). Classical irregular scattering and its quantum-mechanical implications, *Phys. Rev. Lett.* **60**, 477–480.

Blümel, R. and Smilansky, U. (1990a). Quantum mechanical suppression of chaos, *Physics World* **3(2)**, 30–34.

Blümel, R. and Smilansky, U. (1990b). Ionization of hydrogen Rydberg atoms in strong monochromatic and bichromatic microwave fields, *J. Opt. Soc. Am.* **B7**, 664–679.

Blümel, R. and Smilansky, U. (1992). Symmetry breaking and localization in quantum chaotic systems, *Phys. Rev. Lett.* **69**, 217–220.

Blümel, R., Fishman, S. and Smilansky, U. (1986). Excitation of molecular rotation by periodic microwave pulses. A testing ground for Anderson localization, *J. Chem. Phys.* **84**, 2604–2614.

Blümel, R., Kappler, C., Quint, W. and Walther, H. (1989a). Chaos and order of laser-cooled ions in a Paul trap, *Phys. Rev.* **A40**, 808–823.

Blümel, R., Jaeckel, G. and Smilansky, U. (1989b). Ionization of high-n H atoms by bichromatic microwave fields, *Phys. Rev.* **A39**, 450–453.

Blümel, R., Buchleitner, A., Graham, R., Sirko, L., Smilansky, U. and Walther, H. (1991). Dynamical localization in the microwave interaction of Rydberg atoms: The influence of noise, *Phys. Rev.* **A44**, 4521–4540.

Blümel, R., Davidson, I.H., Reinhardt, W.P., Lin, H. and Sharnoff, M. (1992). Quasilinear ridge structures in water surface waves, *Phys. Rev.* **A45**, 2641–2644.

Bogomolny, E.B. and Leboeuf, P. (1994). Statistical properties of the zeros of zeta functions – beyond the Riemann case, *Nonlinearity* **7**, 1155–1167.

Bogomolny, E.B., Georgeot, B., Giannoni, M.-J. and Schmit, C. (1995). Quantum chaos on constant negative curvature surfaces, *Chaos Solitons Fractals* **5**, 1311–1323.

Bohigas, O., Giannoni, M.J. and Schmit, C. (1984). Characterization of chaotic quantum spectra and universality of level fluctuation laws, *Phys. Rev. Lett.* **52**, 1–4.

Bohigas, O., Legrand, O., Schmit, C. and Sornette, D. (1991). Comment on spectral statistics in elastodynamics, *J. Acoust. Soc. Am.* **89**, 1456–1458.

Bohigas, O., Tomsovic, S. and Ullmo, D. (1993). Manifestations of classical phase-space structures in quantum mechanics, *Phys. Rep.* **223**, 43–133.

Bohr, N. (1913a). On the constitution of atoms and molecules, *Phil. Mag.* **26**, 1–25.

Bohr, N. (1913b). On the constitution of atoms and molecules. II., *Phil. Mag.* **26**, 476–502.

Bohr, N. (1936). Neutron capture and nuclear constitution, *Nature* **137**, 344–348.

Bohr, A. and Mottelson, B.R. (1969). *Nuclear Structure* (W. A. Benjamin, New York).

Borel, E. (1914). *Introduction Géométrique á Quelques Théories Physiques* (Gauthier-Villars, Paris).

Born, M. (1969). *Physics In My Generation* (Springer, New York).

Born, M. and Heisenberg, W. (1923). Die Elektronenbahnen im angeregten Heliumatom, *Z. Phys.* **16**, 229–243.

Born, M. and Oppenheimer, J.R. (1927). Zur Quantentheorie der Molekeln, *Ann. Physik* **4-84**, 457–484.

Braun, M. (1975). *Differential Equations and Their Applications* (Springer, New York).

Brenig, W. (1975). *Statistische Theorie der Wärme* (Springer, Berlin).

Breuer, H.P. and Holthaus, M. (1989). Adiabatic processes in the ionization of highly excited hydrogen atoms, *Z. Phys.* D **11**, 1–14.

Breuer, H.P., Dietz, K. and Holthaus, M. (1988). Strong laser fields interacting with matter I, *Z. Phys.* D **10**, 13–26.

Brillouin, L. (1926). Remarques sur la méchanique ondulatoire, *J. Phys. Radium* **7**, 353–368.

Brillouin, L. (1960). Poincaré and the sortcomings of the Hamilton-Jacobi method for classical or quantized mechanics, *Arch. Rational Mech. Analysis* **5(1)**, 76–94.

Buchleitner, A. (1993). Atomes de Rydberg en champ micro-onde: Regularité et chaos, Ph. D. thesis, University of Paris.

Buchleitner, A. and Delande, D. (1993). Dynamical localization in more than one dimension, *Phys. Rev. Lett.* **70**, 33–36.

Buchleitner, A., Grémaud, B. and Delande, D. (1994). Wavefunctions of atomic resonances, *J. Phys.* **B27**, 2663–2679.

Buchleitner, A., Delande, D. and Gay, J.-C. (1995). Microwave ionization of three-dimensional hydrogen atoms in a realistic numerical experiment, *J. Opt. Soc. Am.* **B12**, 505–519.

Bulgac, A. (1991). Level crossings, adiabatic approximation, and beyond, *Phys. Rev. Lett.* **67**, 965–967.

Bürgers, A. (1995). Doppelt angeregte S-Zustände im Heliumatom, Ph. D. thesis, Universität Freiburg.

Bürgers, A. and Wintgen, D. (1994). Inhibited autoionization of planetary atom states, *J. Phys.* **B27**, L131–L135.

Bürgers, A., Wintgen, D. and Rost, J.-M. (1995). Highly doubly excited S states of the helium atom, *J. Phys.* **B28**, 3163–3183.

Burgers, J.M. (1916). Adiabatische Invarianten bij mechanische systemen, I, II, III, *Verlag van Gewone Vergaderingen de Wisen Natuurkundige Afdeeling Konieklijke Aked. Wetenschappente Amsterdam* **25**, pp. 848, 918–922, 1055–1061.

Campbell, D.K. (1987). Nonlinear science, from paradigms to practicalities, in *Los Alamos Science, Vol. 15*, ed. N.G. Cooper (Los Alamos Laboratory, Los Alamos).

Casati, G. and Chirikov, B.V. (1995). *Quantum Chaos* (Cambridge University Press, Cambridge).

Casati, G., Chirikov, B.V., Izraelev, F.M. and Ford, J. (1979). Stochastic behavior of a quantum pendulum under a periodic perturbation, in *Stochastic Behavior in Classical and Quantum Hamiltonian Systems*, eds. G. Casati and J. Ford (Springer, New York).

Casati, G., Chirikov, B.V., Shepelyansky, D.L. and Guarneri, I. (1986). New photoelectric ionization peak in the hydrogen atom, *Phys. Rev. Lett.* **57**, 823–826.

Casati, G., Chirikov, B.V., Guarneri, I. and Shepelyansky, D.L. (1987). Relevance of classical chaos in quantum mechanics: The hydrogen atom in a monochromatic field, *Phys. Rep.* **154**, 77–123.

Chirikov, B.V. (1979). A universal instability of many-dimensional oscillator systems, *Phys. Rep.* **52**, 263–379.

Chirikov, B.V. (1991). Chaotic quantum systems, in *Mathematical Physics X*, ed. K. Schmudgen (Springer, New York).

Chirikov, B.V. and Shepelyansky, D.L. (1984). Correlation properties of dynamical chaos in Hamiltonian systems, *Physica* **D13**, 395–400.

Chirikov, B.V., Izrailev, F.M. and Shepelyansky, D.L. (1981). Dynamical stochasticity in classical and quantum mechanics, *Sov. Sci. Rev.* **C2**, 209–267.

Chu, S.I. and Reinhardt, W.P. (1977). Intense field multiphoton ionization via complex dressed states: Application to the H atom, *Phys. Rev. Lett.* **39**, 1195–1198.

Cole, M.W. (1974). Electronic surface states of liquid helium, *Rev. Mod. Phys.* **46**, 451–464.

Collins, G.P. (1995). Gaseous Bose-Einstein condensate finally observed, *Physics Today* **48(8)**, 17–20.

Cooley, J.W. and Tukey, J.W. (1965). An algorithm for the machine calculation of complex Fourier series, *Math. Comput.* **19**, 297–301.

Cooper, F., Dawson, J.F., Meredith, D. and Shepard, H. (1994). Semiquantum chaos, *Phys. Rev. Lett.* **72**, 1337–1340.

Courtney, M., Jiao, H., Spellmeyer, N., Kleppner, D., Gao, J. and Delos, J.B. (1995). Closed orbit bifurcations in continuum Stark spectra, *Phys. Rev. Lett.* **74**, 1538–1541.

Cvitanović, P. and Eckhardt, B. (1989). Periodic orbit quantization of chaotic systems, *Phys. Rev. Lett.* **63**, 823–826.

d'Abro, A. (1951). *The Rise of the New Physics* (Dover, New York).

Dahlqvist, P. and Russberg, G. (1990). Existence of stable orbits in the x^2y^2 potential, *Phys. Rev. Lett.* **65**, 2837–2838.

Dando, P.A., Monteiro, T.S., Delande, D. and Taylor, K.T. (1995). Beyond periodic orbits: An example in nonhydrogenic atoms, *Phys. Rev. Lett.* **74**, 1099–1102.

Dauben, J.W. (1979). *Georg Cantor, His Mathematics and Philosophy of the Infinite* (Princeton University Press, Princeton).

Davis, M.J. and Gray, S.K. (1986). Unimolecular reactions and phase space bottlenecks, *J. Chem. Phys.* **84**, 5389–5411.

Devaney, R.L. (1992). *A First Course in Chaotic Dynamical Systems* (Addison-Wesley, Reading, MA).

Dirac, P.A.M. (1925). The adiabatic invariance of the quantum integrals, *Proc. Roy. Soc.* **107**, 725–734.

Dittrich, T. and Smilansky, U. (1991a). Spectral properties of systems with dynamical localization: I. The local spectrum, *Nonlinearity* **4**, 59–84.

Dittrich, T. and Smilansky, U. (1991b). Spectral properties of systems with dynamical localization: II. Finite sample approach, *Nonlinearity* **4**, 85–101.

Domke, M., Schulz, K., Remmers, G., Gutiérrez, A., Kaindl, G. and Wintgen, D. (1995). Interferences in photoexcited double-excitation series of He, *Phys. Rev.* **A51**, R4309–R4312.

Doron, E., Smilansky, U. and Frenkel, A. (1990). Experimental demonstration of chaotic scattering of microwaves, *Phys. Rev. Lett.* **65**, 3072–3075.

Draeger, M., Handke, G., Ihra, W. and Friedrich, H. (1994). One- and two-electron excitations of helium in the s-wave model, *Phys. Rev.* **A50**, 3793–3808.

Dunham, W. (1990). *Journey Through Genius. The Great Theorems of Mathematics* (John Wiley, New York).

Dyson, F.J. (1962a). Statistical theory of the energy levels of complex systems I, *J. Math. Phys.* **3**, 140–156.

Dyson, F.J. (1962b). Statistical theory of the energy levels of complex systems II, *J. Math. Phys.* **3**, 157–165.

Dyson, F.J. (1962c). Statistical theory of the energy levels of complex systems III, *J. Math. Phys.* **3**, 166–175.

Eckhardt, B. (1987). Fractal properties of scattering singularities, *J. Phys.* **A20**, 5971–5979.

Eckhardt, B. and Jung, C. (1986). Regular and irregular potential scattering, *J. Phys.* **A19**, L829–L833.

Edmonds, A.R. (1970). The theory of the quadratic Zeeman effect, *J. Phys. (Paris), Colloq.* **31**, **C4**, 71–74.

Edmonds, A.R. (1973). Studies of the quadratic Zeeman effect I. Application of the sturmian functions, *J. Phys.* **B6**, 1603–1615.

Eidson, J. and Fox, R.F. (1986). Quantum chaos in a two-level system in a semiclassical radiation field, *Phys. Rev.* **A34**, 3288–3292.

Einstein, A. (1917). Zum Quantensatz von Sommerfeld und Epstein, *Verh. Deutsch. Phys. Ges.* **19**, 82–92.

Elander, N. and Brändas, E. (1989). The Lertorpet symposium view on a generalized inner product, in *Resonances, Lecture Notes in Physics 325*, eds. E. Brändas and N. Elander (Springer, Berlin).

Ericson, T. (1960). Fluctuations of nuclear cross sections in the "continuum" region, *Phys. Rev. Lett.* **5**, 430–431.

Ericson, T. (1963). A theory of fluctuations in nuclear cross sections, *Ann. Phys.* **23**, 390–414.

Esser, B. and Schanz, H. (1995). Excitonic – vibronic coupled dimers: a dynamic approach, *Zeit. Phys.* **B96**, 553–562.

Ezra, G.S., Richter, K., Tanner, G. and Wintgen, D. (1991). Semiclassical cycle expansion for the helium atom, *J. Phys.* **B24**, L413–L420.

Farmer, J.D., Ott, E. and Yorke, J.A. (1983). The dimension of chaotic attractors, *Physica D* **7**, 153–180.

Farrelly, D. and Howard, J.E. (1994). Double-well dynamics of two ions in the Paul and Penning traps, *Phys. Rev.* **A49**, 1494–1497.

Feigenbaum, M.J. (1978). Quantitative universality for a class of nonlinear transformations, *J. Stat. Phys.* **19**, 25–52.

Feigenbaum, M.J. (1979). The universal metric properties of nonlinear transformations, *J. Stat. Phys.* **21**, 669–706.

Feshbach, H. (1962). A unified theory of nuclear reactions. II, *Ann. Phys. (N.Y.)* **19**, 287–313.

Feynman, R.P. (1948). Space-time approach to non-relativistic quantum mechanics, *Rev. Mod. Phys.* **20**, 367–387.

Feynman, R.P., Leighton, R.B. and Sands, M. (1965). *The Feynman Lectures on Physics, Vol. III* (Addison-Wesley, Reading).

Fishman, S., Grempel, D.R. and Prange, R.E. (1982). Chaos, quantum recurrences and Anderson localization, *Phys. Rev. Lett.* **49**, 509–512.

Ford, J. (1983). How random is a coin toss?, *Phys. Today* **36(4)**, 40–47.

Ford, J. (1989). What is chaos, that we should be mindful of it, in *The New Physics*, ed. P. Davies (Cambridge University Press, Cambridge).

Ford, J., Stoddard, S.D. and Turner, J.S. (1973). On the integrability of the Toda lattice, *Prog. Theor. Phys.* **50**, 1547–1560.

Franck, J. and Hertz, G. (1913). Messung der Ionisierungsspannung in verschiedenen Gasen, *Verh. d. Deutsch. Phys. Ges.* **15**, 34–44.

Franck, J. and Reiche, F. (1920). Über Helium und Parhelium, *Z. Phys.* **1**, 154–160.

French, A.P. and Taylor, E.F. (1978). *An Introduction to Quantum Physics* (W. W. Norton, New York).

Friedrich, H. (1990). *Theoretical Atomic Physics* (Springer, Berlin).

Friedrich, H. and Wintgen, D. (1989). The hydrogen atom in a uniform magnetic field: An example of chaos, *Phys. Rep.* **183**, 37–79.

Galvez, E.J., Sauer, B.E., Moorman, L., Koch, P.M. and Richards, D. (1988). Microwave ionization of H atoms: breakdown of classical dynamics for high frequencies, *Phys. Rev. Lett.* **61**, 2011–2014.

Garstang, R.H. (1977). Atoms in high magnetic fields, *Rep. Prog. Phys.* **40**, 105–154.

Garton, W.R.S. and Tomkins, F.S. (1969). Diamagnetic Zeeman effect and magnetic configuration mixing in long spectral series of Ba I, *Astrophys. J.* **158**, 839–845.

Gaspard, P. and Rice, S.A. (1989a). Scattering from a classically chaotic repellor, *J. Chem. Phys.* **90**, 2225–2241.

Gaspard, P. and Rice, S.A. (1989b). Semiclassical quantization of the scattering from a classically chaotic repellor, *J. Chem. Phys.* **90**, 2242–2254.

Gaspard, P. and Rice, S.A. (1989c). Exact quantization of the scattering from a classically chaotic repellor, *J. Chem. Phys.* **90**, 2255–2262.

Gay, J.-C. (1992). *Irregular Atomic Systems and Quantum Chaos* (Gordon and Breach, Philadelphia).

Ginibre, J. (1965). Statistical ensembles of complex, quaternion and real matrices, *J. Math. Phys.* **6**, 440–449.

Gleick, J. (1987). *Chaos* (Penguin Books, New York).

Goldstein, H. (1976). *Klassische Mechanik* (Akademische Verlagsgesellschaft, Wiesbaden).

Gordon, R.G. (1969). New method for constructing wavefunctions for bound states and scattering, *J. Chem. Phys.* **51**, 14–25.

Gradshteyn, I.S. and Ryzhik, I.M. (1994). *Table of Integrals, Series, and Products* (Academic Press, Boston).

Grebogi, C., McDonald, S.W., Ott, E. and Yorke, J.A. (1985). Exterior dimension of fat fractals, *Phys. Lett.* **A110**, 1–4.

Grempel, D.R., Prange, R.E. and Fishman, S. (1984). Quantum dynamics of a nonintegrable system, *Phys. Rev.* **A29**, 1639–1647.

Grimes, C.C. and Brown, T.R. (1974). Direct spectroscopic observation of electrons in image-potential states outside liquid helium, *Phys. Rev. Lett.* **32**, 280–283.

Gutzwiller, M.C. (1971). Periodic orbits and classical quantization conditions, *J. Math. Phys.* **12**, 343–358.

Gutzwiller, M.C. (1983). Stochastic behavior in quantum scattering, *Physica* **D7**, 341–355.

Gutzwiller, M.C. (1990). *Chaos in Classical and Quantum Mechanics* (Springer, New York).

Haake, F. (1991). *Quantum Signatures of Chaos* (Springer, Berlin).

Haffmans, A., Blümel, R., Koch, P.M. and Sirko, L. (1994). Prediction of a new peak in two-frequency microwave "ionization" of excited hydrogen atoms, *Phys. Rev. Lett.* **73**, 248–251.

Hammel, S.M., Yorke, J.A. and Grebogi, C. (1987). Do numerical orbits of chaotic dynamical processes represent true orbits?, *J. Complexity* **3**, 136–145.

Handke, G. (1994). Fractal dimensions in the phase space of two-electron atoms, *Phys. Rev.* **A50**, R3561–R3564.

Handke, G., Draeger, M. and Friedrich, H. (1993). Classical dynamics of s-wave helium, *Physica* **A197**, 113–129.

Harris, P.G., Bryant, H.C., Mohagheghi, A.H., Reeder, R.A., Tang, C.Y., Donahue, J.B. and Quick, C.R. (1990). Observation of doubly excited resonances in the H$^-$ ion, *Phys. Rev.* **A42**, 6443–6465.

Hausdorff, F. (1919). Dimension und äußeres Maß, *Mathematische Annalen* **79**, 157–179.

Heath, T.L. (1956). *The Thirteen Books of Euclid's Elements* (Dover, New York).

Heisenberg, W. (1925). Über quantentheoretische Umdeutung kinematischer und mechanischer Beziehungen, *Z. Phys.* **33**, 879–893.

Heisenberg, W. (1926). Über die Spektra von Atomsystemen mit zwei Elektronen, *Z. Phys.* **39**, 499–518.

Heisenberg, W. (1969). *Der Teil und das Ganze* (R. Piper & Co., Munich).

Heller, E.J. and Tomsovic, S. (1993). Postmodern quantum mechanics, *Phys. Today* **46(7)**, 38–46.

Hennig, D. and Esser, B. (1992). Transfer dynamics of a quasiparticle in a nonlinear dimer coupled to an intersite vibration: Chaos on the Bloch sphere, *Phys. Rev.* **A46**, 4569–4576.

Hénon, M. (1974). Integrals of the Toda lattice, *Phys. Rev.* **B9**, 1921–1923.

Hillermeier, C.F., Blümel, R. and Smilansky, U. (1992). Ionization of H Rydberg atoms: Fractals and power-law decay, *Phys. Rev.* **A45**, 3486–3502.

Hirschfelder, J.O., Curtiss, C.F. and Bird, R.B. (1954). *Molecular Theory of Gases and Liquids* (Wiley, New York).

Ho, Y.K. (1981). Complex-coordinate calculations for doubly excited states of two-electron atoms, *Phys. Rev.* **A23**, 2137–2149.

Ho, Y.K. (1983). The method of complex coordinate rotation and its applications to atomic collision processes, *Phys. Rep.* **99**, 1–68.

Hogg, T. and Huberman, B.A. (1982). Recurrence phenomena in quantum dynamics, *Phys. Rev. Lett.* **48**, 711–714.

Howard, J.E. (1991). Theory of two-frequency microwave ionization of hydrogen atoms, *Phys.Lett.* **A156**, 286–292.

Howard, J.E. and Farrelly, D. (1993). Integrability of the Paul trap and generalized van der Waals Hamiltonians, *Phys. Lett.* **A178**, 62–72.

Hüpper, B., Main, J. and Wunner, G. (1995). Photoabsorption of nonhydrogenic Rydberg atoms in a magnetic field: effects of core-scattered classical orbits, *Phys. Rev. Lett.* **74**, 2650–2653.

Hylleraas, E.A. (1928). Über den Grundzustand des Heliumatoms, *Z. f. Physik* **48**, 469–494.

Hylleraas, E.A. (1929). Neue Berechnung der Energie des Heliums im Grundzustande, sowie des tiefsten Terms von Ortho-Helium, *Z. Phys* **54**, 347–366.

Ince, E.L. (1956). *Ordinary Differential Equations* (Dover, New York).

Itano, W.M., Wineland, D.J., Hemmati, H., Bergquist, J.C. and Bollinger, J.J. (1983). Time and frequency standards based on charged particle trapping, *IEEE Trans. Nucl. Sci.* **30**, 1521–1523.

Iu, C.H., Welch, G.R., Kash, M.M., Kleppner, D., Delande, D. and Gay, J.C. (1991). Diamagnetic Rydberg atom: Confrontation of calculated and observed spectra, *Phys. Rev. Lett.* **66**, 145–148.

Izrailev, F.M. and Shepelyanskii, D.L. (1979). Quantum resonance for a rotor in a nonlinear periodic field, *Sov. Phys. Dokl.* **24**, 996–998.

Jackson, J.D. (1975). *Classical Electrodynamics* (John Wiley, New York).

Jaeckel, G. (1988). *Rydbergatome in starken Mikrowellenfeldern – Untersuchung eines klassisch chaotischen Systems*, Diploma thesis, Technical University Munich.

Jensen, R.V. (1982). Stochastic ionization of surface-state electrons, *Phys. Rev. Lett.* **49**, 1365–1368.

Jensen, R.V., Susskind, S.M. and Sanders, M.M. (1991). Chaotic ionization of highly excited hydrogen atoms: Comparison of classical and quantum theory with experiment, *Phys. Rep.* **201**, 1–56.

Junker, B.R. (1982). Recent computational developments in the use of complex scaling in resonance phenomena, *Adv. At. Mol. Phys.* **18**, 207–263.

Karney, C.F.F. (1983). Long-time correlations in the stochastic regime, *Physica* **D8**, 360–380.

Keating, J. (1993). The Riemann zeta function and quantum chaology, in *Quantum Chaos*, eds. G. Casati, I. Guarneri and U. Smilansky (North-Holland, Amsterdam).

Keldysh, L.V. (1965). Ionization in the field of a strong electromagnetic wave, *Sov. Phys. JETP* **20**, 1307–1314.

Keller, J.B. (1958). Corrected Bohr-Sommerfeld quantum conditions for nonseparable systems, *Ann. Phys.* **4**, 180–188.

Kellner, G.W. (1927). Die Ionisierungsspannung des Heliums nach der Schrödingerschen Theorie, *Z. f. Physik* **44**, 91–109.

Kim, J.-H. and Ezra, G.S. (1991). Periodic orbits and the classical-quantum correspondence for doubly-excited states of two-electron atoms, in *Proceedings of the Adriatico Conference on Quantum Chaos*, eds. H. Cerdeira *et al.* (World Scientific, Hong Kong).

Kinoshita, T. (1959). Ground state of the helium atom, II., *Phys. Rev.* **115**, 366–374.

Koch, P.M. and van Leeuwen, K.A.H. (1995). The importance of resonances in microwave "ionization" of excited hydrogen atoms, *Phys. Rep.* **255**, 289–403.

Koch, P.M., Moorman, L. and Sauer, B.E. (1992). Microwave ionization of excited hydrogen atoms: experiments versus theories for high scaled frequencies, in *Irregular Atomic Systems and Quantum Chaos*, ed. J.-C. Gay (Gordon and Breach, Philadelphia).

Kramers, H.A. (1923). Über das Modell des Heliumatoms, *Z. Phys.* **13**, 312–341.

Kudrolli, A., Sridhar, S., Pandey, A. and Ramaswamy, R. (1994). Signatures of chaos in quantum billiards: Microwave experiments, *Phys. Rev.* **E49**, R11–R14.

Lagrange, J.L. (1788). *Méchanique Analytique* (Desaint, Paris).

Lai, Y.-C., Ding, M., Grebogi, C. and Blümel, R. (1992a). Algebraic decay and fluctuations of the decay exponent in Hamiltonian systems, *Phys. Rev.* **A46**, 4661–4669.

Lai, Y.-C., Grebogi, C., Blümel, R. and Ding, M. (1992b). Algebraic decay and phase-space metamorphoses in microwave ionization of hydrogen Rydberg atoms, *Phys. Rev.* **A45**, 8284–8287.

Landau, L.D. and Lifschitz, E.M. (1970). *Mechanik* (Vieweg, Braunschweig).

Landau, L.D. and Lifschitz, E.M. (1971). *Quantenmechanik* (Akademie-Verlag, Berlin).

Landau, L.D. and Lifschitz, E.M. (1977). *Klassische Feldtheorie* (Akademie-Verlag, Berlin).

Landé, A. (1919). Das Serienspektrum des Heliums, *Phys. Zeitschr.* **20**, 228–234.

Langmuir, I. (1921). The structure of the helium atom, *Phys. Rev.* **17**, 339–353.

Lapidus, I.R. (1975). One-dimensional models for two-electron systems, *Am. J. Phys.* **43**, 790–792.

Lebowitz, J.L. and Penrose, O. (1973). Modern ergodic theory, *Phys. Today* **26(2)**, 23–29.

Leopold, J.G. and Percival, I.C (1978). Microwave ionization and excitation of Rydberg atoms, *Phys. Rev. Lett.* **41**, 944–947.

Leopold, J.G. and Percival, I.C. (1979). Ionization of highly excited atoms by electric fields III: Microwave ionisation and excitation, *J. Phys.* **B12**, 709–721.

Leopold, J.G. and Percival, I.C. (1980). The semiclassical two-electron atom and the old quantum theory, *J. Phys.* **B13**, 1037–1047.

Lichtenberg, A.J. and Lieberman, M.A. (1983). *Regular and Stochastic Motion* (Springer, New York).

Lichtenberg, A.J. and Lieberman, M.A. (1992). *Regular and Chaotic Dynamics* (Springer, New York).

Lindroth, E. (1994). Calculation of doubly excited states of helium with a finite discrete spectrum, *Phys. Rev.* **A49**, 4473–4480.

Lorenz, E.N. (1963). Deterministic nonperiodic flow, *J. Atmos. Sci.* **20**, 130–141.

Luck, J.M., Orland, H. and Smilansky, U. (1988). On the response of a two-level quantum system to a class of time-dependent perturbations, *J. Stat. Phys.* **53**, 551–564.

McCauley, J.L. (1988). An introduction to nonlinear dynamics and chaos theory, *Physica Scripta* **T20**, 5–57.

McClelland, J.J., Scholten, R.E., Palm, E.C. and Celotta, R.J. (1993). Laser-focused atomic deposition, *Science* **262**, 877–880.

McKean, H.P. (1972). Selberg's trace formula as applied to a compact Riemann surface, *Commun. Pure Appl. Math.* **25**, 225–246.

Madden, R.P. and Codling, K. (1963). New autoionizing atomic energy levels in He, Ne, and Ar, *Phys. Rev. Lett.* **10**, 516–518.

Main, J. and Wunner, G. (1994). Rydberg atoms in external fields as an example of open quantum systems with classical chaos, *J. Phys.* **B27**, 2835–2848.

Main, J., Wiebusch, G. and Welge, K. H. (1992). Spectroscopy of the classically chaotic hydrogen atom in magnetic fields, in *Irregular Atomic Systems and Quantum Chaos*, ed. J.-C. Gay (Gordon and Breach, Philadelphia).

Main, J., Wiebusch, G., Welge, K., Shaw, J. and Delos, J.B. (1994). Recurrence spectroscopy: Observation and interpretation of large-scale structure in the absorption spectra of atoms in magnetic fields, *Phys. Rev.* **A49**, 847–868.

Mandelbrot, B.B. (1975). *Les Objets Fractals: Forme, Hasard et Dimension* (Flammarion, Paris).

Mandelbrot, B.B. (1977). *Fractals: Form, Chance and Dimension* (W. H. Freeman and Co., San Francisco).

Mandelbrot, B.B. (1983). *The Fractal Geometry of Nature* (W. H. Freeman and Co., New York).

Marcus, C.M., Westervelt, R.M., Hopkings, P.F. and Gossard, A.C. (1993). Conductuance fluctuations as quantum chaotic scattering, in *Quantum Dynamics of Chaotic Systems*, eds. J.-M. Yuan, D. H. Feng and G. M. Zaslavsky (Gordon and Breach, Amsterdam).

May, R.M. (1974). Simple mathematical models with very complicated dynamics, *Nature* **261**, 459–467.

May, R.M. (1980). Cree-Ojibwa hunting and the hare-lynx cycle, *Nature* **286**, 108–109.

May, R.M. (1986). When two and two do not make four: nonlinear phenomena in ecology (the Croonian Lecture), *Proc. R. Soc. Lond. B* **228**, 241–266.

May, R.M. (1987). Chaos and the dynamics of biological populations, in *Dynamical Chaos*, eds. M.V. Berry, I.C. Percival and N.O. Weiss (Princeton University Press, Princeton).

Mayer-Kuckuk, T. (1977). *Atomphysik* (Teubner, Stuttgart).

Meerson, B.I., Oks, E.A. and Sasarov, P.V. (1979). Stochastic instability of an oscillator and the ionization of highly-excited atoms under the action of electromagnetic radiation, *JETP Lett.* **29**, 72–75.

Mehta, M.L. (1991). *Random Matrices* (Academic Press, Boston).

Meiss, J.D. and Ott, E. (1985). Markov-tree model of intrinsic transport in Hamiltonian systems, *Phys. Rev. Lett.* **55**, 2741–2744.

Meissner, H. and Schmidt, G. (1986). A simple experiment for studying the transition from order to chaos, *Am. J. Phys.* **54**, 800–804.

Messiah, A. (1979). *Quantenmechanik* (Walter de Gruyter, Berlin).

Miller, W.H. (1975). Semiclassical quantization of nonseparable systems: A new look at periodic orbit theory, *J. Chem. Phys.* **63**, 996–999.

Milne, W.E. (1970). *Numerical Solution of Differential Equations* (Dover, New York).

Moore, M. and Blümel, R. (1993). Quantum manifestations of order and chaos in the Paul trap, *Phys. Rev.* **A48**, 3082–3091.

Moorman, L. and Koch, P.M. (1992). Microwave ionization of Rydberg atoms, in *Quantum Non-Integrability*, eds. D. H. Feng and J. M. Yuan (World Scientific, Singapore).

Müller, J., Yang, X. and Burgdörfer, J. (1994). Calculation of resonances in doubly excited helium using the stabilization method, *Phys. Rev.* **A49**, 2470–2475.

Naeye, R. (1994). Chaos squared, *Discover Magazine*, **March 1994**, p. 28.

Nörenberg, W. and Weidenmüller, H.A. (1976). *Introduction to the Theory of Heavy-Ion Collisions* (Springer, Berlin).

Nogami, Y., Valliéres, M. and van Dijk, W. (1976). Hartree-Fock approximation for the one-dimensional "helium atom", *Am. J. Phys.* **44**, 886–888.

Noid, D.W., Gray, S.K. and Rice, S.A. (1986). Fractal behavior in classical collisional energy transfer, *J. Chem. Phys.* **84**, 2649–2652.

Odlyzko, A.M. (1989). The 10^{20}-th zero of the Riemann zeta function and 70 million of its neighbors, Preprint, AT&T Bell Laboratories.

Ott, E. (1993). *Chaos in Dynamical Systems* (Cambridge University Press, Cambridge).

Pais, A. (1991). *Niels Bohr's Times, in Physics, Philosophy, and Polity* (Oxford University Press, Oxford).

Paul, W. (1990). Electromagnetic traps for charged and neutral particles, *Rev. Mod. Phys.* **62**, 531–540.

Paul, W., Osberghaus, O. and Fischer, E. (1958). Ein Ionenkäfig, *Forschungsberichte des Wirtschafts und Verkehrsministeriums Nordrhein Westfalen* **415**.

Pechukas, Ph. (1983). Distribution of energy eigenvalues in the irregular spectrum, *Phys. Rev. Lett.* **51**, 943–946.

Peitgen, H.-O. and Richter, P.H. (1986). *The Beauty of Fractals* (Springer, New York).

Peitgen, H.-O., Jürgens, H. and Saupe, D. (1993). *Chaos and Fractals, New Frontiers of Science* (Springer, New York).

Pekeris, C.L. (1958). Ground state of two-electron atoms, *Phys. Rev.* **112**, 1649–1658.

Pekeris, C.L. (1959). 1 ^{1}S and 2 ^{3}S states of helium, *Phys. Rev.* **115**, 1216–1221.

Percival, I.C. (1973). Regular and irregular spectra, *J. Phys.* **B6**, L229–L232.

Percival, I.C. (1977). Planetary atoms, *Proc. R. Soc.* **A353**, 289–297.

Percival, I.C. (1987). Chaos in hamiltonian systems, in *Dynamical Chaos*, eds. M.V. Berry, I.C. Percival and N.O. Weiss (Princeton University Press, Princeton).

Peterson, I. (1993). *Newton's Clock* (W. H. Freeman, New York).

Pichard, J.-L., Sanquer, M., Slevin, K. and Debray, P. (1990). Broken symmetries and localization lengths in Anderson insulators: theory and experiment, *Phys. Rev. Lett.* **65**, 1812–1815.

Poincaré, H. (1892). *Les Méthodes Nouvelles de la Méchanique Céleste* (Gauthier-Villars, Paris).

Poincaré, H. (1902). *La Science et l'Hypothése* (Flammarion, Paris).

Poincaré, H.(1993). *New Methods of Celestial Mechanics*, edited and introduced by D. L. Goroff (American Institute of Physics, New York).

Pomeau, Y., Dorizzi, B. and Grammaticos, B. (1986). Chaotic Rabi oscillations under quasiperiodic perturbation, *Phys. Rev. Lett.* **56**, 581–684.

Porter, C.E., ed. (1965). *Statistical Theories of Spectra* (Academic Press, New York).

Prange, R.E., Ott, E., Antonsen, T.M., Georgeot, B. and Blümel, R. (1996). Smoothed density of states for problems with ray splitting, *Phys. Rev.* **E53**, 207–213.

Primack, H. and Smilansky, U. (1995). Quantization of the three-dimensional Sinai billiard, *Phys. Rev. Lett.* **74**, 4831–4834.

Purkert, W. and Ilgauds, H.J. (1987). *Georg Cantor* (Birkhäuser Verlag, Basel).

Rankin, C.C. and Miller, W.H. (1971). Classical S-matrix for linear reactive collision of H + Cl_2, *J. Chem. Phys.* **55**, 3150-3156.

Rechester, A.B. and White, R.B. (1980). Calculation of turbulent diffusion for the Chirikov-Taylor model, *Phys. Rev. Lett.* **44**, 1586–1589.

Reichl, L.E. (1992). *The Transition to Chaos* (Springer, New York).

Reinhardt, W.P. (1982). Complex coordinates in the theory of atomic and molecular structure and dynamics, *Ann. Rev. Phys. Chem.* **33**, 223–255.

Richter, K. (1991). Semiklassik von Zwei-Elektronen-Atomen, Ph. D. thesis, Albert-Ludwigs-Universität, Freiburg.

Richter, K. and Wintgen, D. (1990a). Stable planetary atom configurations, *Phys. Rev. Lett.* **65**, 1965.

Richter, K. and Wintgen, D. (1990b). Analysis of classical motion on the Wannier ridge, *J. Phys.* **B23**, L197–L201.

Richter, K., Rost, J.-M., Thürwächter, R., Briggs, J.S., Wintgen, D. and Solov'ev, E.A. (1991). New State of binding of antiprotons in atoms, *Phys. Rev. Lett.* **66**, 149–152.

Ring, P and Schuck, P. (1980). *The Nuclear Many-Body Problem* (Springer, New York).

Rosenzweig, N. and Porter, C.E. (1960). "Repulsion of energy levels" in complex atomic spectra, *Phys. Rev.* **120**, 1698–1714.

Rost, J.-M. (1994). Two-electron escape near threshold: A classical process?, *Phys. Rev. Lett.* **72**, 1998–2001.

Rost, J.-M. (1995). Threshold ionization of atoms by electron and positron impact, *J. Phys.* **B28**, 3003–3026.

Sagdeev, R.Z., Usikov, D.A. and Zaslavsky, G.M. (1988). *Nonlinear Physics* (Harwood Academic, Chur).

Samuelides, M., Fleckinger, R., Touziller, L. and Bellissard, J. (1986). Instabilities of the quantum rotator and transition in the quasi-energy spectrum, *Europhys. Lett.* **1**, 203–208.

Saville, G.F. and Goodkind, J.M. (1994). Computation of tunneling rates in time-dependent electric fields: Electrons on the surface of liquid helium, a one-dimensional hydrogen atom, *Phys. Rev.* **A50**, 2059–2067.

Saville, G.F., Goodkind, J.M. and Platzman, P.M. (1993). Single-electron tunneling from bound states on the surface of liquid helium, *Phys. Rev. Lett.* **70**, 1517–1520.

Schack, R. (1995). Comment on "Exponential sensitivity and chaos in quantum systems", *Phys. Rev. Lett.* **75**, 581.

Scharf, R. (1989). Kicked rotator for a spin-1/2 particle, *J. Phys.* **A22**, 4223–4242.

Schroeder, M. (1991). *Fractals, Chaos, Power Laws* (Freeman, New York).

Schrödinger, E. (1926). Quantisierung als Eigenwertproblem, *Ann. Physik* **79**, 361–376.

Schuster, H.-G. (1988). *Deterministic Chaos* (VCH, New York).

Seaton, M.J. (1983). Quantum defect theory, *Rep. Prog. Phys.* **46**, 167–257.

Selberg, A. (1956). Harmonic analysis and discontinuous groups in weakly symmetric Riemannian spaces with applications to Dirichlet series, *J. Ind. Math. Soc.* **20**, 47–87.

Shepelyansky, D.L. (1985). Quantum diffusion limitation at excitation of Rydberg atom in variable field, in *Chaotic Behaviour in Quantum Systems*, ed. G. Casati (Plenum, New York).

Shinbrot, T., Grebogi, C., Wisdom, J. and Yorke, J.A. (1992). Chaos in a double pendulum, *Am. J. Phys.* **60**, 491–499.

Silva, J.R. and Canuto, S. (1984). On the spherical quadratic Zeeman problem in hydrogen, *Phys. Lett.* **A101**, 326–330.

So, P., Anlage, S.M., Ott, E. and Oerter, R.N. (1995). Wave chaos experiments with and without time reversal symmetry: GUE and GOE statistics, *Phys. Rev. Lett.* **74**, 2662–2665.

Stöckmann, H.-J. and Stein, J. (1990). "Quantum" chaos in billiards studied by microwave absorption, *Phys. Rev. Lett.* **64**, 2215–2218.

Stoer, J. and Bulirsch, R. (1978). *Einführung in die Numerische Mathematik* (Springer, Berlin).

Stoffregen, U., Stein, J., Stöckmann, H.-J., Kuś, M. and Haake, F. (1995). Microwave billiards with broken time reversal symmetry, *Phys. Rev. Lett.* **74**, 2666–2669.

Stroud, A.H. and Secrest, D. (1966). *Gaussian Quadrature Formulas* (Prentice Hall, Englewood Cliffs, N.J.).

Susskind, S.M. and Jensen, R.V. (1988). Numerical calculations of the ionization of one-dimensional hydrogen atoms using hydrogenic and sturmian basis functions, *Phys. Rev.* **A38**, 711–728.

Symon, K.R. (1971). *Mechanics* (Addison-Wesley, Reading).

Tabor, M. (1989). *Chaos and Integrability in Nonlinear Dynamics* (John Wiley, New York).

Thaha, M. and Blümel, R. (1994). Nonuniversality of the localization length in a quantum chaotic system, *Phys. Rev. Lett.* **72**, 72–75.

Thaha, M., Blümel, R. and Smilansky, U. (1993). Symmetry breaking and localization in quantum chaotic systems, *Phys. Rev.* **E48**, 1764–1781.

Timp, G., Behringer, R.E., Tennant, D.M., Cunningham, J.E., Prentiss, M. and Berggren, K.K. (1992). Using light as a lens for submicron, neutral-atom lithography, *Phys. Rev. Lett.* **69**, 1636–1639.

Townes, C.H. and Schawlow, A.L. (1975). *Microwave Spectroscopy* (Dover, New York).

Ulam, S.M. and von Neumann, J. (1947). On combination of stochastic and deterministic processes, *Bull. Am. Math. Soc.* **53**, 1120.

Umberger, D.K., Mayer-Kress, G. and Jen, E. (1986). Hausdorff dimensions for sets with broken scaling symmetry, in *Dimensions and Entropies in Chaotic Systems*, ed. G. Mayer-Kress (Springer, New York).

van Leeuwen, K.A.H, Oppen, von, G., Renwick, S., Bowlin, J.B., Koch, P.M., Jensen, R.V., Rath, O., Richards, D. and Leopold, J.G. (1985). Microwave ionization of hydrogen Rydberg atoms: experiment versus classical dynamics, *Phys. Rev. Lett.* **55**, 2231–2234.

Van Vleck, J.H. (1922). The normal helium atom and its relation to the quantum theory, *Phil. Mag.* **44**, 842–869.

Vattay, G., Wirzba, A. and Rosenqvist, P.E. (1994). Periodic orbit theory of diffraction, *Phys. Rev. Lett.* **73**, 2304–2307.

von Klarenbosch, A., Geerinck, K.K., Klaassen, T.O., Wenckebach, W.Th. and Foxon, C.T. (1990). Ionization energies and lifetime broadening of autoionizing states of the hydrogen atom in strong magnetic fields: theory vs. experiment, *Europhys. Lett.* **13**, 237–242.

Watanabe, S. (1987). Kummer-function representation of ridge travelling waves, *Phys. Rev.* **A36**, 1566–1574.

Weaver, R.L. (1989). Spectral statistics in elastodynamics, *J. Acoust. Soc. Am.* **85**, 1005–1013.

Weigert, S. (1993). Quantum chaos in the configurational quantum cat map, *Phys. Rev.* **A48**, 1780–1798.

Whittaker, E.T. and Watson, G.N. (1927). *A Course of Modern Analysis* (Cambridge University Press, Cambridge).

Wineland, D.J., Itano, W.M., Bergquist, J.C., Bollinger, J.J. and Hemmati, H. (1984). Frequency standard research using stored ions, *Prog. Quant. Electr.* **8**, 139–142.

Wintgen, D. (1987). Connection between long-range correlations in quantum spectra and classical periodic orbits, *Phys. Rev. Lett.* **58**, 1589–1592.

Wintgen, D. (1988). Semiclassical path-integral quantization of nonintegrable Hamiltonian systems, *Phys. Rev. Lett.* **61**, 1803–1806.

Wintgen, D. (1993). Quantisiertes Chaos, *Phys. Blätter* **49**, 641–644.

Wintgen, D. and Delande, D. (1993). Double photoexcitation of $^1P^o$ states in helium, *J. Phys.* **B26**, L399–L405.

Wintgen, D. and Friedrich, H. (1986). Regularity and irregularity in spectra of the magnetized hydrogen atom, *Phys. Rev. Lett.* **57**, 571–574.

Wintgen, D. and Friedrich, H. (1992). The status of semiclassical methods for chaotic systems, in *Irregular Atomic Systems and Quantum Chaos,* ed. J.-C. Gay (Gordon and Breach, Philadelphia).

Wintgen, D., Richter, K. and Tanner, G. (1993). The semi-classical helium atom, in *Quantum Chaos*, eds. G. Casati, I. Guarneri and U. Smilansky (North-Holland, Amsterdam).

Wunner, G. (1989). Gibt es Chaos in der Quantenmechanik?, *Phys. Blätter* **45**, 139–145.

Yang, X. and Burgdörfer, J. (1991). Molecular dynamics approach to the statistical properties of energy levels, *Phys. Rev. Lett.* **66**, 982–985.

Yukawa, T. (1985). New approach to the statistical properties of energy levels, *Phys. Rev. Lett.* **54**, 1883–1886.

Zaslavsky, G.M.I (1985). *Chaos in Dynamic Systems* (Harwood Academic, Chur).

Index

absolute space, 22
absolute time, 22
accelerator modes, 126
acoustic chaos, 97
acoustics, 85, 287
action, 27, 65–7, 71, 104, 105, 128,
129, 154, 157, 178, 191, 193, 198,
199, 201, 202, 204, 206, 207, 242,
248, 249, 252, 254, 266–8, 270,
271, 277, 286
    classical, 103, 104, 191, 242
    critical, 202
action-angle variables, 154, 155, 160,
199, 206, 208, 248
action integral, 66
adiabaticity parameter, 184
algebraic decay, 213, 214, 295
algebraic irrational, 33, 34
algebraic number, 33, 34
alphabet, 60–2, 250
aluminium block, 96
Anderson localization, 134, 144, 146,
182, 289
astrophysics, 290, 291
atomic beam, 159, 183, 184, 190
atomic clock, 82
atomic decay, 5, 51, 56
atomic lithography, 239
atomic nucleus, 86, 95
atomic oven, 183

atomic physics, 1, 4–6, 11, 12, 21, 36,
52–4, 84–6, 95, 104–6, 116–18,
123, 126, 127, 130, 146, 149–51,
153, 167, 181, 182, 195, 197,
214–17, 219, 233, 234, 238, 240,
242, 243, 277, 284, 288, 290, 295
atomic spectroscopy, 217
atomic spectrum, 2, 86, 284
atomic states, 86, 105, 183, 185
attractor, 14, 37
auto-correlation function, 26, 27, 237
auto-correlation test, 26
autonomous systems, 5, 6, 70, 79, 84,
102, 112, 154, 160, 179, 264, 290
avoided crossing, 2

baker's map, 49–51, 114
ballistic electrons, 2
balloon, 106
Bayfield, J.E., 181, 183–5, 288
Bayfield-Koch experiment, 181
Berry, M.V., 84, 285, 286
Bessel functions, 132, 162, 179, 193
bifurcation, 16, 17, 41, 292
bifurcation diagram, 17, 40
bifurcation point, 17, 42, 292
bifurcation scenario, 16
billiard, 11, 12, 97
binary code, 295
binary digit, 42, 43, 45

313